# Applied Mathematical Sciences
Volume 44

# Applied Mathematical Sciences

*(continued following index)*

A. Pazy

# Semigroups of Linear Operators and Applications to Partial Differential Equations

Springer-Verlag
New York Berlin Heidelberg London Paris
Tokyo Hong Kong Barcelona Budapest

Amnon Pazy
Council for Higher Education
Planning and Budgeting Committee
Jerusalem 91040
Israel

*Editors*

F. John
Courant Institute of
  Mathematical Sciences
New York University
New York, NY 10012
USA

J.E. Marsden
Department of
  Mathematics
University of California
Berkeley, CA 94720
USA

L. Sirovich
Division of Applied
  Mathematics
Brown University
Providence, RI 02912
USA

Mathematics Subject Classification: H7D05, 35F10, 35F25, 35G25

Library of Congress Cataloging in Publication Data
Pazy, A.
  Semigroups of linear operators and applications to
partial differential equations.
  (Applied mathematical sciences; v. 44)
  Includes bibliographical references and index.
  1. Differential equations, Partial. 2. Initial
value problems. 3. Semigroups of operators.
I. Title.  II. Series: Applied mathematical sciences
(Springer-Verlag, New York Inc.); v.44.

Printed on acid-free paper.

Production coordinated by Brian Howe and managed by Francine Sikorski; manufacturing
supervised by Robert Paella.
Media conversion by Science Typographers, Inc., Medford, NY.
Printed and bound by Edwards Brothers, Inc., Ann Arbor, MI.
Printed in the United States of America.

9 8 7 6 5 4 3 2 (Corrected second printing)

ISBN 0-387-90845-5 Springer-Verlag New York Berlin Heidelberg Tokyo
ISBN 3-540-90845-5 Springer-Verlag Berlin Heidelberg New York Tokyo

# Preface to the Second Printing

This second printing of the book contains a few minor changes and corrections. It is a pleasure for me to thank Peter Hess, Gunter Lumer, R. de Roo, and Hans Sager for drawing my attention to many misprints and some errors.

I am especially indebted to Shinnosuke Oharu, who went through the whole book and recommended many valuable clarifications, modifications, and corrections.

A. Pazy

# Preface to the First Printing

The aim of this book is to give a simple and self-contained presentation of the theory of semigroups of bounded linear operators and its applications to partial differential equations.

The book is a corrected and expanded version of a set of lecture notes which I wrote at the University of Maryland in 1972–1973. The first three chapters present a short account of the abstract theory of semigroups of bounded linear operators. Chapters 4 and 5 give a somewhat more detailed study of the abstract Cauchy problem for autonomous and nonautonomous linear initial value problems, while Chapter 6 is devoted to some abstract nonlinear initial value problems. The first six chapters are self-contained and the only prerequisite needed is some elementary knowledge of functional analysis. Chapters 7 and 8 present applications of the abstract theory to concrete initial value problems for linear and nonlinear partial differential equations. Some of the auxiliary results from the theory of partial differential equations used in these chapters are stated without proof. References where the proofs can be found are given in the bibliographical notes to these chapters.

I am indebted to many good friends who read the lecture notes on which this book is based, corrected errors, and suggested improvements. In particular I would like to express my thanks to H. Brezis, M.G. Crandall, and P. Rabinowitz for their valuable advice, and to Danit Sharon for the tedious work of typing the manuscript.

A. PAZY

# Contents

# CHAPTER 1

# Generation and Representation

## 1.1. Uniformly Continuous Semigroups of Bounded Linear Operators

**Definition 1.1.** Let $X$ be a Banach space. A one parameter family $T(t)$, $0 \leq t < \infty$, of bounded linear operators from $X$ into $X$ is a *semigroup of bounded linear operators on $X$* if

(i) $T(0) = I$, ($I$ is the identity operator on $X$).
(ii) $T(t + s) = T(t)T(s)$ for every $t$, $s \geq 0$ (the semigroup property).

A semigroup of bounded linear operators, $T(t)$, is *uniformly continuous* if

$$\lim_{t \downarrow 0} \| T(t) - I \| = 0. \tag{1.1}$$

The linear operator $A$ defined by

$$D(A) = \left\{ x \in X : \lim_{t \downarrow 0} \frac{T(t)x - x}{t} \text{ exists} \right\} \tag{1.2}$$

and

$$Ax = \lim_{t \downarrow 0} \frac{T(t)x - x}{t} = \left. \frac{d^+ T(t)x}{dt} \right|_{t=0} \quad \text{for} \quad x \in D(A) \tag{1.3}$$

is the *infinitesimal generator* of the semigroup $T(t)$, $D(A)$ is the domain of $A$.

This section is devoted to the study of uniformly continuous semigroups of bounded linear operators. From the definition it is clear that if $T(t)$ is a uniformly continuous semigroup of bounded linear operators then

$$\lim_{s \to t} \| T(s) - T(t) \| = 0. \tag{1.4}$$

**Theorem 1.2.** *A linear operator $A$ is the infinitesimal generator of a uniformly continuous semigroup if and only if $A$ is a bounded linear operator.*

PROOF. Let $A$ be a bounded linear operator on $X$ and set

$$T(t) = e^{tA} = \sum_{n=0}^{\infty} \frac{(tA)^n}{n!}. \tag{1.5}$$

The right-hand side of (1.5) converges in norm for every $t \geq 0$ and defines, for each such $t$, a bounded linear operator $T(t)$. It is clear that $T(0) = I$ and a straightforward computation with the power series shows that $T(t + s) = T(t)T(s)$. Estimating the power series yields

$$\|T(t) - I\| \leq t\|A\|e^{t\|A\|}$$

and

$$\left\| \frac{T(t) - I}{t} - A \right\| \leq \|A\| \cdot \max_{0 \leq s \leq t} \|T(s) - I\|$$

which imply that $T(t)$ is a uniformly continuous semigroup of bounded linear operators on $X$ and that $A$ is its infinitesimal generator.

Let $T(t)$ be a uniformly continuous semigroup of bounded linear operators on $X$. Fix $\rho > 0$, small enough, such that $\|I - \rho^{-1}\int_0^\rho T(s)\,ds\| < 1$. This implies that $\rho^{-1}\int_0^\rho T(s)\,ds$ is invertible and therefore $\int_0^\rho T(s)\,ds$ is invertible. Now,

$$h^{-1}(T(h) - I)\int_0^\rho T(s)\,ds = h^{-1}\left( \int_0^\rho T(s+h)\,ds - \int_0^\rho T(s)\,ds \right)$$

$$= h^{-1}\left( \int_\rho^{\rho+h} T(s)\,ds - \int_0^h T(s)\,ds \right)$$

and therefore

$$h^{-1}(T(h) - I) = \left( h^{-1}\int_\rho^{\rho+h} T(s)\,ds - h^{-1}\int_0^h T(s)\,ds \right)\left( \int_0^\rho T(s)\,ds \right)^{-1} \tag{1.6}$$

Letting $h \downarrow 0$ in (1.6) shows that $h^{-1}(T(h) - I)$ converges in norm and therefore strongly to the bounded linear operator $(T(\rho) - I)(\int_0^\rho T(s)\,ds)^{-1}$ which is the infinitesimal generator of $T(t)$. $\square$

From Definition 1.1 it is clear that a semigroup $T(t)$ has a unique infinitesimal generator. If $T(t)$ is uniformly continuous its infinitesimal generator is a bounded linear operator. On the other hand, every bounded linear operator $A$ is the infinitesimal generator of a uniformly continuous semigroup $T(t)$. Is this semigroup unique? The affirmative answer to this question is given next.

**Theorem 1.3.** *Let $T(t)$ and $S(t)$ be uniformly continuous semigroups of bounded linear operators. If*

$$\lim_{t \downarrow 0} \frac{T(t) - I}{t} = A = \lim_{t \downarrow 0} \frac{S(t) - I}{t} \tag{1.7}$$

*then $T(t) = S(t)$ for $t \geq 0$.*

PROOF. We will show that given $T > 0$, $S(t) = T(t)$ for $0 \leq t \leq T$. Let $T > 0$ be fixed, since $t \to \|T(t)\|$ and $t \to \|S(t)\|$ are continuous there is a constant $C$ such that $\|T(t)\| \, \|S(s)\| \leq C$ for $0 \leq s, t \leq T$. Given $\varepsilon > 0$ it follows from (1.7) that there is a $\delta > 0$ such that

$$h^{-1}\|T(h) - S(h)\| < \varepsilon/TC \qquad \text{for} \quad 0 \leq h \leq \delta. \tag{1.8}$$

Let $0 \leq t \leq T$ and choose $n \geq 1$ such that $t/n < \delta$. From the semigroup property and (1.8) it then follows that

$$\|T(t) - S(t)\| = \left\| T\left(n\frac{t}{n}\right) - S\left(n\frac{t}{n}\right) \right\|$$

$$\leq \sum_{k=0}^{n-1} \left\| T\left((n-k)\frac{t}{n}\right) S\left(\frac{kt}{n}\right) - T\left((n-k-1)\frac{t}{n}\right) S\left(\frac{(k+1)t}{n}\right) \right\|$$

$$\leq \sum_{k=0}^{n-1} \left\| T\left((n-k-1)\frac{t}{n}\right) \right\| \left\| T\left(\frac{t}{n}\right) - S\left(\frac{t}{n}\right) \right\| \left\| S\left(\frac{kt}{n}\right) \right\| \leq Cn\frac{\varepsilon}{TC}\frac{t}{n} \leq \varepsilon.$$

Since $\varepsilon > 0$ was arbitrary $T(t) = S(t)$ for $0 \leq t \leq T$ and the proof is complete. $\qquad\square$

**Corollary 1.4.** *Let $T(t)$ be a uniformly continuous semigroup of bounded linear operators. Then*

a) *There exists a constant $\omega \geq 0$ such that $\|T(t)\| \leq e^{\omega t}$.*
b) *There exists a unique bounded linear operator $A$ such that $T(t) = e^{tA}$.*
c) *The operator $A$ in part (b) is the infinitesimal generator of $T(t)$.*
d) *$t \to T(t)$ is differentiable in norm and*

$$\frac{dT(t)}{dt} = AT(t) = T(t)A \tag{1.9}$$

PROOF. All the assertions of Corollary 1.4 follow easily from (b). To prove (b) note that the infinitesimal generator of $T(t)$ is a bounded linear operator $A$. $A$ is also the infinitesimal generator of $e^{tA}$ defined by (1.5) and therefore, by Theorem 1.3, $T(t) = e^{tA}$. $\qquad\square$

## 1.2. Strongly Continuous Semigroups of Bounded Linear Operators

Throughout this section $X$ will be a Banach space.

**Definition 2.1.** A semigroup $T(t)$, $0 \leq t < \infty$, of bounded linear operators on $X$ is a *strongly continuous* semigroup of bounded linear operators if

$$\lim_{t \downarrow 0} T(t)x = x \qquad \text{for every} \quad x \in X. \tag{2.1}$$

A strongly continuous semigroup of bounded linear operators on $X$ will be called a *semigroup of class $C_0$* or simply a $C_0$ *semigroup*.

**Theorem 2.2.** *Let $T(t)$ be a $C_0$ semigroup. There exist constants $\omega \geq 0$ and $M \geq 1$ such that*

$$\|T(t)\| \leq Me^{\omega t} \qquad for \quad 0 \leq t < \infty. \tag{2.2}.$$

PROOF. We show first that there is an $\eta > 0$ such that $\|T(t)\|$ is bounded for $0 \leq t \leq \eta$. If this is false then there is a sequence $\{t_n\}$ satisfying $t_n \geq 0$, $\lim_{n \to \infty} t_n = 0$ and $\|T(t_n)\| \geq n$. From the uniform boundedness theorem it then follows that for some $x \in X$, $\|T(t_n)x\|$ is unbounded contrary to (2.1). Thus, $\|T(t)\| \leq M$ for $0 \leq t \leq \eta$. Since $\|T(0)\| = 1$, $M \geq 1$. Let $\omega = \eta^{-1} \log M \geq 0$. Given $t \geq 0$ we have $t = n\eta + \delta$ where $0 \leq \delta < \eta$ and therefore by the semigroup property

$$\|T(t)\| = \|T(\delta)T(\eta)^n\| \leq M^{n+1} \leq MM^{t/\eta} = Me^{\omega t}. \qquad \square$$

**Corollary 2.3.** *If $T(t)$ is a $C_0$ semigroup then for every $x \in X$, $t \to T(t)x$ is a continuous function from $\mathbb{R}_0^+$ (the nonnegative real line) into $X$.*

PROOF. Let $t, h \geq 0$. The continuity of $t \to T(t)x$ follows from

$$\|T(t+h)x - T(t)x\| \leq \|T(t)\| \, \|T(h)x - x\| \leq Me^{\omega t}\|T(h)x - x\|$$

and for $t \geq h \geq 0$

$$\|T(t-h)x - T(t)x\| \leq \|T(t-h)\| \, \|x - T(h)x\|$$

$$\leq Me^{\omega t}\|x - T(h)x\|. \qquad \square$$

**Theorem 2.4.** *Let $T(t)$ be a $C_0$ semigroup and let $A$ be its infinitesimal generator. Then*

a) *For $x \in X$,*

$$\lim_{h \to 0} \frac{1}{h} \int_t^{t+h} T(s)x \, ds = T(t)x. \tag{2.3}$$

b) *For $x \in X$, $\int_0^t T(s)x \, ds \in D(A)$ and*

$$A\left(\int_0^t T(s)x \, ds\right) = T(t)x - x. \tag{2.4}$$

c) *For $x \in D(A)$, $T(t)x \in D(A)$ and*

$$\frac{d}{dt} T(t)x = AT(t)x = T(t)Ax. \tag{2.5}$$

d) *For $x \in D(A)$,*

$$T(t)x - T(s)x = \int_s^t T(\tau)Ax \, d\tau = \int_s^t AT(\tau)x \, d\tau. \tag{2.6}$$

PROOF. Part (a) follows directly from the continuity of $t \to T(t)x$. To prove (b) let $x \in X$ and $h > 0$. Then,

$$\frac{T(h) - I}{h} \int_0^t T(s)x \, ds = \frac{1}{h} \int_0^t (T(s+h)x - T(s)x) \, ds$$

$$= \frac{1}{h} \int_t^{t+h} T(s)x \, ds - \frac{1}{h} \int_0^h T(s)x \, ds$$

and as $h \downarrow 0$ the right-hand side tends to $T(t)x - x$, which proves (b). To prove (c) let $x \in D(A)$ and $h > 0$. Then

$$\frac{T(h) - I}{h} T(t)x = T(t)\left(\frac{T(h) - I}{h}\right)x \to T(t)Ax \qquad \text{as} \quad h \downarrow 0.$$

$$\tag{2.7}$$

Thus, $T(t)x \in D(A)$ and $AT(t)x = T(t)Ax$. (2.7) implies also that

$$\frac{d^+}{dt} T(t)x = AT(t)x = T(t)Ax,$$

i.e., that the right derivative of $T(t)x$ is $T(t)Ax$. To prove (2.5) we have to show that for $t > 0$, the left derivative of $T(t)x$ exists and equals $T(t)Ax$. This follows from,

$$\lim_{h \downarrow 0} \left[ \frac{T(t)x - T(t-h)x}{h} - T(t)Ax \right]$$

$$= \lim_{h \downarrow 0} T(t-h)\left[ \frac{T(h)x - x}{h} - Ax \right] + \lim_{h \downarrow 0} (T(t-h)Ax - T(t)Ax),$$

and the fact that both terms on the right-hand side are zero, the first since $x \in D(A)$ and $\|T(t-h)\|$ is bounded on $0 \le h \le t$ and the second by the strong continuity of $T(t)$. This concludes the proof of (c). Part (d) is obtained by integration of (2.5) from $s$ to $t$. □

**Corollary 2.5.** *If $A$ is the infinitesimal generator of a $C_0$ semigroup $T(t)$ then $D(A)$, the domain of $A$, is dense in $X$ and $A$ is a closed linear operator.*

PROOF. For every $x \in X$ set $x_t = 1/t \int_0^t T(s)x \, ds$. By part (b) of Theorem 2.4, $x_t \in D(A)$ for $t > 0$ and by part (a) of the same theorem $x_t \to x$ as $t \downarrow 0$. Thus $\overline{D(A)}$, the closure of $D(A)$, equals $X$. The linearity of $A$ is evident. To prove its closedness let $x_n \in D(A)$, $x_n \to x$ and $Ax_n \to y$ as $n \to \infty$. From part (d) of Theorem 2.4 we have

$$T(t)x_n - x_n = \int_0^t T(s)Ax_n \, ds. \tag{2.8}$$

The integrand on the right-hand side of (2.8) converges to $T(s)y$ uniformly on bounded intervals. Consequently letting $n \to \infty$ in (2.8) yields

$$T(t)x - x = \int_0^t T(s)y \, ds. \tag{2.9}$$

Dividing (2.9) by $t > 0$ and letting $t \downarrow 0$, we see, using part (a) of Theorem 2.4, that $x \in D(A)$ and $Ax = y$. $\square$

**Theorem 2.6.** *Let $T(t)$ and $S(t)$ be $C_0$ semigroups of bounded linear operators with infinitesimal generators $A$ and $B$ respectively. If $A = B$ then $T(t) = S(t)$ for $t \geq 0$.*

PROOF. Let $x \in D(A) = D(B)$. From Theorem 2.4 (c) it follows easily that the function $s \to T(t - s)S(s)x$ is differentiable and that

$$\frac{d}{ds}T(t - s)S(s)x = -AT(t - s)S(s)x + T(t - s)BS(s)x$$

$$= -T(t - s)AS(s)x + T(t - s)BS(s)x = 0.$$

Therefore $s \to T(t - s)S(s)x$ is constant and in particular its values at $s = 0$ and $s = t$ are the same, i.e., $T(t)x = S(t)x$. This holds for every $x \in D(A)$ and since, by Corollary 2.5, $D(A)$ is dense in $X$ and $T(t), S(t)$ are bounded, $T(t)x = S(t)x$ for every $x \in X$. $\square$

If $A$ is the infinitesimal generator of a $C_0$ semigroup then by Corollary 2.5, $\overline{D(A)} = X$. Actually, a much stronger result is true. Indeed we have,

**Theorem 2.7.** *Let $A$ be the infinitesimal generator of the $C_0$ semigroup $T(t)$. If $D(A^n)$ is the domain of $A^n$, then $\bigcap_{n=1}^{\infty} D(A^n)$ is dense in $X$.*

PROOF. Let $\mathcal{D}$ be the set of all infinitely differentiable compactly supported complex valued functions on $]0, \infty[$. For $x \in X$ and $\varphi \in \mathcal{D}$ set

$$y = x(\varphi) = \int_0^{\infty} \varphi(s)T(s)x \, ds. \tag{2.10}$$

If $h > 0$ then

$$\frac{T(h) - I}{h}y = \frac{1}{h}\int_0^{\infty} \varphi(s)[T(s + h)x - T(s)x] \, ds$$

$$= \int_0^{\infty} \frac{1}{h}[\varphi(s - h) - \varphi(s)]T(s)x \, ds. \tag{2.11}$$

The integrand on the right-hand side of (2.11) converges as $h \downarrow 0$ to $-\varphi'(s)T(s)x$ uniformly on $[0, \infty[$. Therefore $y \in D(A)$ and

$$Ay = \lim_{h \downarrow 0} \frac{T(h) - I}{h} y = -\int_0^\infty \varphi'(s)T(s)x \, ds.$$

Clearly, if $\varphi \in \mathcal{D}$ then $\varphi^{(n)}$, the $n$-th derivative of $\varphi$, is also in $\mathcal{D}$ for $n = 1, 2, \ldots$. Thus, repeating the previous argument we find that $y \in D(A^n)$

$$A^n y = (-1)^n \int_0^\infty \varphi^{(n)}(s)T(s)x \, ds \qquad \text{for} \quad n = 1, 2, \ldots$$

and consequently $y \in \bigcap_{n=1}^\infty D(A^n)$. Let $Y$ be the linear span of $\{x(\varphi) : x \in X, \varphi \in \mathcal{D}\}$. $Y$ is clearly a linear manifold. From what we have proved so far it follows that $Y \subseteq \bigcap_{n=1}^\infty D(A^n)$. To conclude the proof we will show that $Y$ is dense in $X$. If $Y$ is not dense in $X$, then by Hahn-Banach's theorem there is a functional $x^* \in X^*$, $x^* \neq 0$ such that $x^*(y) = 0$ for every $y \in Y$ and therefore

$$\int_0^\infty \varphi(s)x^*(T(s)x) \, ds = x^*\left(\int_0^\infty \varphi(s)T(s)x \, ds\right) = 0 \qquad (2.12)$$

for every $x \in X$, $\varphi \in \mathcal{D}$. This implies that for $x \in X$ the continuous function $s \to x^*(T(s)x)$ must vanish identically on $[0, \infty[$ since otherwise, it would have been possible to choose $\varphi \in \mathcal{D}$ such that the left-hand side of (2.12) does not vanish. Thus in particular for $s = 0$, $x^*(x) = 0$. This holds for every $x \in X$ and therefore $x^* = 0$ contrary to the choice of $x^*$. □

We conclude this section with a simple application of Theorem 2.4.

**Lemma 2.8.** *Let $A$ be the infinitesimal generator of a $C_0$ semigroup $T(t)$ satisfying $\|T(t)\| \leq M$ for $t \geq 0$. If $x \in D(A^2)$ then*

$$\|Ax\|^2 \leq 4M^2 \|A^2 x\| \, \|x\|. \qquad (2.13)$$

PROOF. Using (2.6) it is easy to check that for $x \in D(A^2)$

$$T(t)x - x = tAx + \int_0^t (t - s)T(s)A^2 x \, ds.$$

Therefore,

$$\|Ax\| \leq t^{-1}(\|T(t)x\| + \|x\|) + t^{-1} \int_0^t (t - s)\|T(s)A^2 x\| ds$$

$$\leq \frac{2M}{t}\|x\| + \frac{Mt}{2}\|A^2 x\|. \qquad (2.14)$$

Here we used that $M \geq 1$ (since $\|T(0)\| = 1$). If $A^2 x = 0$ then (2.14) implies $Ax = 0$ and (2.13) is satisfied. If $A^2 x \neq 0$ we substitute $t = 2\|x\|^{1/2}\|A^2 x\|^{-1/2}$ in (2.14) and (2.13) follows. □

EXAMPLE 2.9. Let $X$ be the Banach space of bounded uniformly continuous functions on $]-\infty, \infty[$ with the supremum norm. For $f \in X$ we define

$$(T(t)f)(s) = f(t + s).$$

It is easy to check that $T(t)$ is a $C_0$ semigroup satisfying $\|T(t)\| \le 1$ for $t \ge 0$. The infinitesimal generator of $T(t)$ is defined on $D(A) = \{f : f \in X,$ $f'$ exists, $f' \in X\}$ and $(Af)(s) = f'(s)$ for $f \in D(A)$. From Lemma 2.8 we obtain Landau's inequality

$$(\sup|f'(s)|)^2 \le 4(\sup|f''(s)|)(\sup|f(s)|) \tag{2.15}$$

where the sup are taken over $]-\infty, \infty[$. Example 2.9 can be easily modified to the case where $X = L^p(-\infty, \infty)$, $1 < p < \infty$.

## 1.3. The Hille-Yosida Theorem

Let $T(t)$ be a $C_0$ semigroup. From Theorem 2.2 it follows that there are constants $\omega \ge 0$ and $M \ge 1$ such that $\|T(t)\| \le Me^{\omega t}$ for $t \ge 0$. If $\omega = 0$, $T(t)$ is called *uniformly bounded* and if moreover $M = 1$ it is called a $C_0$ *semigroup of contractions*. This section is devoted to the characterization of the infinitesimal generators of $C_0$ semigroups of contractions. Conditions on the behavior of the resolvent of an operator $A$, which are necessary and sufficient for $A$ to be the infinitesimal generator of a $C_0$ semigroup of contractions, are given.

Recall that if $A$ is a linear, not necessarily bounded, operator in $X$, the resolvent set $\rho(A)$ of $A$ is the set of all complex numbers $\lambda$ for which $\lambda I - A$ is invertible, i.e., $(\lambda I - A)^{-1}$ is a bounded linear operator in $X$. The family $R(\lambda : A) = (\lambda I - A)^{-1}$, $\lambda \in \rho(A)$ of bounded linear operators is called the resolvent of $A$.

**Theorem 3.1** (Hille-Yosida). *A linear (unbounded) operator $A$ is the infinitesimal generator of a $C_0$ semigroup of contractions $T(t)$, $t \ge 0$ if and only if*

(i) *$A$ is closed and $\overline{D(A)} = X$.*
(ii) *The resolvent set $\rho(A)$ of $A$ contains $\mathbb{R}^+$ and for every $\lambda > 0$*

$$\|R(\lambda : A)\| \le \frac{1}{\lambda}. \tag{3.1}$$

PROOF OF THEOREM 3.1 (Necessity). If $A$ is the infinitesimal generator of a $C_0$ semigroup then it is closed and $\overline{D(A)} = X$ by Corollary 2.5. For $\lambda > 0$ and $x \in X$ let

$$R(\lambda)x = \int_0^\infty e^{-\lambda t} T(t)x \, dt. \tag{3.2}$$

Since $t \to T(t)x$ is continuous and uniformly bounded the integral exists as an improper Riemann integral and defines a bounded linear operator $R(\lambda)$ satisfying

$$\|R(\lambda)x\| \le \int_0^\infty e^{-\lambda t}\|T(t)x\| \, dt \le \frac{1}{\lambda}\|x\|. \tag{3.3}$$

Furthermore, for $h > 0$

$$\frac{T(h) - I}{h} R(\lambda)x = \frac{1}{h} \int_0^\infty e^{-\lambda t}(T(t+h)x - T(t)x)\, dt$$

$$= \frac{e^{\lambda h} - 1}{h} \int_0^\infty e^{-\lambda t}T(t)x\, dt - \frac{e^{\lambda h}}{h} \int_0^h e^{-\lambda t}T(t)x\, dt.$$

$$(3.4)$$

As $h \downarrow 0$, the right-hand side of (3.4) converges to $\lambda R(\lambda)x - x$. This implies that for every $x \in X$ and $\lambda > 0$, $R(\lambda)x \in D(A)$ and $AR(\lambda) = \lambda R(\lambda) - I$, or

$$(\lambda I - A)R(\lambda) = I.\qquad (3.5)$$

For $x \in D(A)$ we have

$$R(\lambda)Ax = \int_0^\infty e^{-\lambda t}T(t)Ax\, dt = \int_0^\infty e^{-\lambda t}AT(t)x\, dt \qquad \text{where was}$$
$$\text{closedness of}$$
$$A \text{ used.}$$
$$= A\left(\int_0^\infty e^{-\lambda t}T(t)x\, dt\right) = AR(\lambda)x.\qquad (3.6)$$

Here we used Theorem 2.4 (c) and the closedness of $A$. From (3.5) and (3.6) it follows that

$$R(\lambda)(\lambda I - A)x = x \qquad \text{for} \quad x \in D(A).\qquad (3.7)$$

Thus, $R(\lambda)$ is the inverse of $\lambda I - A$, it exists for all $\lambda > 0$ and satisfies the desired estimate (3.1). Conditions (i) and (ii) are therefore necessary.   $\square$

In order to prove that the conditions (i) and (ii) are sufficient for $A$ to be the infinitesimal generator of a $C_0$ semigroup of contractions we will need some lemmas.

**Lemma 3.2.** *Let $A$ satisfy the conditions* (i) *and* (ii) *of Theorem* 3.1 *and let* $R(\lambda : A) = (\lambda I - A)^{-1}$. *Then*                                  $(Proof\ 2/3/97)$
$$(\lambda I - A)^{-1}$$
$$\lim_{\lambda \to \infty} \lambda R(\lambda : A)x = x \qquad \text{for} \quad x \in X.\qquad (3.8)$$
$$?$$

PROOF. Suppose first that $x \in D(A)$. Then

$$\|\lambda R(\lambda : A)x - x\| = \|AR(\lambda : A)x\| \qquad \lambda(\lambda I - A)^{-1}x - x = (\lambda(\lambda I - A)^{-1} - I)x$$
$$= (\lambda I - A)(\lambda I - (\lambda I - A))x$$
$$= R_\lambda(A)Ax$$
$$= \|R(\lambda : A)Ax\| \le \frac{1}{\lambda}\|Ax\| \to 0 \qquad \text{as} \quad \lambda \to \infty.$$

But $D(A)$ is dense in $X$ and $\|\lambda R(\lambda : A)\| \le 1$. Therefore $\lambda R(\lambda : A)x \to x$ as $\lambda \to \infty$ for every $x \in X$.   $\nexists \|R(\lambda)$   $\square$

We now define, for every $\lambda > 0$, the *Yosida approximation* of $A$ by

$$A_\lambda = \lambda AR(\lambda : A) = \lambda^2 R(\lambda : A) - \lambda I.\qquad (3.9)$$

$A_\lambda$ is an approximation of $A$ in the following sense:

**Lemma 3.3.** *Let $A$ satisfy the conditions* (i) *and* (ii) *of Theorem* 3.1. *If $A_\lambda$ is the Yosida approximation of $A$, then*

$$\lim_{\lambda \to \infty} A_\lambda x = Ax \quad for \quad x \in D(A). \tag{3.10}$$

PROOF. For $x \in D(A)$ we have by Lemma 3.2 and the definition of $A_\lambda$ that

$$\lim_{\lambda \to \infty} A_\lambda x = \lim_{\lambda \to \infty} \lambda R(\lambda : A) Ax = Ax. \qquad \square$$

**Lemma 3.4.** *Let $A$ satisfy the conditions* (i) *and* (ii) *of Theorem* 3.1. *If $A_\lambda$ is the Yosida approximation of $A$, then $A_\lambda$ is the infinitesimal generator of a uniformly continuous semigroup of contractions $e^{tA_\lambda}$. Furthermore, for every $x \in X$, $\lambda, \mu > 0$ we have*

$$\|e^{tA_\lambda}x - e^{tA_\mu}x\| \le t\|A_\lambda x - A_\mu x\|. \tag{3.11}$$

PROOF. From (3.9) it is clear that $A_\lambda$ is a bounded linear operator and thus is the infinitesimal generator of a uniformly continuous semigroup $e^{tA_\lambda}$ of bounded linear operators (see e.g., Theorem 1.2). Also,

$$\|e^{tA_\lambda}\| = e^{-t\lambda}\|e^{t\lambda^2 R(\lambda:A)}\| \le e^{-t\lambda}e^{t\lambda^2\|R(\lambda:A)\|} \le 1 \tag{3.12}$$

and therefore $e^{tA_\lambda}$ is a semigroup of contractions. It is clear from the definitions that $e^{tA_\lambda}$, $e^{tA_\mu}$, $A_\lambda$ and $A_\mu$ commute with each other. Consequently

$$\|e^{tA_\lambda}x - e^{tA_\mu}x\| = \left\| \int_0^1 \frac{d}{ds} \left( e^{tsA_\lambda} e^{t(1-s)A_\mu} x \right) ds \right\|$$

$$\le \int_0^1 t\|e^{tsA_\lambda} e^{t(1-s)A_\mu} (A_\lambda x - A_\mu x)\| ds \le t\|A_\lambda x - A_\mu x\|.$$

$$\square$$

PROOF OF THEOREM 3.1 (Sufficiency). Let $x \in D(A)$. Then

$$\|e^{tA_\lambda}x - e^{tA_\mu}x\| \le t\|A_\lambda x - A_\mu x\| \le t\|A_\lambda x - Ax\| + t\|Ax - A_\mu x\|. \tag{3.13}$$

From (3.13) and Lemma 3.3 it follows that for $x \in D(A)$, $e^{tA_\lambda}x$ converges as $\lambda \to \infty$ and the convergence is uniform on bounded intervals. Since $D(A)$ is dense in $X$ and $\|e^{tA_\lambda}\| \le 1$, it follows that

$$\lim_{\lambda \to \infty} e^{tA_\lambda}x = T(t)x \quad for \ every \quad x \in X. \tag{3.14}$$

The limit in (3.14) is again uniform on bounded intervals. From (3.14) it follows readily that the limit $T(t)$ satisfies the semigroup property, that $T(0) = I$ and that $\|T(t)\| \le 1$. Also $t \to T(t)x$ is continuous for $t \ge 0$ as a uniform limit of the continuous functions $t \to e^{tA_\lambda}x$. Thus $T(t)$ is a $C_0$

semigroup of contractions on $X$. To conclude the proof we will show that $A$ is the infinitesimal generator of $T(t)$. Let $x \in D(A)$. Then using (3.14) and Theorem 2.4 we have

$$T(t)x - x = \lim_{\lambda \to \infty} \left( e^{tA_\lambda}x - x \right) = \lim_{\lambda \to \infty} \int_0^t e^{sA_\lambda} A_\lambda x \, ds = \int_0^t T(s) Ax \, ds.$$

(3.15)

The last equality follows from the uniform convergence of $e^{tA_\lambda} A_\lambda x$ to $T(t) Ax$ on bounded intervals. Let $B$ be the infinitesimal generator of $T(t)$ and let $x \in D(A)$. Dividing (3.15) by $t > 0$ and letting $t \downarrow 0$ we see that $x \in D(B)$ and that $Bx = Ax$. Thus $B \supseteq A$. Since $B$ is the infinitesimal generator of $T(t)$, it follows from the necessary conditions that $1 \in \rho(B)$. On the other hand, we assume (assumption (ii)) that $1 \in \rho(A)$. Since $B \supseteq A$, $(I - B)D(A) = (I - A)D(A) = X$ which implies $D(B) = (I - B)^{-1}X = D(A)$ and therefore $A = B$.  $\square$

Theorem 3.1 and its proof have some simple consequences which we now state.

**Corollary 3.5.** *Let $A$ be the infinitesimal generator of a $C_0$ semigroup of contractions $T(t)$. If $A_\lambda$ is the Yosida approximation of $A$, then*

$$T(t)x = \lim_{\lambda \to \infty} e^{tA_\lambda}x \quad for \quad x \in X.$$

(3.16)

PROOF. From the proof of Theorem 3.1 it follows that the right-hand side of (3.16) defines a $C_0$ semigroup of contractions, $S(t)$, whose infinitesimal generator is $A$. From Theorem 2.6 it then follows that $T(t) = S(t)$.  $\square$

**Corollary 3.6.** *Let $A$ be the infinitesimal generator of a $C_0$ semigroup of contractions $T(t)$. The resolvent set of $A$ contains the open right half-plane, i.e., $\rho(A) \supseteq \{\lambda : \operatorname{Re} \lambda > 0\}$ and for such $\lambda$*

$$\|R(\lambda : A)\| \le \frac{1}{\operatorname{Re} \lambda}.$$

(3.17)

PROOF. The operator $R(\lambda)x = \int_0^\infty e^{-\lambda t} T(t)x \, dt$ is well-defined for $\lambda$ satisfying $\operatorname{Re} \lambda > 0$. In the proof of the necessary part of Theorem 3.1 it was shown that $R(\lambda) = (\lambda I - A)^{-1}$ and therefore $\rho(A) \supseteq \{\lambda : \operatorname{Re} \lambda > 0\}$. The estimate (3.17) for $R(\lambda)$ is obvious.  $\square$

The following example shows that the resolvent set of the infinitesimal generator of a $C_0$ semigroup of contractions need not contain more than the open right half-plane.

EXAMPLE 3.7. Let $X = BU(0, \infty)$, that is, the space of all bounded uniformly continuous functions on $[0, \infty[$. Define

$$(T(t)f)(s) = f(t + s).$$

(3.18)

$T(t)$ is a $C_0$ semigroup of contractions on $X$. Its infinitesimal generator $A$ is given by

$$D(A) = \{f : f \text{ and } f' \in X\} \tag{3.19}$$

and

$$(Af)(s) = f'(s) \quad \text{for} \quad f \in D(A). \tag{3.20}$$

From Corollary 3.6 we know that $\rho(A) \supseteq \{\lambda : \operatorname{Re} \lambda > 0\}$. For every complex $\lambda$ the equation $(\lambda - A)\varphi_\lambda = 0$ has the nontrivial solution $\varphi_\lambda(s) = e^{\lambda s}$. If $\operatorname{Re} \lambda \le 0$, $\varphi_\lambda \in X$ and therefore the closed left half-plane is in the spectrum $\sigma(A)$ of $A$.

Let $T(t)$ be a $C_0$ semigroup satisfying $\|T(t)\| \le e^{\omega t}$ (for some $\omega \ge 0$). Consider $S(t) = e^{-\omega t} T(t)$. $S(t)$ is obviously a $C_0$ semigroup of contractions. If $A$ is the infinitesimal generator of $T(t)$ then $A - \omega I$ is the infinitesimal generator of $S(t)$. On the other hand if $A$ is the infinitesimal generator of a $C_0$ semigroup of contractions $S(t)$, then $A + \omega I$ is the infinitesimal generator of a $C_0$ semigroup $T(t)$ satisfying $\|T(t)\| \le e^{\omega t}$. Indeed, $T(t) = e^{\omega t} S(t)$. These remarks lead us to the characterization of the infinitesimal generators of $C_0$ semigroups satisfying $\|T(t)\| \le e^{\omega t}$.

**Corollary 3.8.** *A linear operator $A$ is the infinitesimal generator of a $C_0$ semigroup satisfying $\|T(t)\| \le e^{\omega t}$ if and only if*

(i) *$A$ is closed and $\overline{D(A)} = X$*
(ii) *The resolvent set $\rho(A)$ of $A$ contains the ray $\{\lambda : \operatorname{Im} \lambda = 0, \lambda > \omega\}$ and for such $\lambda$*

$$\|R(\lambda : A)\| \le \frac{1}{\lambda - \omega}. \tag{3.21}$$

We conclude this section with a result that is often useful in proving that a given operator $A$ satisfies the sufficient conditions of the Hille-Yosida theorem (Theorem 3.1) and thus is the infinitesimal generator of a $C_0$ semigroup of contractions.

Let $X$ be a Banach space and let $X^*$ be its dual. We denote the value of $x^* \in X^*$ at $x \in X$ by $\langle x^*, x \rangle$ or $\langle x, x^* \rangle$. If $A$ is a linear operator in $X$ its *numerical range* $S(A)$ is the set

$$S(A) = \{\langle x^*, Ax \rangle : x \in D(A), \|x\| = 1,$$
$$x^* \in X^*, \|x^*\| = 1, \langle x^*, x \rangle = 1\}. \tag{3.22}$$

**Theorem 3.9.** *Let $A$ be a closed linear operator with dense domain $D(A)$ in $X$. Let $S(A)$ be the numerical range of $A$ and let $\Sigma$ be the complement of $\overline{S(A)}$ in $\mathbb{C}$. If $\lambda \in \Sigma$ then $\lambda I - A$ is one-to-one and has closed range. Moreover, if $\Sigma_0$ is a component of $\Sigma$ satisfying $\rho(A) \cap \Sigma_0 \ne \varnothing$ then the spectrum of $A$ is* *contained in the complement $S_0$ of $\Sigma_0$ and*

$$\|R(\lambda:A)\| \leq \frac{1}{d(\lambda:\overline{S(A)})} \tag{3.23}$$

*where $d(\lambda:\overline{S(A)})$ is the distance of $\lambda$ from $\overline{S(A)}$.*

PROOF. Let $\lambda \in \Sigma$. If $x \in D(A)$, $\|x\| = 1$, $x^* \in X^*$, $\|x^*\| = 1$ and $\langle x^*, x \rangle = 1$ then

$$0 < d(\lambda:\overline{S(A)}) \leq |\lambda - \langle x^*, Ax \rangle| = |\langle x^*, \lambda x - Ax \rangle| \leq \|\lambda x - Ax\| \tag{3.24}$$

and therefore $\lambda I - A$ is one-to-one and has closed range. If moreover $\lambda \in \rho(A)$ then (3.24) implies (3.23) and

$$d(\lambda:\overline{S(A)}) \leq \|R(\lambda:A)\|^{-1}. \tag{3.25}$$

It remains to show that if $\Sigma_0$ is a component of $\Sigma$ which has a nonempty intersection with the resolvent set $\rho(A)$ of $A$ then $\sigma(A) \subseteq S_0$. To this end consider the set $\rho(A) \cap \Sigma_0$. This set is obviously open in $\Sigma_0$. But it is also closed in $\Sigma_0$ since $\lambda_n \in \rho(A) \cap \Sigma_0$ and $\lambda_n \to \lambda \in \Sigma_0$ imply for $n$ large enough that $d(\lambda_n:\overline{S(A)}) > \frac{1}{2}d(\lambda:\overline{S(A)}) > 0$ and consequently for $n$ large enough $|\lambda - \lambda_n| < d(\lambda_n:\overline{S(A)})$. From (3.25) it then follows that for large $n$, $\lambda$ is in a ball of radius less than $\|R(\lambda_n:A)\|^{-1}$ centered at $\lambda_n$ which implies that $\lambda \in \rho(A)$ and therefore $\rho(A) \cap \Sigma_0$ is closed in $\Sigma_0$. The connectedness of $\Sigma_0$ then implies that $\rho(A) \cap \Sigma_0 = \Sigma_0$ or $\rho(A) \supseteq \Sigma_0$ which is equivalent to $\sigma(A) \subseteq S_0$ and the proof is complete.     □

## 1.4. The Lumer Phillips Theorem

In the previous section we saw the Hille-Yosida characterization of the infinitesimal generator of a $C_0$ semigroup of contractions. In this section we will see a different characterization of such infinitesimal generators. In order to state and prove the result we need some preliminaries.

Let $X$ be a Banach space and let $X^*$ be its dual. We denote the value of $x^* \in X^*$ at $x \in X$ by $\langle x^*, x \rangle$ or $\langle x, x^* \rangle$. For every $x \in X$ we define the duality set $F(x) \subseteq X^*$ by

$$F(x) = \{x^* : x^* \in X^* \quad \text{and} \quad \langle x^*, x \rangle = \|x\|^2 = \|x^*\|^2\}. \tag{4.1}$$

From the Hahn-Banach theorem it follows that $F(x) \neq \varnothing$ for every $x \in X$.

**Definition 4.1.** A linear operator $A$ is *dissipative* if for every $x \in D(A)$ there is a $x^* \in F(x)$ such that $\mathrm{Re}\langle Ax, x^* \rangle \leq 0$.

A useful characterization of dissipative operators is given next.

**Theorem 4.2.** *A linear operator $A$ is dissipative if and only if*

$$\|(\lambda I - A)x\| \geq \lambda\|x\| \qquad \text{for all} \quad x \in D(A) \quad \text{and} \quad \lambda > 0. \quad (4.2)$$

PROOF. Let $A$ be dissipative, $\lambda > 0$ and $x \in D(A)$. If $x^* \in F(x)$ and $\text{Re}\langle Ax, x^*\rangle \leq 0$ then

$$\|\lambda x - Ax\|\,\|x\| \geq |\langle \lambda x - Ax, x^*\rangle| \geq \text{Re}\langle \lambda x - Ax, x^*\rangle \geq \lambda\|x\|^2$$

and (4.2) follows at once. Conversely, let $x \in D(A)$ and assume that $\lambda\|x\| \leq \|\lambda x - Ax\|$ for all $\lambda > 0$. If $y_\lambda^* \in F(\lambda x - Ax)$ and $z_\lambda^* = y_\lambda^*/\|y_\lambda^*\|$ then $\|z_\lambda^*\| = 1$ and

$$\lambda\|x\| \leq \|\lambda x - Ax\| = \langle \lambda x - Ax, z_\lambda^*\rangle$$
$$= \lambda\,\text{Re}\langle x, z_\lambda^*\rangle - \text{Re}\langle Ax, z_\lambda^*\rangle \leq \lambda\|x\| - \text{Re}\langle Ax, z_\lambda^*\rangle$$

for every $\lambda > 0$. Therefore

$$\text{Re}\langle Ax, z_\lambda^*\rangle \leq 0 \qquad \text{and} \qquad \text{Re}\langle x, z_\lambda^*\rangle \geq \|x\| - \frac{1}{\lambda}\|Ax\|. \quad (4.3)$$

Since the unit ball of $X^*$ is compact in the weak-star topology of $X^*$ the net $z_\lambda^*$, $\lambda \to \infty$, has a weak-star cluster point $z^* \in X^*$, $\|z^*\| \leq 1$. From (4.3) it follows that $\text{Re}\langle Ax, z^*\rangle \leq 0$ and $\text{Re}\langle x, z^*\rangle \geq \|x\|$. But $\text{Re}\langle x, z^*\rangle \leq |\langle x, z^*\rangle| \leq \|x\|$ and therefore $\langle x, z^*\rangle = \|x\|$. Taking $x^* = \|x\|z^*$ we have $x^* \in F(x)$ and $\text{Re}\langle Ax, x^*\rangle \leq 0$. Thus for every $x \in D(A)$ there is an $x^* \in F(x)$ such that $\text{Re}\langle Ax, x^*\rangle \leq 0$ and $A$ is dissipative.         $\square$

**Theorem 4.3** (Lumer-Phillips). *Let $A$ be a linear operator with dense domain $D(A)$ in $X$.*

(a) *If $A$ is dissipative and there is a $\lambda_0 > 0$ such that the range, $R(\lambda_0 I - A)$, of $\lambda_0 I - A$ is $X$, then $A$ is the infinitesimal generator of a $C_0$ semigroup of contractions on $X$.*

(b) *If $A$ is the infinitesimal generator of a $C_0$ semigroup of contractions on $X$ then $R(\lambda I - A) = X$ for all $\lambda > 0$ and $A$ is dissipative. Moreover, for every $x \in D(A)$ and every $x^* \in F(x)$, $\text{Re}\langle Ax, x^*\rangle \leq 0$.*

PROOF. Let $\lambda > 0$, the dissipativeness of $A$ implies by Theorem 4.2 that

$$\|\lambda x - Ax\| \geq \lambda\|x\| \qquad \text{for every} \quad \lambda > 0 \quad \text{and} \quad x \in D(A). \quad (4.4)$$

Since $R(\lambda_0 I - A) = X$, it follows from (4.4) with $\lambda = \lambda_0$ that $(\lambda_0 I - A)^{-1}$ is a bounded linear operator and thus closed. But then $\lambda_0 I - A$ is closed and therefore also $A$ is closed. If $R(\lambda I - A) = X$ for every $\lambda > 0$ then $\rho(A) \supseteq ]0, \infty[$ and $\|R(\lambda : A)\| \leq \lambda^{-1}$ by (4.4). It then follows from the Hille-Yosida theorem that $A$ is the infinitesimal generator of a $C_0$ semigroup of contractions on $X$.

To complete the proof of (a) it remains to show that $R(\lambda I - A) = X$ for all $\lambda > 0$. Consider the set

$$\Lambda = \{\lambda : 0 < \lambda < \infty \qquad \text{and} \qquad R(\lambda I - A) = X\}.$$

Let $\lambda \in \Lambda$. By (4.4), $\lambda \in \rho(A)$. Since $\rho(A)$ is open, a neighborhood of $\lambda$ is in $\rho(A)$. The intersection of this neighborhood with the real line is clearly in $\Lambda$ and therefore $\Lambda$ is open. On the other hand, let $\lambda_n \in \Lambda$, $\lambda_n \to \lambda > 0$. For every $y \in X$ there exists an $x_n \in D(A)$ such that

$$\lambda_n x_n - A x_n = y. \tag{4.5}$$

From (4.4) it follows that $\|x_n\| \leq \lambda_n^{-1}\|y\| \leq C$ for some $C > 0$. Now,

$$\lambda_m \|x_n - x_m\| \leq \|\lambda_m(x_n - x_m) - A(x_n - x_m)\|$$
$$= |\lambda_n - \lambda_m| \, \|x_n\| \leq C|\lambda_n - \lambda_m|. \tag{4.6}$$

Therefore $\{x_n\}$ is a Cauchy sequence. Let $x_n \to x$. Then by (4.5) $A x_n \to \lambda x - y$. Since $A$ is closed, $x \in D(A)$ and $\lambda x - A x = y$. Therefore $R(\lambda I - A) = X$ and $\lambda \in \Lambda$. Thus $\Lambda$ is also closed in $]0, \infty[$ and since $\lambda_0 \in \Lambda$ by assumption $\Lambda \neq \varnothing$ and therefore $\Lambda = ]0, \infty[$. This completes the proof of (a).

If $A$ is the infinitesimal generator of a $C_0$ semigroup of contractions, $T(t)$, on $X$, then by the Hille-Yosida theorem $\rho(A) \supseteq ]0, \infty[$ and therefore $R(\lambda I - A) = X$ for all $\lambda > 0$. Furthermore, if $x \in D(A)$, $x^* \in F(x)$ then

$$|\langle T(t)x, x^* \rangle| \leq \|T(t)x\| \, \|x^*\| \leq \|x\|^2$$

and therefore,

$$\mathrm{Re}\langle T(t)x - x, x^* \rangle = \mathrm{Re}\langle T(t)x, x^* \rangle - \|x\|^2 \leq 0. \tag{4.7}$$

Dividing (4.7) by $t > 0$ and letting $t \downarrow 0$ yields

$$\mathrm{Re}\langle A x, x^* \rangle \leq 0. \tag{4.8}$$

This holds for every $x^* \in F(x)$ and the proof is complete.  □

**Corollary 4.4.** *Let $A$ be a densely defined closed linear operator. If both $A$ and $A^*$ are dissipative, then $A$ is the infinitesimal generator of a $C_0$ semigroup of contractions on $X$.*

PROOF. By Theorem 4.3(a) it suffices to prove that $R(I - A) = X$. Since $A$ is dissipative and closed $R(I - A)$ is a closed subspace of $X$. If $R(I - A) \neq X$ then there exists $x^* \in X^*$, $x^* \neq 0$ such that $\langle x^*, x - A x \rangle = 0$ for $x \in D(A)$. This implies $x^* - A^* x^* = 0$. Since $A^*$ is also dissipative it follows from Theorem 4.2. that $x^* = 0$, contradicting the construction of $x^*$.  □

We conclude this section with some properties of dissipative operators.

**Theorem 4.5.** *Let $A$ be a dissipative operator in $X$.*

(a) *If for some $\lambda_0 > 0$, $R(\lambda_0 I - A) = X$ then $R(\lambda I - A) = X$ for all $\lambda > 0$.*
(b) *If $A$ is closable then $\bar{A}$, the closure of $A$, is also dissipative.*
(c) *If $\overline{D(A)} = X$ then $A$ is closable.*

PROOF. The assertion (a) was proved in the proof of part (a) of Theorem 4.3. To prove (b) let $x \in D(\bar{A})$, $y = \bar{A}x$. Then there is a sequence $\{x_n\}$ $x_n \in D(A)$, such that $x_n \to x$ and $Ax_n \to y = \bar{A}x$. From Theorem 4.2 it follows that $\|\lambda x_n - Ax_n\| \geq \lambda\|x_n\|$ for $\lambda > 0$ and letting $n \to \infty$ we have

$$\|\lambda x - \bar{A}x\| \geq \lambda\|x\| \qquad \text{for} \quad \lambda > 0. \tag{4.9}$$

Since (4.9) holds for every $x \in D(\bar{A})$, $\bar{A}$ is dissipative by Theorem 4.2. To prove (c) assume that $A$ is not closable. Then there is a sequence $\{x_n\}$ such that $x_n \in D(A)$, $x_n \to 0$ and $Ax_n \to y$ with $\|y\| = 1$. From Theorem 4.2 it follows that for every $t > 0$ and $x \in D(A)$

$$\left\|\left(x + t^{-1}x_n\right) - tA\left(x + t^{-1}x_n\right)\right\| \geq \left\|x + t^{-1}x_n\right\|.$$

Letting $n \to \infty$ and then $t \to 0$ yields $\|x - y\| \geq \|x\|$ for every $x \in D(A)$. But this is impossible if $D(A)$ is dense in $X$ and therefore $A$ is closable. $\square$

**Theorem 4.6.** *Let $A$ be dissipative with $R(I - A) = X$. If $X$ is reflexive then $\overline{D(A)} = X$.*

PROOF. Let $x^* \in X^*$ be such that $\langle x^*, x \rangle = 0$ for every $x \in D(A)$. We will show that $x^* = 0$. Since $R(I - A) = X$ it suffices to show that $\langle x^*, x - Ax \rangle = 0$ for every $x \in D(A)$ which is equivalent to $\langle x^*, Ax \rangle = 0$ for every $x \in D(A)$. Let $x \in D(A)$ then by Theorem 4.5 (a) there is an $x_n$ such that $x = x_n - (1/n)Ax_n$. Since $Ax_n = n(x_n - x) \in D(A)$, $x_n \in D(A^2)$ and $Ax = Ax_n - (1/n)A^2x_n$ or $Ax_n = (I - (1/n)A)^{-1}Ax$. From Theorem 4.2 it follows that $\|(I - (1/n)A)^{-1}\| \leq 1$ and therefore $\|Ax_n\| \leq \|Ax\|$. Also, $\|x_n - x\| \leq 1/n\|Ax_n\| \leq 1/n\|Ax\|$ and therefore $x_n \to x$. Since $\|Ax_n\| \leq C$ and $X$ is reflexive there is a subsequence $Ax_{n_k}$ of $Ax_n$ such that $Ax_{n_k} \to y$ weakly. Since $A$ is closed (see Theorem 4.3 (a)) it follows that $y = Ax$. Finally, since $\langle x^*, z \rangle = 0$ for every $z \in D(A)$, we have

$$\langle x^*, Ax_{n_k} \rangle = n_k \langle x^*, x_{n_k} - x \rangle = 0. \tag{4.10}$$

Letting $n_k \to \infty$ in (4.10) yields $\langle x^*, Ax \rangle = 0$. This holds for every $x \in D(A)$ and therefore $x^* = 0$ and $\overline{D(A)} = X$. $\square$

The next example shows that Theorem 4.6 is not true for general Banach spaces.

EXAMPLE 4.7. Let $X = C([0, 1])$, i.e., the continuous functions on $[0, 1]$ with the sup norm. Let $D(A) = \{u : u \in C^1([0, 1])$ and $u(0) = 0\}$ and $Au = -u'$ for $u \in D(A)$. For every $f \in X$ the equation $\lambda u - Au = f$ has a solution $u$ given by

$$u(x) = \int_0^x e^{\lambda(\xi - x)}f(\xi) \, d\xi. \tag{4.11}$$

This shows that $R(\lambda I - A) = X$. From (4.11) it also follows that

$$\lambda |u(x)| \leq (1 - e^{-\lambda x})\|f\| \leq \|\lambda u - Au\|. \qquad (4.12)$$

Taking the sup over $x \in [0, 1]$ of the left-hand side of (4.12) we find that $\lambda \|u\| \leq \|\lambda u - Au\|$ and therefore $A$ is dissipative by Theorem 4.2. But $\overline{D(A)} = \{u : u \in X \text{ and } u(0) = 0\} \neq X = C([0, 1])$.

## 1.5. The Characterization of the Infinitesimal Generators of $C_0$ Semigroups

In the previous two sections we gave two different characterizations of the infinitesimal generators of $C_0$ semigroups of contractions. We saw at the end of Section 1.3 that these characterizations yield characterizations of the infinitesimal generators of $C_0$ semigroups of bounded operators satisfying $\|T(t)\| \leq e^{\omega t}$. We turn now to the characterization of the infinitesimal generators of general $C_0$ semigroups of bounded operators. From Theorem 2.2 it follows that for such semigroups there exist real constants $M \geq 1$ and $\omega$ such that $\|T(t)\| \leq Me^{\omega t}$. Using arguments similar to those used at the end of Section 1.3, we show that in order to characterize the infinitesimal generator in the general case it suffices to characterize the infinitesimal generators of uniformly bounded $C_0$ semigroups. This will be done by renorming the Banach space $X$ so that the uniformly bounded $C_0$ semigroup becomes, in the new norm, a $C_0$ semigroup of contractions and then using the previously proved characterizations of the infinitesimal generators of $C_0$ semigroups of contractions.

We start with a renorming lemma.

**Lemma 5.1.** *Let $A$ be a linear operator for which $\rho(A) \supset ]0, \infty[$. If*

$$\|\lambda^n R(\lambda : A)^n\| \leq M \qquad \text{for} \quad n = 1, 2, \ldots, \lambda > 0, \qquad (5.1)$$

*then there exists a norm $|\cdot|$ on $X$ which is equivalent to the original norm $\|\cdot\|$ on $X$ and satisfies:*

$$\|x\| \leq |x| \leq M\|x\| \qquad \text{for} \quad x \in X \qquad (5.2)$$

*and*

$$|\lambda R(\lambda : A)x| \leq |x| \qquad \text{for} \quad x \in X, \quad \lambda > 0. \qquad (5.3)$$

PROOF. Let $\mu > 0$ and

$$\|x\|_\mu = \sup_{n \geq 0} \|\mu^n R(\mu : A)^n x\|. \qquad (5.4)$$

Then obviously,

$$\|x\| \leq \|x\|_\mu \leq M\|x\| \tag{5.5}$$

and

$$\|\mu R(\mu : A)\|_\mu \leq 1. \tag{5.6}$$

We claim that

$$\|\lambda R(\lambda : A)\|_\mu \leq 1 \qquad \text{for} \quad 0 < \lambda \leq \mu. \tag{5.7}$$

Indeed, if $y = R(\lambda : A)x$ then $y = R(\mu : A)(x + (\mu - \lambda)y)$ and by (5.6),

$$\|y\|_\mu \leq \frac{1}{\mu}\|x\|_\mu + \left(1 - \frac{\lambda}{\mu}\right)\|y\|_\mu$$

whence $\lambda\|y\|_\mu \leq \|x\|_\mu$ as claimed. From (5.5) and (5.7) it follows that

$$\|\lambda^n R(\lambda : A)^n x\| \leq \|\lambda^n R(\lambda : A)^n x\|_\mu \leq \|x\|_\mu \qquad \text{for} \quad 0 < \lambda \leq \mu. \tag{5.8}$$

Taking the sup over $n \geq 0$ on the left-hand side of (5.8) implies that $\|x\|_\lambda \leq \|x\|_\mu$ for $0 < \lambda \leq \mu$. Finally, we define

$$|x| = \lim_{\mu \to \infty} \|x\|_\mu. \tag{5.9}$$

Then, (5.2) follows from (5.5). Taking $n = 1$ in (5.8) we have

$$\|\lambda R(\lambda : A)x\|_\mu \leq \|x\|_\mu$$

and (5.3) follows upon letting $\mu \to \infty$.                                                            $\square$

Lemma 5.1 is closely related to the following observation. Let $\{B_\gamma\}, \gamma \in \Gamma$ be a family of uniformly bounded commuting linear operators. Then there exists an equivalent norm on $X$ for which all the $B_\gamma$ are contractions, if and only if there is a constant $M$ such that

$$\|B_{\gamma_1} B_{\gamma_2} \cdots B_{\gamma_m} x\| \leq M\|x\| \tag{5.10}$$

for every finite subset $\{\gamma_1, \gamma_2, \ldots, \gamma_m\}$ of $\Gamma$. Indeed, it is clear that if there is such an equivalent norm then (5.10) is satisfied. On the other hand if (5.10) is satisfied we define

$$|x| = \sup\|B_{\gamma_1} B_{\gamma_2} \cdots B_{\gamma_m} x\|, \tag{5.11}$$

where the sup is taken over all finite subsets of $\Gamma$ (including the empty set), and $|\cdot|$ is the desired equivalent norm. The weaker condition

$$\|B_\gamma^n x\| \leq M\|x\| \qquad \text{for every} \quad \gamma \in \Gamma \quad \text{and} \quad n \geq 0 \tag{5.12}$$

is not sufficient, in general, to insure the existence of an equivalent norm on $X$ for which all $B_\gamma$ are contractions. In the special case where $\Gamma = \mathbb{R}^+$ and $B_\gamma = R(\gamma : A)$ for some fixed linear operator $A$, the previous lemma shows that the weaker condition (5.12) suffices to insure such an equivalent norm.

**Theorem 5.2.** *A linear operator $A$ is the infinitesimal generator of a $C_0$ semigroup $T(t)$, satisfying $\|T(t)\| \leq M$ ($M \geq 1$), if and only if*

(i) *$A$ is closed and $D(A)$ is dense in $X$.*
(ii) *The resolvent set $\rho(A)$ of $A$ contains $\mathbb{R}^+$ and*

$$\|R(\lambda : A)^n\| \leq M/\lambda^n \quad for \quad \lambda > 0, \quad n = 1, 2, \ldots. \quad (5.13)$$

PROOF. Let $T(t)$ be a $C_0$ semigroup on a Banach space $X$ and let $A$ be its infinitesimal generator. If the norm in $X$ is changed to an equivalent norm, $T(t)$ stays a $C_0$ semigroup on $X$ with the new norm. The infinitesimal generator $A$ does not change nor does the fact that $A$ is closed and densely defined change when we pass to an equivalent norm on $X$. All these are topological properties which are independent of the particular equivalent norm with which $X$ is endowed.

Let $A$ be the infinitesimal generator of a $C_0$ semigroup satisfying $\|T(t)\| \leq M$. Define,

$$|x| = \sup_{t \geq 0} \|T(t)x\|. \quad (5.14)$$

Then

$$\|x\| \leq |x| \leq M\|x\| \quad (5.15)$$

and therefore $|\cdot|$ is a norm on $X$ which is equivalent to the original norm $\|\cdot\|$ on $X$. Furthermore,

$$|T(t)x| = \sup_{s \geq 0} \|T(s)T(t)x\| \leq \sup_{s \geq 0} \|T(s)x\| = |x| \quad (5.16)$$

and $T(t)$ is a $C_0$ semigroup of contractions on $X$ endowed with the norm $|\cdot|$. It follows from the Hille-Yosida theorem and the remarks at the beginning of the proof, that $A$ is closed and densely defined and that $|R(\lambda : A)| \leq \lambda^{-1}$ for $\lambda > 0$. Therefore by (5.15) and (5.16) we have

$$\|R(\lambda : A)^n x\| \leq |R(\lambda : A)^n x| \leq \lambda^{-n}|x| \leq M\lambda^{-n}\|x\|$$

and the conditions (i) and (ii) are necessary.

Let the conditions (i) and (ii) be satisfied. By Lemma 5.1 there exists a norm $|\cdot|$ on $X$ satisfying (5.2) and (5.3). Considering $X$ with this norm, $A$ is a closed densely defined operator with $\rho(A) \supset ]0, \infty[$ and $|R(\lambda : A)| \leq \lambda^{-1}$ for $\lambda > 0$. Thus by the Hille-Yosida theorem, $A$ is the infinitesimal generator of a $C_0$ semigroup of contractions on $X$ endowed with the norm $|\cdot|$. Returning to the original norm, $A$ is again the infinitesimal generator of $T(t)$ and,

$$\|T(t)x\| \leq |T(t)x| \leq |x| \leq M\|x\|$$

so $\|T(t)\| \leq M$ as required. The conditions (i) and (ii) are therefore also sufficient. □

If $T(t)$ is a general $C_0$ semigroup on $X$ then, by Theorem 2.2, there are constants $M \geq 1$ and $\omega$ such that

$$\|T(t)\| \leq Me^{\omega t}. \tag{5.17}$$

Consider the $C_0$ semigroup $S(t) = e^{-\omega t}T(t)$ then $\|S(t)\| \leq M$ and $A$ is the infinitesimal generator of $T(t)$ if and only if $A - \omega I$ is the infinitesimal generator of $S(t)$. Using these remarks together with Theorem 5.2 we obtain

**Theorem 5.3.** *A linear operator $A$ is the infinitesimal generator of a $C_0$ semigroup $T(t)$ satisfying $\|T(t)\| \leq Me^{\omega t}$, if and only if*

(i) *$A$ is closed and $D(A)$ is dense in $X$.*
(ii) *The resolvent set $\rho(A)$ of $A$ contains the ray $]\omega, \infty[$ and*

$$\|R(\lambda : A)^n\| \leq M/(\lambda - \omega)^n \quad for \quad \lambda > \omega, \quad n = 1, 2, \ldots. \tag{5.18}$$

**Remark 5.4.** The condition that every real $\lambda$, $\lambda > \omega$, is in the resolvent set of $A$ together with the estimate (5.18) imply that every complex $\lambda$ satisfying $\operatorname{Re} \lambda > \omega$ is in the resolvent set of $A$ and

$$\|R(\lambda : A)^n\| \leq M/(\operatorname{Re} \lambda - \omega)^n \quad for \operatorname{Re} \lambda > \omega, \quad n = 1, 2, \ldots. \tag{5.19}$$

PROOF. We define

$$R(\lambda)x = \int_0^\infty e^{-\lambda t}T(t)x \, dt.$$

Since $\|T(t)\| \leq Me^{\omega t}$, $R(\lambda)$ is well-defined for every $\lambda$ satisfying $\operatorname{Re} \lambda > \omega$. An argument, identical to the argument used in the proof of Theorem 3.1, shows that $R(\lambda) = R(\lambda : A)$. To prove (5.19) we assume that $\operatorname{Re} \lambda > \omega$, then

$$\frac{d}{d\lambda}R(\lambda : A)x = \frac{d}{d\lambda}\int_0^\infty e^{-\lambda t}T(t)x \, dt = -\int_0^\infty te^{-\lambda t}T(t)x \, dt.$$

Proceeding by induction we obtain

$$\frac{d^n}{d\lambda^n}R(\lambda : A)x = (-1)^n \int_0^\infty t^n e^{-\lambda t}T(t)x \, dt. \tag{5.20}$$

On the other hand, from the resolvent identity

$$R(\lambda : A) - R(\mu : A) = (\mu - \lambda)R(\lambda : A)R(\mu : A)$$

it follows that for every $\lambda \in \rho(A)$, $\lambda \to R(\lambda : A)$ is holomorphic and

$$\frac{d}{d\lambda}R(\lambda : A) = -R(\lambda : A)^2. \tag{5.21}$$

Proceeding again by induction we find

$$\frac{d^n R(\lambda : A)}{d\lambda^n} = (-1)^n n! R(\lambda : A)^{n+1}. \tag{5.22}$$

Comparing (5.20) and (5.22) yields

$$R(\lambda : A)^n x = \frac{1}{(n-1)!} \int_0^\infty t^{n-1} e^{-\lambda t} T(t) x \, dt \qquad (5.23)$$

whence

$$\|R(\lambda : A)^n x\| \le \frac{M}{(n-1)!} \int_0^\infty t^{n-1} e^{(\omega - \mathrm{Re}\,\lambda)t} \|x\| dt = \frac{M}{(\mathrm{Re}\,\lambda - \omega)^n} \|x\|.$$

$\square$

We conclude this section by extending the representation formula of Corollary 3.5 to the general case.

**Theorem 5.5.** *Let $A$ be the infinitesimal generator of a $C_0$ semigroup $T(t)$ on $X$. If $A_\lambda$ is the Yosida approximation of $A$, i.e., $A_\lambda = \lambda A R(\lambda : A)$ then*

$$T(t)x = \lim_{\lambda \to \infty} e^{tA_\lambda} x. \qquad (5.24)$$

PROOF. We start with the case where $\|T(t)\| \le M$. In the proof of Theorem 5.2 we exhibited a norm $\|| \cdot \||$ on $X$ which is equivalent to the original norm $\| \cdot \|$ on $X$ and for which $T(t)$ is a $C_0$ semigroup of contractions. From Corollary 3.5 it then follows that $\||e^{tA_\lambda} x - T(t)x\|| \to 0$ as $\lambda \to \infty$ for every $x \in X$. Since $\|| \cdot \||$ is equivalent to $\| \cdot \|$ (5.24) holds in $X$. In the general case where $\|T(t)\| \le Me^{\omega t}$ we have for $\omega \le 0$, $\|T(t)\| \le M$ and therefore by what we have just proved, the result holds. It remains to prove the result for $\omega > 0$. Let $\omega > 0$ and note that $\lambda \to \|e^{tA_\lambda}\|$ is bounded for $\lambda > 2\omega$. Indeed,

$$\|e^{tA_\lambda}\| = e^{-\lambda t} \|e^{\lambda^2 R(\lambda : A)t}\|$$

$$\le e^{-\lambda t} \sum_{k=0}^\infty \frac{\lambda^{2k} t^k \|R(\lambda : A)^k\|}{k!} \le Me^{(\lambda\omega/\lambda - \omega)t} \le Me^{2\omega t}.$$

$$(5.25)$$

Next we consider the uniformly bounded semigroup $S(t) = e^{-\omega t} T(t)$ whose infinitesimal generator is $A - \omega I$. From the first part of the proof we have

$$T(t)x = \lim_{\lambda \to \infty} e^{t(A - \omega I)_\lambda + \omega t} x \qquad \text{for} \quad x \in X. \qquad (5.26)$$

A simple computation shows that

$$(A - \omega I)_\lambda + \omega I = A_{\lambda + \omega} + H(\lambda)$$

where

$$H(\lambda) = 2\omega I - \omega(\omega + 2\lambda) R(\lambda + \omega : A)$$

$$= \omega[\omega R(\lambda + \omega : A) - 2A R(\lambda + \omega : A)].$$

It is easy to check that $\|H(\lambda)\| \le 2\omega + (2\omega + \lambda^{-1}\omega^2)M$ and that for

$x \in D(A) \; \|H(\lambda)x\| \le M\lambda^{-1}(\omega^2\|x\| + 2\omega\|Ax\|) \to 0$ as $\lambda \to \infty$. Therefore $H(\lambda)x \to 0$ as $\lambda \to \infty$ for every $x \in X$. Since

$$\|e^{tH(\lambda)}x - x\| \le te^{t\|H(\lambda)\|}\|H(\lambda)x\|$$

we have

$$\lim_{\lambda \to \infty} e^{tH(\lambda)}x = x \qquad \text{for} \quad x \in X. \tag{5.27}$$

Finally, since $H(\lambda)$ and $A_{\lambda+\omega}$ commute we have

$$\|e^{tA_\lambda}x - T(t)x\| \le \|e^{tA_\lambda + tH(\lambda-\omega)}x - T(t)x\| + \|e^{tA_\lambda}\| \|e^{tH(\lambda-\omega)}x - x\|. \tag{5.28}$$

As $\lambda \to \infty$ the first term on the right-hand side tends to zero by (5.26) while
the second term tends to zero by (5.25) and (5.27). Therefore

$$\lim_{\lambda \to \infty} e^{tA_\lambda}x = T(t)x \qquad \text{for} \quad x \in X$$

and the proof is complete.                                                            □


## 1.6. Groups of Bounded Operators

**Definition 6.1.** A one parameter family $T(t)$, $-\infty < t < \infty$, of bounded
linear operators on a Banach space $X$ is a $C_0$ *group of bounded operators* if it
satisfies

(i) $T(0) = I$,
(ii) $T(t + s) = T(t)T(s)$ for $-\infty < t, s < \infty$.
(iii) $\lim_{t \to 0} T(t)x = x$ for $x \in X$.

**Definition 6.2.** The infinitesimal generator $A$ of a group $T(t)$ is defined by

$$Ax = \lim_{t \to 0} \frac{T(t)x - x}{t} \tag{6.1}$$

whenever the limit exists; the domain of $A$ is the set of all elements $x \in X$
for which the limit (6.1) exists.

Note that in (6.1) $t \to 0$ from both sides and not only $t \to 0^+$ as in the
case of the infinitesimal generator of a $C_0$ semigroup.

Let $T(t)$ be a $C_0$ group of bounded operators. It is clear from the
definitions that for $t \ge 0$, $T(t)$ is a $C_0$ semigroup of bounded operators
whose infinitesimal generator is $A$. Moreover, for $t \ge 0$, $S(t) = T(-t)$ is
also a $C_0$ semigroup of bounded operators with the infinitesimal generator
$-A$. Thus if $T(t)$ is a $C_0$ group of bounded operators on $X$, both $A$ and $-A$
are infinitesimal generators of $C_0$ semigroups which are denoted by $T_+(t)$
and $T_-(t)$ respectively. Conversely, if $A$ and $-A$ are the infinitesimal
generators of $C_0$ semigroups $T_+(t)$ and $T_-(t)$ then we will see that $A$ is the

infinitesimal generator of a $C_0$ group $T(t)$ given by

$$T(t) = \begin{cases} T_+(t) & \text{for } t \geq 0 \\ T_-(-t) & \text{for } t \leq 0 \end{cases}. \tag{6.2}$$

**Theorem 6.3.** *A is the infinitesimal generator of a $C_0$ group of bounded operators $T(t)$ satisfying $\|T(t)\| \leq Me^{\omega|t|}$ if and only if*

(i) *A is closed and $\overline{D(A)} = X$.*
(ii) *Every real $\lambda$, $|\lambda| > \omega$, is in the resolvent set $\rho(A)$ of A and for such $\lambda$*

$$\|R(\lambda:A)^n\| \leq M(|\lambda| - \omega)^{-n}, \qquad n = 1,2,\dots . \tag{6.3}$$

PROOF. The necessity of the conditions follows from the fact that both $A$ and $-A$ are the infinitesimal generators of $C_0$ semigroups of bounded operators satisfying the estimate $\|T(t)\| \leq Me^{\omega t}$. Since $A$ is the infinitesimal generator of such a semigroup it follows from Theorem 5.3 that $A$ is closed , $\overline{D(A)} = X$ and (6.3) is satisfied for $\lambda > \omega$. Moreover, since $-A$ is also the infinitesimal generator of such a semigroup and clearly $R(\lambda:A) = -R(-\lambda:-A)$ it follows that $\sigma(-A) = -\sigma(A)$ and that (6.3) is satisfied for $-\lambda < -\omega$. The conditions (i) and (ii) are therefore necessary.

If the conditions (i) and (ii) are satisfied it follows from Theorem 5.3 that $A$ and $-A$ are the infinitesimal generators of $C_0$ semigroups $T_+(t)$ and $T_-(t)$ respectively and that $\|T_\pm(t)\| \leq Me^{\omega t}$. The semigroups $T_+(t)$ and $T_-(t)$ commute since clearly $e^{tA_\lambda}$ and $e^{-tA_\mu}$, where $A_\nu$ is the Yosida approximation of $A$, commute and by Theorem 5.5, $T_+(t)x = \lim_{\lambda \to \infty} e^{tA_\lambda}x$ and $T_-(t)x = \lim_{\mu \to \infty} e^{-tA_\mu}x$. If $W(t) = T_+(t)T_-(t)$ then $W(t)$ is a $C_0$ semigroup of bounded operators for $t \geq 0$. For $x \in D(A) = D(-A)$ we have

$$\frac{W(t)x - x}{t} = T_-(t)\frac{T_+(t)x - x}{t} + \frac{T_-(t)x - x}{t}$$
$$\to Ax - Ax = 0 \qquad \text{as } t \downarrow 0. \tag{6.4}$$

Therefore, for $x \in D(A)$ we have $W(t)x = x$. Since $D(A)$ is dense in $X$ and $W(t)$ is bounded we have $W(t) = I$ or $T_-(t) = (T_+(t))^{-1}$. Defining

$$T(t) = \begin{cases} T_+(t) & \text{for } t \geq 0 \\ T_-(-t) & \text{for } t \leq 0 \end{cases} \tag{6.5}$$

we obtain a $C_0$ group of bounded operators satisfying $\|T(t)\| \leq Me^{\omega|t|}$. The conditions (i) and (ii) are therefore sufficient and the proof is complete.
□

**Lemma 6.4.** *Let $T(t)$ be a $C_0$ semigroup of bounded operators. If for every $t > 0$, $T(t)^{-1}$ exists and is a bounded operator then $S(t) = T(t)^{-1}$ is a $C_0$ semigroup of bounded operators whose infinitesimal generator is $-A$.*

*Moreover if*

$$U(t) = \begin{cases} T(t) & \text{for } t \geq 0 \\ T(-t)^{-1} & \text{for } t \leq 0 \end{cases}$$

*then $U(t)$ is a $C_0$ group of bounded operators.*

PROOF. The semigroup property for $S(t)$ is obvious since

$$S(t+s) = T(t+s)^{-1} = (T(t)T(s))^{-1} = T(s)^{-1}T(t)^{-1} = S(s)S(t).$$

We prove the strong continuity of $S(t)$. For $s > 0$ the range of $T(s)$ is all of $X$. Let $x \in X$ and let $s > 1$. There exists a $y \in X$ such that $T(s)y = x$. For $t < 1$ we then have

$$\|T(t)^{-1}x - x\| = \|T(t)^{-1}T(t)T(s-t)y - T(s)y\|$$
$$= \|T(s-t)y - T(s)y\| \to 0 \qquad \text{as } t \downarrow 0.$$

Therefore, $S(t)$ is strongly continuous. Finally, for $x \in D(A)$ we have

$$\lim_{t \downarrow 0} \frac{T(t)^{-1}x - x}{t} = \lim_{t \downarrow 0} T(t) \frac{T(t)^{-1}x - x}{t} = \lim_{t \downarrow 0} \frac{x - T(t)x}{t} = -Ax$$

and therefore $-A$ is the infinitesimal generator of $T(t)^{-1}$. The rest of the proof is obvious.                                                                        □

**Theorem 6.5.** *Let $T(t)$ be a $C_0$ semigroup of bounded operators. If $0 \in \rho(T(t_0))$ for some $t_0 > 0$ then $0 \in \rho(T(t))$ for all $t > 0$ and $T(t)$ can be embedded in a $C_0$ group.*

PROOF. In view of Lemma 6.4 it is sufficient to show that $0 \in \rho(T(t))$ for all $t > 0$. Since $0 \in \rho(T(t_0))$, $T(t_0)^n = T(nt_0)$ is one-to-one for every $n \geq 1$. Let $T(t)x = 0$. Choosing $n$ such that $nt_0 > t$ we have $T(nt_0)x = T(nt_0 - t)T(t)x = 0$ which implies $x = 0$. Thus $T(t)$ is one-to-one for every $t > 0$. We show next that $R(T(t)) = X$ for every $t > 0$. This is clear for $t \leq t_0$ since by the semigroup property $R(T(t)) \supset R(T(t_0))$ for $t \leq t_0$. For $t > t_0$ let $t = kt_0 + t_1$ with $0 \leq t_1 < t_0$. Then $T(t) = T(t_0)^k T(t_1)$ and therefore, again, $R(T(t)) = X$. Thus $T(t)$ is one-to-one and $R(T(t)) = X$ for every $t > 0$ and by the closed graph theorem $0 \in \rho(T(t))$ for all $t > 0$.   □

**Theorem 6.6.** *Let $T(t)$ be a $C_0$ semigroup of bounded operators. If for some $s_0 > 0$, $T(s_0) - I$ is compact, then $T(t)$ is invertible for every $t > 0$ and $T(t)$ can be embedded in a $C_0$ group.*

PROOF. In view of Theorem 6.5 it suffices to prove that $T(s_0)$ is invertible. If $T(s_0)$ is not invertible then $0 \in \sigma(T(s_0))$ but by our assumption $T(s_0) - I$ is compact and so $0$ is an eigenvalue of $T(s_0)$ with finite multiplicity. Let $x \neq 0$ be such that $T(s_0)x = 0$. Set $s_1 = s_0/2$ then $T(s_1)T(s_1)x = T(s_0)x = 0$ and $0$ is an eigenvalue of $T(s_1)$. Proceeding by induction we define a sequence $s_n \downarrow 0$ such that $0$ is an eigenvalue of $T(s_n)$. If $N(T(t))$ is the null space of $T(t)$ then clearly $N(T(s)) \subset N(T(t))$ for $s \leq t$. Let $Q_n = N(T(s_n)) \cap \{x: \|x\| = 1\}$. $Q_n$ is a decreasing sequence of closed nonempty subsets of $X$. Since $N(T(s_0))$ has finite dimension, $Q_0$ is compact and consequently $\cap_{n=0}^{\infty} Q_n \neq \varnothing$. If $x \in \cap_{n=0}^{\infty} Q_n$ then

$$\| T(s_n)x - x \| = \|x\| = 1 \qquad \text{for all } s_n. \tag{6.6}$$

But $s_n \rightarrow 0$ as $n \rightarrow \infty$ and therefore (6.6) contradicts the strong continuity of $T(t)$. This contradiction shows that $T(s_0)$ must be invertible and the proof is complete.                                                    □

## 1.7. The Inversion of the Laplace Transform

One of the fundamental problems in the theory of semigroups of operators is the relation between the semigroup and its infinitesimal generator. Given a semigroup $T(t)$ one obtains its infinitesimal generator, by definition, as

$$Ax = \lim_{t \downarrow 0} \frac{T(t)x - x}{t} \qquad \text{for } x \in D(A).$$

A different way of obtaining $A$, or rather the resolvent of $A$, is by Remark 5.4. There we showed that if $\| T(t) \| \leq M e^{\omega t}$ then

$$R(\lambda : A)x = \int_0^{\infty} e^{-\lambda t} T(t)x \, dt \qquad \text{for } x \in X, \text{Re } \lambda > \omega. \tag{7.1}$$

From the point of view of applications to partial differential equations it is more interesting to obtain $T(t)$ from its infinitesimal generator. The reason for this is that for $x \in D(A)$, $T(t)x$ is the solution of the initial value problem

$$\frac{du}{dt} - Au = 0, \qquad u(0) = x.$$

This section and the next one are dedicated to the problem of representing $T(t)$ in terms of its infinitesimal generator. One way of doing this has already been exhibited in Theorem 5.5. Here we will use a different method. If $T(t)$ satisfies $\| T(t) \| \leq M e^{\omega t}$ then the resolvent of $A$ satisfies (7.1), i.e., the resolvent of $A$ is the Laplace transform of the semigroup. We therefore expect to obtain the semigroup $T(t)$ from the resolvent of $A$ by inverting the Laplace transform. This will be done in this section. We start with some preliminaries.

**Lemma 7.1.** Let $B$ be a bounded linear operator. If $\gamma > \|B\|$ then

$$e^{tB} = \frac{1}{2\pi i} \int_{\gamma - i\infty}^{\gamma + i\infty} e^{\lambda t} R(\lambda : B) \, d\lambda. \tag{7.2}$$

*The convergence in (7.2) is in the uniform operator topology and uniformly in $t$ on bounded intervals.*

PROOF. Let $\gamma > \|B\|$. Choose $r$ such that $\gamma > r > \|B\|$ and let $C_r$ be the circle of radius $r$ centered at the origin. For $|\lambda| > r$ we have

$$R(\lambda : B) = \sum_{k=0}^{\infty} \frac{B^k}{\lambda^{k+1}} \qquad (7.3)$$

where the convergence is in the uniform operator topology uniformly for $|\lambda| \geq r$. Multiplying (7.3) by $(1/2\pi i)e^{\lambda t}$ and integrating over $C_r$ term by term yields.

$$e^{tB} = \frac{1}{2\pi i} \int_{C_r} e^{\lambda t} R(\lambda : B) \, d\lambda. \qquad (7.4)$$

Here we used the identities

$$\frac{1}{2\pi i} \int_{C_r} \lambda^{-k-1} e^{\lambda t} \, d\lambda = \frac{t^k}{k!} \qquad \text{for} \quad k = 0, 1, 2, \dots . \qquad (7.5)$$

Since outside $C_r$ the integrand of (7.4) is analytic and $\|R(\lambda : B)\| \leq C|\lambda|^{-1}$, we can shift the path of integration from $C_r$ to the line Re $z = \gamma$, using Cauchy's theorem. $\qquad\square$

**Lemma 7.2.** Let $A$ be the infinitesimal generator of a $C_0$ semigroup $T(t)$ satisfying $\|T(t)\| \leq Me^{\omega t}$. Let $\mu$ be real, $\mu > \omega \geq 0$, and let

$$A_\mu = \mu A R(\mu : A) = \mu^2 R(\mu : A) - \mu I \qquad (7.6)$$

be the Yosida approximation of $A$. Then for Re $\lambda > \omega\mu/(\mu - \omega)$ we have

$$R(\lambda : A_\mu) = (\lambda + \mu)^{-1}(\mu I - A) R\left(\frac{\mu\lambda}{\mu + \lambda} : A\right) \qquad (7.7)$$

and

$$\|R(\lambda : A_\mu)\| \leq M\left(\text{Re } \lambda - \frac{\omega\mu}{\mu - \omega}\right)^{-1}. \qquad (7.8)$$

For Re $\lambda > \varepsilon + \omega\mu/(\mu - \omega)$ and $\mu > 2\omega$, there is a constant $C$ depending only on $M$ and $\varepsilon$ such that for every $x \in D(A)$

$$\|R(\lambda : A_\mu)x\| \leq \frac{C}{|\lambda|}(\|x\| + \|Ax\|). \qquad (7.9)$$

PROOF. Multiplying the right-hand side of (7.7) from the right or from the left by $\lambda I - A_\mu$ and using the commutativity of $A$ and its resolvent, one obtains the identity, thus proving (7.7). To prove (7.8) we note that $A_\mu$ is the infinitesimal generator of $e^{tA_\mu}$ and that by (5.25)

$$\|e^{tA_\mu}\| \leq M \exp\left\{ t\left(\frac{\omega\mu}{\mu - \omega}\right)\right\}$$

which implies (7.8) by Theorem 5.3. Finally for Re $\lambda > \varepsilon + \omega\mu/(\mu - \omega)$ it follows from (7.8) that $\|R(\lambda : A_\mu)\| \le M\varepsilon^{-1}$. If $x \in D(A)$ and $\mu > 2\omega$ then,

$$\|A_\mu x\| = \|\mu R(\mu : A) Ax\| \le \frac{\mu M}{\mu - \omega} \|Ax\| \le 2M\|Ax\|$$

and therefore

$$\|R(\lambda : A_\mu)x\| = \left\| \frac{x}{\lambda} + \frac{R(\lambda : A_\mu)A_\mu x}{\lambda} \right\| \le \frac{1}{|\lambda|}\left( \|x\| + \frac{2M^2}{\varepsilon}\|Ax\| \right)$$

$$\le \frac{C}{|\lambda|}(\|x\| + \|Ax\|). \qquad \square$$

**Lemma 7.3.** *Let $A$ be as in Lemma 7.2, $\lambda = \gamma + i\eta$ where $\gamma > \omega + \varepsilon$ is fixed. For every $x \in X$ we have*

$$\lim_{\mu \to \infty} R(\lambda : A_\mu)x = R(\lambda : A)x \qquad (7.10)$$

*and for every $Y > 0$, the limit is uniform in $\eta$ for $|\eta| \le Y$.*

PROOF. Set $\nu = \mu\lambda/(\mu + \lambda)$. From (7.7) we then have for $\mu$ large enough

$$R(\lambda : A_\mu) - R(\lambda : A)$$

$$= (\mu + \lambda)^{-1}[(\mu I - A)R(\nu : A) - (\mu + \lambda)R(\lambda : A)]$$

$$= (\mu + \lambda)^{-1}(\mu I - A)R(\nu : A)[(\lambda I - A)(\mu I - A)$$

$$- (\mu + \lambda)(\nu I - A)]R(\mu : A)R(\lambda : A)$$

$$= (\mu + \lambda)^{-1}(\mu I - A)R(\nu : A)A^2 R(\mu : A)R(\lambda : A)$$

$$= (\mu + \lambda)^{-1}A^2 R(\nu : A)R(\lambda : A).$$

For $\gamma > \omega + \varepsilon$ Theorem 5.3 implies $\|R(\lambda : A)\| \le M\varepsilon^{-1}$. Given $Y > 0$, we can find $\mu_0$ depending on $Y$ and $\gamma$ such that if $\lambda = \gamma + i\eta$, $|\eta| \le Y$ and $\mu > \mu_0$ then $\operatorname{Re} \mu\lambda/(\mu + \lambda) > \omega + \varepsilon/2$. Thus, for $\mu > \mu_0$ we have $\|R(\nu : A)\| \le 2M\varepsilon^{-1}$. Therefore if $x \in D(A^2)$ and $\mu > \mu_0$, we have

$$\|R(\lambda : A_\mu)x - R(\lambda : A)x\| \le \frac{1}{|\mu + \lambda|}\|R(\nu : A)\|\,\|R(\lambda : A)\|\,\|A^2 x\|$$

$$\le \frac{1}{\mu}\frac{2M^2}{\varepsilon^2}\|A^2 x\|$$

and (7.10) follows for $x \in D(A^2)$. Since $D(A^2)$ is dense in $X$ (Theorem 2.7), and since by Lemma 7.2, $\|R(\lambda : A_\mu)\|$ is uniformly bounded for $\operatorname{Re} \lambda > \omega + \varepsilon$ provided that $\mu > \omega + \omega^2/\varepsilon$ and by Theorem 5.3 the same is true for $\|R(\lambda : A)\|$, (7.10) follows for every $x \in X$. $\qquad \square$

**Theorem 7.4.** *Let $A$ be the infinitesimal generator of a $C_0$ semigroup $T(t)$ satisfying $\|T(t)\| \leq Me^{\omega t}$ and let $\gamma > \max(0, \omega)$. If $x \in D(A)$ then*

$$\int_0^t T(s)x\,ds = \frac{1}{2\pi i} \int_{\gamma - i\infty}^{\gamma + i\infty} e^{\lambda t} R(\lambda : A)x\,\frac{d\lambda}{\lambda}, \tag{7.11}$$

*and the integral on the right converges uniformly in $t$ for $t$ in bounded intervals.*

PROOF. Let $\mu > 0$ be fixed and let $\delta > \|A_\mu\|$. Set

$$\rho_k(s) = \frac{1}{2\pi i} \int_{\delta - ik}^{\delta + ik} e^{\lambda s} R(\lambda : A_\mu)x\,d\lambda. \tag{7.12}$$

Integrating both sides of (7.12) from 0 to $t$ and interchanging the order of integration we find

$$\int_0^t \rho_k(s)\,ds = \frac{1}{2\pi i} \int_{\delta - ik}^{\delta + ik} e^{\lambda t} R(\lambda : A_\mu)x\,\frac{d\lambda}{\lambda} - \frac{1}{2\pi i} \int_{\delta - ik}^{\delta + ik} R(\lambda : A_\mu)x\,\frac{d\lambda}{\lambda}. \tag{7.13}$$

Letting $k \to \infty$ it follows from Lemma 7.1 that $\rho_k(t) \to e^{tA_\mu}x$ uniformly on $0 \leq t \leq T$. Also,

$$\lim_{k \to \infty} \int_{\delta - ik}^{\delta + ik} R(\lambda : A_\mu)x\,\frac{d\lambda}{\lambda} = 0. \tag{7.14}$$

This can be seen by integrating $\lambda^{-1} R(\lambda : A_\mu)x$ on the path $\Gamma_k$ composed of $\Gamma_k^{(1)} = \{\gamma + i\eta : -k \leq \eta \leq k\}$ and the semi circle $\Gamma_k^{(2)} = \{\gamma + ke^{i\varphi} : -\pi/2 \leq \varphi \leq \pi/2\}$. From Cauchy's theorem the integral around $\Gamma_k$ is zero. As $k \to \infty$ the integral along $\Gamma_k^{(2)}$ tends to zero since $\|R(\lambda : A_\mu)\| \leq C_\mu |\lambda|^{-1}$ for $|\lambda| \geq \delta$. Therefore passing to the limit as $k \to \infty$ in (7.13) we find

$$\int_0^t e^{sA_\mu}x\,ds = \frac{1}{2\pi i} \int_{\delta - i\infty}^{\delta + i\infty} e^{\lambda t} R(\lambda : A_\mu)x\,\frac{d\lambda}{\lambda}. \tag{7.15}$$

If $\gamma > \max(\omega, 0)$ it is clear from Lemma 7.2 that there is a $\mu_0 > 0$ such that for $\mu \geq \mu_0$ $\{\lambda : \operatorname{Re}\lambda \geq \gamma\}$ is in $\rho(A_\mu)$ and for $x \in D(A)$

$$\|R(\lambda : A_\mu)x\| \leq \frac{C}{|\lambda|}(\|x\| + \|Ax\|) \tag{7.16}$$

where $C$ depends only on $M$ and $\gamma$. Therefore for $\mu \geq \mu_0$ we can shift the path of integration in (7.15) from $\operatorname{Re}\lambda = \delta$ to $\operatorname{Re}\lambda = \gamma$ and obtain

$$\int_0^t e^{sA_\mu}x\,ds = \frac{1}{2\pi i} \int_{\gamma - i\infty}^{\gamma + i\infty} e^{\lambda t} R(\lambda : A_\mu)x\,\frac{d\lambda}{\lambda}. \tag{7.17}$$

From (7.16) it follows that for $x \in D(A)$ the integral

$$\int_{-\infty}^{\infty} e^{\gamma t}\|R(\gamma + i\eta : A_\mu)x\|\,\frac{d\eta}{\sqrt{\gamma^2 + \eta^2}} \tag{7.18}$$

converges uniformly for $\mu \geq \mu_0$ and $t$ on bounded intervals. For $x \in D(A)$

the integral

$$\int_{-\infty}^{\infty} e^{\gamma t} \|R(\gamma + i\eta : A)x\| \frac{d\eta}{\sqrt{\gamma^2 + \eta^2}} \qquad (7.19)$$

also converges uniformly in $t$ on bounded intervals since for $\operatorname{Re} \lambda > \omega$
$\|R(\lambda : A)x\| \leq C|\lambda|^{-1}(\|x\| + \|Ax\|)$. Finally, using Theorem 5.5, it is
clear that as $\mu \to \infty$ the left-hand side of (7.17) converges to $\int_0^t T(s)x \, ds$
whereas by Lemma 7.3, (7.18) and (7.19) the right-hand side converges to
the right-hand side of (7.11) and the proof is complete.                    $\square$

**Corollary 7.5.** *Let $A$ be the infinitesimal generator of a $C_0$ semigroup $T(t)$
satisfying $\|T(t)\| \leq M e^{\omega t}$. Let $\gamma > \max(0, \omega)$. If $x \in D(A^2)$, then*

$$T(t)x = \frac{1}{2\pi i} \int_{\gamma - i\infty}^{\gamma + i\infty} e^{\lambda t} R(\lambda : A)x \, d\lambda \qquad (7.20)$$

*and for every $\delta > 0$, the integral converges uniformly in $t$ for $t \in [\delta, 1/\delta]$.*

PROOF. If $x \in D(A^2)$, then $Ax \in D(A)$. Using Theorem 7.4 for $Ax$ we find

$$T(t)x - x = \int_0^t T(s)Ax \, ds = \frac{1}{2\pi i} \int_{\gamma - i\infty}^{\gamma + i\infty} e^{\lambda t} R(\lambda : A)Ax \frac{d\lambda}{\lambda}$$

$$= \frac{1}{2\pi i} \int_{\gamma - i\infty}^{\gamma + i\infty} e^{\lambda t} \left( R(\lambda : A)x - \frac{x}{\lambda} \right) d\lambda. \qquad (7.21)$$

But

$$\frac{1}{2\pi i} \int_{\gamma - i\infty}^{\gamma + i\infty} e^{\lambda t} x \frac{d\lambda}{\lambda} = x \qquad \text{for} \quad t > 0 \qquad (7.22)$$

and (7.22) converges uniformly in $t$ for $t \in [\delta, 1/\delta]$. Combining (7.21) and
(7.22) gives (7.20).                                                         $\square$

**Corollary 7.6.** *Let $A$ be the infinitesimal generator of a $C_0$ semigroup $T(t)$
satisfying $\|T(t)\| \leq M e^{\omega t}$. Let $\gamma > \max(0, \omega)$. For every $x \in X$ we have*

$$\int_0^t (t - s)T(s)x \, ds = \frac{1}{2\pi i} \int_{\gamma - i\infty}^{\gamma + i\infty} e^{\lambda t} R(\lambda : A)x \frac{d\lambda}{\lambda^2} \qquad (7.23)$$

*and the convergence is uniform in $t$ on bounded intervals.*

PROOF. Integrating (7.11) from 0 to $t$ we obtain

$$\int_0^t (t - s)T(s)x \, ds = \frac{1}{2\pi i} \int_0^t \int_{\gamma - i\infty}^{\gamma + i\infty} e^{\lambda s} R(\lambda : A)x \frac{d\lambda}{\lambda} \, ds$$

$$= \frac{1}{2\pi i} \int_{\gamma - i\infty}^{\gamma + i\infty} (e^{\lambda t} - 1) R(\lambda : A) x \frac{d\lambda}{\lambda^2}.$$

But

$$\frac{1}{2\pi i}\int_{\gamma-i\infty}^{\gamma+i\infty}R(\lambda:A)x\frac{d\lambda}{\lambda^2}=0$$

and therefore (7.23) follows for $x \in D(A)$. The right-hand side of (7.23) converges in the uniform operator topology and therefore defines a bounded linear operator. Since $D(A)$ is dense in $X$, (7.23) holds for every $x \in X$.  □

We conclude this section with an important sufficient (but not necessary) condition for an operator $A$ to be the infinitesimal generator of a $C_0$ semigroup. In contrast to the conditions of Theorem 5.2 and 5.3, the conditions of Theorem 7.7 below, are often rather easy to check for concrete examples.

**Theorem 7.7.** *Let $A$ be a densely defined operator in $X$ satisfying the following conditions.*

(i) *For some $0 < \delta < \pi/2$, $\rho(A) \supset \Sigma_\delta = \{\lambda: |\arg \lambda| < \pi/2 + \delta\} \cup \{0\}$.*
(ii) *There exists a constant $M$ such that*

$$\|R(\lambda:A)\| \le \frac{M}{|\lambda|} \quad for \quad \lambda \in \Sigma_\delta, \quad \lambda \ne 0. \quad (7.24)$$

*Then, $A$ is the infinitesimal generator of a $C_0$ semigroup $T(t)$ satisfying $\|T(t)\| \le C$ for some constant $C$. Moreover*

$$T(t) = \frac{1}{2\pi i}\int_\Gamma e^{\lambda t}R(\lambda:A)\,d\lambda \quad (7.25)$$

*where $\Gamma$ is a smooth curve in $\Sigma_\delta$ running from $\infty e^{-i\vartheta}$ to $\infty e^{i\vartheta}$ for $\pi/2 < \vartheta < \pi/2 + \delta$. The integral (7.25) converges for $t > 0$ in the uniform operator topology.*

PROOF. Set

$$U(t) = \frac{1}{2\pi i}\int_\Gamma e^{\mu t}R(\mu:A)\,d\mu. \quad (7.26)$$

From (7.24) it follows easily that for $t > 0$ the integral in (7.26) converges in the uniform topology. Moreover, since $R(\lambda:A)$ is analytic in $\Sigma_\delta$ we may shift the path of integration in (7.26) to $\Gamma_t$ where $\Gamma_t = \Gamma_1 \cup \Gamma_2 \cup \Gamma_3$ and $\Gamma_1 = \{re^{-i\vartheta}:t^{-1} \le r < \infty\}$, $\Gamma_2 = \{t^{-1}e^{i\varphi}: -\vartheta \le \varphi \le \vartheta\}$ and $\Gamma_3 = \{re^{i\vartheta}:t^{-1} \le r < \infty\}$ without changing the value of the integral. But,

$$\left\|\frac{1}{2\pi i}\int_{\Gamma_3} e^{\mu t}R(\mu:A)\,d\mu\right\| \le \frac{1}{2\pi}\int_{t^{-1}}^\infty e^{-rt\sin(\vartheta-\pi/2)}Mr^{-1}\,dr$$

$$= \frac{M}{2\pi}\int_{\sin(\vartheta-\pi/2)}^\infty e^{-s}\frac{ds}{s} \le C_1.$$

The integral on $\Gamma_1$ is estimated similarly and on $\Gamma_2$ we have

$$\left\| \frac{1}{2\pi i} \int_{\Gamma_2} e^{\mu t} R(\mu : A) \, d\mu \right\| < \frac{M}{2\pi} \int_{-\vartheta}^{\vartheta} e^{\cos \varphi} \, d\varphi \leq C_2.$$

Therefore there is a constant $C$ such that $\|U(t)\| \leq C$ for $0 < t < \infty$. Next we show that for $\lambda > 0$

$$R(\lambda : A) = \int_0^\infty e^{-\lambda t} U(t) \, dt. \tag{7.27}$$

To this end we multiply (7.26) by $e^{-\lambda t}$ and integrate from 0 to $T$. Using Fubini's theorem and the residue theorem we find

$$\int_0^T e^{-\lambda t} U(t) \, dt = \frac{1}{2\pi i} \int_\Gamma \frac{1}{\mu - \lambda} (e^{(\mu - \lambda)T} - 1) R(\mu : A) \, d\mu$$

$$= R(\lambda : A) + \frac{1}{2\pi i} \int_\Gamma e^{(\mu - \lambda)T} \frac{R(\mu : A)}{\mu - \lambda} \, d\mu. \tag{7.28}$$

But,

$$\left\| \int_\Gamma e^{T(\mu - \lambda)} \frac{R(\mu : A)}{\mu - \lambda} \, d\mu \right\| \leq M e^{-T\lambda} \int_\Gamma \frac{|d\mu|}{|\mu| \, |\lambda - \mu|} \to 0 \qquad \text{as} \quad T \to \infty.$$

Therefore, passing to the limit as $T \to \infty$ in (7.28) we obtain (7.27). Since $\|U(t)\| \leq C$ we can differentiate (7.27) $n - 1$ times under the integral sign to find

$$\frac{d^{n-1}}{d\lambda^{n-1}} R(\lambda : A) = (-1)^{n-1} \int_0^\infty t^{n-1} e^{-\lambda t} U(t) \, dt.$$

Since by (5.22)

$$\frac{d^{n-1}}{d\lambda^{n-1}} R(\lambda : A) = (-1)^{n-1} (n-1)! R(\lambda : A)^n$$

we obtain

$$\|R(\lambda : A)^n\| = \left\| \frac{1}{(n-1)!} \int_0^\infty t^{n-1} e^{-\lambda t} U(t) \, dt \right\|$$

$$\leq C \frac{1}{(n-1)!} \int_0^\infty t^{n-1} e^{-\lambda t} \, dt = \frac{C}{\lambda^n}. \tag{7.29}$$

Therefore by Theorem 5.2, $A$ is the infinitesimal generator of a $C_0$ semigroup $T(t)$ satisfying $\|T(t)\| \leq C$. It remains to prove (7.25). Let $x \in D(A^2)$. From Corollary 7.5 it follows that

$$T(t)x = \frac{1}{2\pi i} \int_{\gamma - i\infty}^{\gamma + i\infty} e^{\lambda t} R(\lambda : A) x \, d\lambda. \tag{7.30}$$

Using (7.24) we can shift the path of integration in (7.30) to $\Gamma$ and so

$$T(t)x = \frac{1}{2\pi i} \int_\Gamma e^{\lambda t} R(\lambda : A) x \, d\lambda \qquad (7.31)$$

holds for every $x \in D(A^2)$. Since by the first part of the proof the integral $\int_\Gamma e^{\lambda t} R(\lambda : A) \, d\lambda$ converges in the uniform operator topology and since $D(A^2)$ is dense in $X$ (Theorem 2.7) it follows that (7.31) holds for every $x \in X$ whence the result.                                                         $\square$

## 1.8. Two Exponential Formulas

As we have already mentioned a $C_0$ semigroup $T(t)$ is equal in some sense to $e^{tA}$ where $A$ is the infinitesimal generator of $T(t)$. Equality holds if $A$ is a bounded linear operator. In the case where $A$ is unbounded Theorem 5.5 gives one possible interpretation to the sense in which $T(t)$ "equals" $e^{tA}$. In this section we give two more results of the same nature.

**Theorem 8.1.** *Let $T(t)$ be a $C_0$ semigroup on $X$. If*

$$A(h)x = \frac{T(h)x - x}{h} \qquad (8.1)$$

*then for every $x \in X$ we have*

$$T(t)x = \lim_{h \downarrow 0} e^{tA(h)}x \qquad (8.2)$$

*and the limit is uniform in $t$ on any bounded interval $[0, T]$.*

PROOF. Let $\|T(t)\| \le Me^{\omega t}$ with $\omega \ge 0$ and let $A$ be the infinitesimal generator of $T(t)$. Since for every $h > 0$ $A(h)$ is bounded, $e^{tA(h)}$ is well-defined. Furthermore, since $A(h)$ and $T(t)$ commute, so do $e^{tA(h)}$ and $T(t)$. Also,

$$\|e^{tA(h)}\| \le e^{-t/h} \sum_{k=0}^{\infty} \left(\frac{t}{h}\right)^k \frac{\|T(hk)\|}{k!} \le M \exp\left\{\frac{t}{h}(e^{\omega h} - 1)\right\}.$$

Therefore, for $0 < h \le 1$ we have

$$\|e^{tA(h)}\| \le Me^{t(e^\omega - 1)}.$$

It is easy to verify that for $x \in D(A)$, $e^{(t-s)A(h)}T(s)x$ is differentiable in $s$ and that

$$\frac{d}{ds}\left(e^{(t-s)A(h)}T(s)x\right) = -A(h)e^{(t-s)A(h)}T(s)x + e^{(t-s)A(h)}AT(s)x$$

$$= e^{(t-s)A(h)}T(s)(Ax - A(h)x).$$

Consequently, for $0 < h \le 1$ and $x \in D(A)$ we have

$$\|T(t)x - e^{tA(h)}x\| = \left\| \int_0^t \frac{d}{ds} \left( e^{(t-s)A(h)} T(s)x \right) ds \right\|$$

$$\le \int_0^t \|e^{(t-s)A(h)}\| \, \|T(s)\| \, \|Ax - A(h)x\| ds$$

$$\le tM^2 e^{t(e^\omega + \omega - 1)} \|Ax - A(h)x\|. \tag{8.3}$$

Letting $h \downarrow 0$ in (8.3) yields (8.2) for $x \in D(A)$. Since both $\|e^{tA(h)}\|$ and $\|T(t)\|$ are uniformly bounded on any finite $t$-interval and since $D(A)$ is dense in $X$, (8.2) holds for every $x \in X$. $\qquad\qquad\square$

EXAMPLE 8.2. Let $X = BU(\mathbb{R})$, i.e., $X$ is the space of uniformly continuous bounded functions on $\mathbb{R}$. Let,

$$(T(t)f)(x) = f(x + t) \qquad \text{for} \quad -\infty < x < \infty, \quad 0 \le t < \infty. \tag{8.4}$$

$T(t)$ is a $C_0$ semigroup of contractions on $X$. Its infinitesimal generator $A$ has the domain

$$D(A) = \{ f : f \in X, \, f' \text{ exists and } \, f' \in X \}$$

and on $D(A)$, $Af = f'$. For this semigroup we have

$$(A(h)f)(x) = \frac{f(x + h) - f(x)}{h} = (\Delta_h f)(x).$$

It is easy to verify that

$$(A(h)^k f)(x) = \frac{1}{h^k} \sum_{m=0}^k (-1)^{k-m} \binom{k}{m} f(x + mh) = (\Delta_h^k f)(x).$$

Using Theorem 8.1 we obtain

$$f(x + t) = \lim_{h \downarrow 0} \sum_{k=0}^\infty \frac{t^k}{k!} (\Delta_h^k f)(x). \tag{8.5}$$

The limit in (8.5) exists uniformly with respect to $x$ in $\mathbb{R}$ and uniformly with respect to $t$ on any finite interval. Formula (8.5) is a generalization of Taylor's formula for an $f$ which is merely continuous. Note that if $f$ has $k$ continuous derivatives then

$$\lim_{h \downarrow 0} (\Delta_h^k f)(x) = f^{(k)}(x).$$

**Theorem 8.3** (The exponential formula). *Let $T(t)$ be a $C_0$ semigroup on $X$. If $A$ is the infinitesimal generator of $T(t)$ then*

$$T(t)x = \lim_{n \to \infty} \left( I - \frac{t}{n} A \right)^{-n} x = \lim_{n \to \infty} \left[ \frac{n}{t} R\left( \frac{n}{t} : A \right) \right]^n x \qquad \text{for} \quad x \in X \tag{8.6}$$

*and the limit is uniform in $t$ on any bounded interval.*

PROOF. Assume that $\|T(t)\| \leq Me^{\omega t}$. We have seen that for $\operatorname{Re} \lambda > \omega$, $R(\lambda : A)$ is analytic in $\lambda$ and

$$R(\lambda : A)x = \int_0^\infty e^{-\lambda s}T(s)x\,ds \qquad \text{for} \quad x \in X. \qquad (8.7)$$

Differentiating (8.7) $n$ times with respect to $\lambda$, substituting $s = vt$ and taking $\lambda = n/t$ we find

$$R\left(\frac{n}{t} : A\right)^{(n)}x = (-1)^n t^{n+1}\int_0^\infty (ve^{-v})^n T(tv)x\,dv.$$

But

$$R(\lambda : A)^{(n)} = (-1)^n n! R(\lambda : A)^{n+1}$$

and therefore

$$\left[\frac{n}{t}R\left(\frac{n}{t} : A\right)\right]^{n+1}x = \frac{n^{n+1}}{n!}\int_0^\infty (ve^{-v})^n T(tv)x\,dv.$$

Noting that

$$\frac{n^{n+1}}{n!}\int_0^\infty (ve^{-v})^n\,dv = 1$$

we obtain

$$\left[\frac{n}{t}R\left(\frac{n}{t}:A\right)\right]^{n+1}x - T(t)x = \frac{n^{n+1}}{n!}\int_0^\infty (ve^{-v})^n[T(vt)x - T(t)x]\,dv.$$

$$(8.8)$$

Given $\varepsilon > 0$, we choose $0 < a < 1 < b < \infty$ such that $t \in [0, t_0]$ implies

$$\|T(tv)x - T(t)x\| < \varepsilon \qquad \text{for} \quad a \leq v \leq b.$$

Then we break the integral on the right-hand side of (8.8) into three integrals $I_1, I_2, I_3$ on the intervals $[0, a]$, $[a, b]$ and $[b, \infty[$ respectively. We have

$$\|I_1\| \leq \frac{n^{n+1}}{n!}(ae^{-a})^n\int_0^a \|T(vt)x - T(t)x\|dv,$$

$$\|I_2\| \leq \varepsilon\frac{n^{n+1}}{n!}\int_a^b (ve^{-v})^n\,dv < \varepsilon$$

$$\|I_3\| = \|\frac{n^{n+1}}{n!}\int_b^\infty (ve^{-v})^n(T(tv)x - T(t)x)\,dv\|.$$

Here we used the fact that $ve^{-v} \geq 0$ is monotonically non decreasing for $0 \leq v \leq 1$ and non increasing on $v \geq 1$. Since furthermore $ve^{-v} < e^{-1}$ for $v \neq 1$, $\|I_1\| \to 0$ uniformly in $t \in [0, t_0]$ as $n \to \infty$. Choosing $n > \omega t$ in $I_3$, we see that the integral in the estimate of $I_3$ converges and that $\|I_3\| \to 0$

uniformly in $t \in [0, t_0]$ as $n \to \infty$. Consequently,

$$\limsup_{n \to \infty} \left\| \left[ \frac{n}{t} R\left( \frac{n}{t} : A \right) \right]^{n+1} x - T(t)x \right\| \leq \varepsilon$$

and since $\varepsilon > 0$ was arbitrary we have

$$\lim_{n \to \infty} \left[ \frac{n}{t} R\left( \frac{n}{t} : A \right) \right]^{n+1} x = T(t)x.$$

But by Lemma 3.2

$$\lim_{n \to \infty} \frac{n}{t} R\left( \frac{n}{t} : A \right) x = x$$

and (8.6) follows.                                                                                    □

**Remark 8.4.** In section 7 we saw that $T(t)$ can be obtained from the resolvent of its infinitesimal generator by inverting the Laplace transform. Theorem 8.3 gives us also an inversion of the Laplace transform which is related to the Post-Widder real inversion formula, namely

$$f(t) = \lim_{k \to \infty} \frac{(-1)^k}{k!} \left( \frac{k}{t} \right)^{k+1} \hat{f}^{(k)}\left( \frac{k}{t} \right)$$

where $\hat{f}$ is the Laplace transform of $f$.

**Remark 8.5.** The formula (8.6) has another interesting interpretation. Let $A$ be the infinitesimal generator of a $C_0$ semigroup $T(t)$. Suppose we want to solve the initial value problem

$$\frac{du}{dt} = Au, \qquad u(0) = x. \tag{8.9}$$

A standard way of doing this is to replace (8.9) by

$$\frac{u_n\left( \frac{jt}{n} \right) - u_n\left( \frac{(j-1)t}{n} \right)}{\frac{t}{n}} = Au_n\left( \frac{jt}{n} \right), \qquad u_n(0) = x \tag{8.10}$$

which is an implicit difference approximation of (8.9). The equations (8.10) can be solved explicitly and their solution $u_n(t)$ is given by

$$u_n(t) = \left( I - \frac{t}{n} A \right)^{-n} x$$

$u_n(t)$ is an approximation of the solution of (8.9) at $t$. Theorem 8.3 implies that as $n \to \infty$, $u_n(t) \to T(t)x$. From what we know already it is not difficult to deduce that if $x \in D(A)$, $T(t)x$ is the unique solution of (8.9). Thus the solutions of the difference equations (8.10) converge to the solution of the differential equation (8.9). If $x \notin D(A)$ then (8.9) need not have a solution at all. The solutions of the difference equations do, nevertheless, converge to $T(t)x$ which should be considered as a generalized solution of (8.9) in this case.


## 1.9. Pseudo Resolvents

We have seen that the characterization of the infinitesimal generator of a $C_0$ semigroup on $X$ is usually done in terms of conditions on the resolvent of $A$ (see e.g. Theorems 3.1 and 5.3). This is not an exceptional situation. Indeed, in the study of unbounded linear operators on $X$ it is often more convenient to deal with their resolvent families which consist of bounded linear operators. This short section is devoted to the characterization of the resolvent family of an operator $A$ in $X$ by means of its main properties.

Let $A$ be a closed and densely defined operator on $X$ and let $R(\lambda : A) = (\lambda I - A)^{-1}$ be its resolvent. If $\mu$ and $\lambda$ are in the resolvent set $\rho(A)$ of $A$, then we have the resolvent identity

$$R(\lambda : A) - R(\mu : A) = (\mu - \lambda) R(\lambda : A) R(\mu : A). \qquad (9.1)$$

This identity motivates our next definition.

**Definition 9.1.** Let $\Delta$ be a subset of the complex plane. A family $J(\lambda)$, $\lambda \in \Delta$, of bounded linear operators on $X$ satisfying

$$J(\lambda) - J(\mu) = (\mu - \lambda) J(\lambda) J(\mu) \qquad \text{for} \quad \lambda, \mu \in \Delta \qquad (9.2)$$

is called a *pseudo resolvent* on $\Delta$.

Our main objective in this section is to determine conditions under which there exists a densely defined closed linear operator $A$ such that $J(\lambda)$ is the resolvent family of $A$.

**Lemma 9.2.** *Let $\Delta$ be a subset of $\mathbb{C}$ (the complex plane). If $J(\lambda)$ is a pseudo resolvent on $\Delta$, then $J(\lambda)J(\mu) = J(\mu)J(\lambda)$. The null space $N(J(\lambda))$ and the range, $R(J(\lambda))$, are independent of $\lambda \in \Delta$. $N(J(\lambda))$ is a closed subspace of $X$.*

PROOF. It is evident from (9.2) that $J(\lambda)$ and $J(\mu)$ commute for $\lambda, \mu \in \Delta$. Also rewriting (9.2) in the form

$$J(\lambda) = J(\mu)[I + (\mu - \lambda)J(\lambda)]$$

it is clear that $R(J(\mu)) \supset R(J(\lambda))$ and, by symmetry, we have equality. Similarly $N(J(\lambda)) = N(J(\mu))$. The closedness of $N(J(\lambda))$ is evident.    □

**Theorem 9.3.** *Let $\Delta$ be a subset of $\mathbb{C}$ and let $J(\lambda)$ be a pseudo resolvent on $\Delta$. Then, $J(\lambda)$ is the resolvent of a unique densely defined closed linear operator $A$ if and only if $N(J(\lambda)) = \{0\}$ and $R(J(\lambda))$ is dense in $X$.*

PROOF. Clearly if $J(\lambda)$ is the resolvent of a densely defined closed operator $A$, we have $N(J(\lambda)) = \{0\}$ and $R(J(\lambda)) = D(A)$ is dense in $X$. Assume now that $N(J(\lambda)) = \{0\}$ and $R(J(\lambda))$ is dense in $X$. From $N(J(\lambda)) = \{0\}$ it follows that $J(\lambda)$ is one-to-one. Let $\lambda_0 \in \Delta$ and define

$$A = \lambda_0 I - J(\lambda_0)^{-1}. \tag{9.3}$$

The operator $A$ thus defined is clearly linear, closed and $D(A) = R(J(\lambda_0))$ is dense in $X$. From (9.3) it is clear that

$$(\lambda_0 I - A)J(\lambda_0) = J(\lambda_0)(\lambda_0 I - A) = I, \tag{9.4}$$

and therefore $J(\lambda_0) = R(\lambda_0 : A)$. If $\lambda \in \Delta$ then

$$\begin{aligned}
(\lambda I - A)J(\lambda) &= ((\lambda - \lambda_0)I + (\lambda_0 I - A))J(\lambda) \\
&= ((\lambda - \lambda_0)I + (\lambda_0 I - A))J(\lambda_0)[I - (\lambda - \lambda_0)J(\lambda)] \\
&= I + (\lambda - \lambda_0)[J(\lambda_0) - J(\lambda) - (\lambda - \lambda_0)J(\lambda)J(\lambda_0)] \\
&= I
\end{aligned}$$

and similarly $J(\lambda)(\lambda I - A) = I$. Therefore $J(\lambda) = R(\lambda : A)$ for every $\lambda \in \Delta$. In particular $A$ is independent of $\lambda_0$ and is uniquely determined by $J(\lambda)$. $\square$

We conclude this section with two useful sufficient conditions for a pseudo resolvent to be a resolvent.

**Theorem 9.4.** *Let $\Delta$ be an unbounded subset of $\mathbb{C}$ and let $J(\lambda)$ be a pseudo resolvent on $\Delta$. If $R(J(\lambda))$ is dense in $X$ and there is a sequence $\lambda_n \in \Delta$ such that $|\lambda_n| \to \infty$ and*

$$\|\lambda_n J(\lambda_n)\| \le M \tag{9.5}$$

*for some constant $M$ then $J(\lambda)$ is the resolvent of a unique densely defined closed linear operator $A$.*

PROOF. From (9.5) it follows that $\|J(\lambda_n)\| \to 0$ as $n \to \infty$. Let $\mu \in \Delta$. From (9.2) we deduce that

$$\|(\lambda_n J(\lambda_n) - I)J(\mu)\| \to 0 \qquad \text{as} \quad n \to \infty. \tag{9.6}$$

Therefore, if $x$ is in the range of $J(\mu)$ we have

$$\lambda_n J(\lambda_n)x \to x \qquad \text{as} \quad n \to \infty. \tag{9.7}$$

Since $R(J(\mu))$ is dense in $X$ and $\lambda_n J(\lambda_n)$ are uniformly bounded, we have (9.7) for every $x \in X$. If $x \in N(J(\lambda))$ then $\lambda_n J(\lambda_n)x = 0$ and from (9.7) we deduce that $x = 0$. Thus $N(J(\lambda)) = \{0\}$ and $J(\lambda)$ is the resolvent of a densely defined closed operator $A$ by Theorem 9.3. $\square$

**Corollary 9.5.** *Let $\Delta$ be an unbounded subset of $\mathbb{C}$ and let $J(\lambda)$ be a pseudo resolvent on $\Delta$. If there is a sequence $\lambda_n \in \Delta$ such that $|\lambda_n| \to \infty$ as $n \to \infty$*

*and*

$$\lim_{n \to \infty} \lambda_n J(\lambda_n)x = x \qquad \text{for all} \quad x \in X \tag{9.8}$$

*then $J(\lambda)$ is the resolvent of a unique densely defined closed operator A.*

PROOF. From the uniform boundedness theorem and (9.8) it follows that (9.5) holds. From Lemma 9.2 we know that $R(J(\lambda))$ is independent of $\lambda \in \Delta$ and therefore (9.8) implies that $R(J(\lambda))$ is dense in $X$. Thus, the conditions of Theorem 9.4 hold and $J(\lambda)$ is the resolvent of an operator $A$.
□

## 1.10. The Dual Semigroup

We start with a few preliminaries. Let $X$ be a Banach space with dual $X^*$. We denote by $\langle x^*, x \rangle$ or $\langle x, x^* \rangle$ the value of $x^* \in X^*$ at $x \in X$. Let $S$ be a linear operator with dense domain, $D(S)$, in $X$. Recall that the adjoint $S^*$ of $S$, is a linear operator from $D(S^*) \subset X^*$ into $X^*$ defined as follows: $D(S^*)$ is the set of all elements $x^* \in X^*$ for which there is a $y^* \in X^*$ such that

$$\langle x^*, Sx \rangle = \langle y^*, x \rangle \qquad \text{for all} \quad x \in D(S) \tag{10.1}$$

and if $x^* \in D(S^*)$ then $y^* = S^*x^*$ where $y^*$ is the element of $X^*$ satisfying (10.1). Note that since $D(S)$ is dense in $X$ there is at most one $y^* \in X^*$ for which (10.1) can hold.

**Lemma 10.1.** *Let $S$ be a bounded operator on $X$ then $S^*$ is a bounded operator on $X^*$ and $\|S\| = \|S^*\|$.*

PROOF. For every $x^* \in X^*$, $\langle x^*, Sx \rangle$ is a bounded linear functional on $X$ and so it determines a unique element $y^* \in X^*$ for which $\langle y^*, x \rangle = \langle x^*, Sx \rangle$ and so $D(S^*) = X^*$. Moreover,

$$\|S^*\| = \sup_{\|x^*\| \le 1} \|S^*x^*\| = \sup_{\|x^*\| \le 1} \sup_{\|x\| \le 1} |\langle S^*x^*, x \rangle|$$

$$= \sup_{\|x\| \le 1} \sup_{\|x^*\| \le 1} |\langle x^*, Sx \rangle| = \sup_{\|x\| \le 1} \|Sx\| = \|S\|. \qquad \square$$

**Lemma 10.2.** *Let $A$ be a linear densely defined operator in $X$. If $\lambda \in \rho(A)$ then $\lambda \in \rho(A^*)$ and*

$$R(\lambda : A^*) = R(\lambda : A)^*. \tag{10.2}$$

PROOF. From the definition of the adjoint we have $(\lambda I - A)^* = \lambda I^* - A^*$ where $I^*$ is the identity in $X^*$. Since $R(\lambda : A)$ is a bounded operator $R(\lambda : A)^*$ is a bounded operator on $X^*$ by Lemma 10.1. We will prove that $R(\lambda : A^*)$ exists and that it equals $R(\lambda : A)^*$. First we show that $\lambda I^* - A^*$ is one-to-one. If for some $x^* \ne 0$, $(\lambda I^* - A^*)x^* = 0$ then $0 = \langle(\lambda I^* - A^*)x^*, x \rangle = \langle(\lambda I - A)x, x^* \rangle$ for all $x \in D(A)$. But since $\lambda \in \rho(A)$,

$R(\lambda I - A) = X$ and therefore $x^* = 0$ and $\lambda I^* - A^*$ is one-to-one. Now if $x \in X$, $x^* \in D(A^*)$ then

$$\langle x^*, x \rangle = \langle x^*, (\lambda I - A)R(\lambda : A)x \rangle = \langle (\lambda I^* - A^*)x^*, R(\lambda : A)x \rangle$$

and therefore

$$R(\lambda : A)^*(\lambda I^* - A^*)x^* = x^* \qquad \text{for every} \quad x^* \in D(A^*). \quad (10.3)$$

On the other hand if $x^* \in X^*$ and $x \in D(A)$ then

$$\langle x^*, x \rangle = \langle x^*, R(\lambda : A)(\lambda I - A)x \rangle = \langle R(\lambda : A)^*x^*, (\lambda I - A)x \rangle$$

which implies

$$(\lambda I^* - A^*)R(\lambda : A)^*x^* = x^* \qquad \text{for every} \quad x^* \in X^*. \quad (10.4)$$

From (10.3) and (10.4) it follows that $\lambda \in \rho(A^*)$ and that $R(\lambda : A^*) = R(\lambda : A)^*$. $\qquad \square$

Let $T(t)$, $t \geq 0$, be a $C_0$ semigroup on $X$. For $t > 0$ let $T(t)^*$ be the adjoint operator of $T(t)$. From the definition of the adjoint operator it is clear that the family $T^*(t)$, $t \geq 0$, of bounded operators on $X^*$, satisfies the semigroup property. This family is therefore called the *adjoint semigroup* of $T(t)$. The adjoint semigroup however, need not be a $C_0$ semigroup on $X^*$ since the mapping $T(t) \to T(t)^*$ does not necessarily conserve the strong continuity of $T(t)$. Before we state and prove the main result of this section concerning the relations between the semigroups $T(t)$, $T(t)^*$ and their infinitesimal generators we need one more definition.

**Definition 10.3.** Let $S$ be a linear operator in $X$ and let $Y$ be a subspace of $X$. The operator $\tilde{S}$ defined by $D(\tilde{S}) = \{x \in D(S) \cap Y : Sx \in Y\}$ and $\tilde{S}x = Sx$ for $x \in D(\tilde{S})$ is called *the part of $S$ in $Y$*.

**Theorem 10.4.** *Let $T(t)$ be a $C_0$ semigroup on $X$ with the infinitesimal generator $A$ and let $T(t)^*$ be its adjoint semigroup. If $A^*$ is the adjoint of $A$ and $Y^*$ is the closure of $D(A^*)$ in $X^*$ then the restriction $T(t)^+$ of $T(t)^*$ to $Y^*$ is a $C_0$ semigroup on $Y^*$. The infinitesimal generator $A^+$ of $T(t)^+$ is the part of $A^*$ in $Y^*$.*

PROOF. Since $A$ is the infinitesimal generator of $T(t)$, there are constants $\omega$ and $M$ such that for all real $\lambda$, $\lambda > \omega$, $\lambda \in \rho(A)$ and

$$\|R(\lambda : A)^n\| \leq \frac{M}{(\lambda - \omega)^n} \qquad n = 1, 2, \ldots . \quad (10.5)$$

This is a consequence of Theorem 5.3. From Lemma 10.2 and Lemma 10.1 it follows that if $\lambda > \omega$, $\lambda \in \rho(A^*)$ and

$$\|R(\lambda : A^*)^n\| \leq \frac{M}{(\lambda - \omega)^n} \qquad n = 1, 2, \ldots . \quad (10.6)$$

Let $J(\lambda)$ be the restriction of $R(\lambda : A^*)$ to $Y^*$. Then obviously we have

$$\|J(\lambda)^n\| \leq \frac{M}{(\lambda - \omega)^n},\tag{10.7}$$

$$J(\lambda) - J(\mu) = (\mu - \lambda)J(\lambda)J(\mu) \qquad \text{for} \quad \lambda, \mu > \omega \tag{10.8}$$

and by Lemma 3.2

$$\lim_{\lambda \to \infty} \lambda J(\lambda)x^* = x^* \qquad \text{for every} \quad x^* \in Y^*. \tag{10.9}$$

From (10.8), (10.9) and Corollary 9.5 it follows that $J(\lambda)$ is a resolvent of a closed densely defined operator $A^+$ in $Y^*$. From (10.7) and Theorem 5.3 it follows that $A^+$ is the infinitesimal generator of a $C_0$ semigroup $T(t)^+$ on $Y^*$. For $x \in X$ and $x^* \in Y^*$ we have by the definitions

$$\left\langle x^*, \left(I - \frac{t}{n}A\right)^{-n} x\right\rangle = \left\langle \left(I - \frac{t}{n}A^+\right)^{-n} x^*, x\right\rangle \qquad n = 1, 2, \ldots .$$

$$\tag{10.10}$$

Letting $n \to \infty$ in (10.10) and using Theorem 8.3 we obtain

$$\langle x^*, T(t)x\rangle = \langle T(t)^+ x^*, x\rangle \tag{10.11}$$

and so for $x^* \in Y^*$, $T(t)^*x^* = T(t)^+x^*$ and $T(t)^+$ is the restriction of $T(t)^*$ to $Y^*$.

To conclude the proof we have to show that $A^+$ is the part of $A^*$ in $Y^*$. Let $x^* \in D(A^*)$ be such that $x^* \in Y^*$ and $A^*x^* \in Y^*$. Then $(\lambda I^* - A^*)x^* \in Y^*$ and

$$(\lambda I^* - A^+)^{-1}(\lambda I^* - A^*)x^* = x^*. \tag{10.12}$$

Therefore $x^* \in D(A^+)$ and applying $\lambda I^* - A^+$ on both sides of (10.12) yields $(\lambda I^* - A^*)x^* = (\lambda I^* - A^+)x^*$ and therefore $A^+x^* = A^*x^*$. Thus $A^+$ is the part of $A^*$ in $Y^*$.                                                            □

In the special case where $X$ is a reflexive Banach space we have,

**Lemma 10.5.** *If $S$ is a densely defined closed operator in $X$ then $D(S^*)$ is dense in $X^*$.*

PROOF. If $D(S^*)$ is not dense in $X^*$ then there is an element $x_0 \in X$ such that $x_0 \neq 0$ and $\langle x^*, x_0\rangle = 0$ for every $x^* \in D(S^*)$. Since $S$ is closed its graph in $X \times X$ is closed and does not contain $(0, x_0)$. From the Hahn-Banach theorem it follows that there are $x_1^*, x_2^* \in X^*$ such that $\langle x_1^*, x\rangle - \langle x_2^*, Sx\rangle = 0$ for every $x \in D(S)$ and $\langle x_1^*, 0\rangle - \langle x_2^*, x_0\rangle \neq 0$. From the second equation it follows that $x_2^* \neq 0$ and that $\langle x_2^*, x_0\rangle \neq 0$. But from the first equation it follows that $x_2^* \in D(S^*)$ which implies $\langle x_2^*, x_0\rangle = 0$, a contradiction. Thus $\overline{D(S^*)} = X^*$.                                                            □

As a consequence of Theorem 10.4 and Lemma 10.5 we have,

**Corollary 10.6.** *Let $X$ be a reflexive Banach space and let $T(t)$ be a $C_0$ semigroup on $X$ with infinitesimal generator $A$. The adjoint semigroup $T(t)^*$ of $T(t)$ is a $C_0$ semigroup on $X^*$ whose infinitesimal generator is $A^*$ the adjoint of $A$.*

We conclude this section with a result in Hilbert space.

**Definition 10.7.** Let $H$ be a Hilbert space with scalar product $(\,,\,)$. An operator $A$ in $H$ is *symmetric* if $\overline{D(A)} = H$ and $A \subset A^*$, that is, $(Ax, y) = (x, Ay)$ for all $x, y \in D(A)$. $A$ is *self-adjoint* if $A = A^*$. A bounded operator $U$ on $H$ is *unitary* if $U^* = U^{-1}$.

We recall that any adjoint operator is closed and that $U$ is unitary if and only if $R(U) = H$ and $U$ is an isometry. Both these facts are easy to prove and are left as exercises to the reader.

**Theorem 10.8** (Stone). *$A$ is the infinitesimal generator of a $C_0$ group of unitary operators on a Hilbert space $H$ if and only if $iA$ is self-adjoint.*

PROOF. If $A$ is the infinitesimal generator of a $C_0$ group of unitary operators $U(t)$, then $A$ is densely defined (Corollary 2.5) and for $x \in D(A)$

$$-Ax = \lim_{t \downarrow 0} t^{-1}(U(-t)x - x) = \lim_{t \downarrow 0} t^{-1}(U(t)^*x - x) = A^*x$$

which implies that $A = -A^*$ and therefore $iA = (iA)^*$ and $iA$ is self-adjoint.

If $iA$ is self adjoint then $A$ is densely defined and $A = -A^*$. Thus for every $x \in D(A)$ we have

$$(Ax, x) = (x, A^*x) = -(x, Ax) = -\overline{(Ax,x)}$$

and therefore $\text{Re}(Ax, x) = 0$ for every $x \in D(A)$, i.e., $A$ is dissipative. Since $A = -A^*$ also $\text{Re}(A^*x, x) = 0$ for every $x \in D(A^*) = D(A)$ and also $A^*$ is dissipative. By the remarks preceding the theorem it follows that $A$ and $A^*$ are closed and since $A^{**} = A$, both $A$ and $A^* = -A$ are the infinitesimal generators of $C_0$ semigroups of contractions on $H$ by Corollary 4.4. If $U_+(t)$ and $U_-(t)$ are the semigroups generated by $A$ and $A^*$ respectively we define

$$U(t) = \begin{cases} U_+(t) & \text{for } t \geq 0 \\ U_-(-t) & \text{for } t \leq 0. \end{cases} \tag{10.13}$$

Then $U(t)$ is a group (see Section 1.6) and since $U(t)^{-1} = U(-t)$, $\|U(t)\| \leq 1$, $\|U(-t)\| \leq 1$ it follows that $R(U(t)) = X$ and $U(t)$ is an isometry for every $t$ and thus $U(t)$ is a group of unitary operators on $H$ as desired.  $\square$

# CHAPTER 2

# Spectral Properties and Regularity

## 2.1. Weak Equals Strong

Let $T(t)$ be a $C_0$ semigroup of bounded linear operators on a Banach space $X$. Let $A$ be its infinitesimal generator as defined in Definition 1.1.1. We consider now the operator

$$\tilde{A}x = w - \lim_{h \downarrow 0} \frac{T(h)x - x}{h} \tag{1.1}$$

where $w - \lim$ denotes the weak limit in $X$. The domain of $\tilde{A}$ is the set of all $x \in X$ for which the weak limit on the right-hand side of (1.1) exists. Since the existence of a limit implies the existence of a weak limit, it is clear that $\tilde{A}$ extends $A$. That this extension is not genuine follows from Theorem 1.3 below. In the proof of this theorem we will need the following real variable results.

**Lemma 1.1.** *Let the real valued function $\omega$ be continuous and differentiable from the right on $[a, b[$. Let $D^+\omega$ be the right derivative of $\omega$. If $\omega(a) = 0$ and $D^+\omega(t) \leq 0$ on $[a, b[$ then $\omega(t) \leq 0$ on $[a, b[$.*

PROOF. Assume first that $D^+\omega(t) < 0$. If the result is false then there is a $t_1 \in ]a, b[$ for which $\omega(t_1) > 0$. Let $t_0 = \inf\{t : \omega(t) > 0\}$. By the continuity of $\omega$, $\omega(t_0) = 0$ and by the definition of $t_0$ we have a sequence $\langle t_n \rangle$ such that $t_n \downarrow t_0$ and $\omega(t_n) > 0$. Therefore,

$$D^+\omega(t_0) = \lim_{t_n \downarrow t_0} \frac{\omega(t_n) - \omega(t_0)}{t_n - t_0} \geq 0$$

in contradiction to our assumption that $D^+\omega(t) < 0$ and thus $\omega(t) \leq 0$ on $[a, b[$.

Returning to the general case where $D^+\omega(t) \leq 0$ we consider for every $\varepsilon > 0$ the function $\omega_\varepsilon(t) = \omega(t) - \varepsilon(t - a)$. For $\omega_\varepsilon(t)$ we have $\omega_\varepsilon(a) = 0$ and $D^+\omega_\varepsilon \leq -\varepsilon < 0$. Therefore, by the first part of the proof, $\omega_\varepsilon(t) \leq 0$ on $[a, b[$, i.e., $\omega(t) \leq \varepsilon(t - a)$. Since $\varepsilon > 0$ is arbitrary, $\omega(t) \leq 0$ on $[a, b[$.  □

**Corollary 1.2.** *Let* $\varphi$ *be continuous and differentiable from the right on* $[a, b[$. *If* $D^+\varphi$ *is continuous on* $[a, b[$ *then* $\varphi$ *is continuously differentiable on* $[a, b[$.

PROOF. Let $\psi = D^+\varphi$ and define $\chi(t) = \varphi(a) + \int_a^t \psi(\tau)\, d\tau$. The function $\chi$ thus defined is clearly continuously differentiable on $[a, b[$. Let $\omega(t) = \chi(t) - \varphi(t)$ then $\omega(a) = 0$ and $D^+\omega(t) = 0$ on $[a, b[$. From Lemma 1.1 it then follows that $\omega(t) \leq 0$ on $[a, b[$. Similarly $-\omega(t)$ also satisfies the conditions of Lemma 1.1 and therefore $\omega(t) \geq 0$. Hence $\omega(t) = 0$ on $[a, b[$, i.e., $\varphi(t) = \chi(t)$ and the proof is complete.  □

**Theorem 1.3.** *Let* $T(t)$ *be a* $C_0$ *semigroup of bounded operators and let* $A$ *be its infinitesimal generator. If* $\tilde{A}$ *is the operator defined by* (1.1) *then* $\tilde{A} = A$.

PROOF. From the definitions of $A$ and $\tilde{A}$ it is clear that $\tilde{A} \supset A$. Let $x \in D(\tilde{A})$. Since bounded linear operators are weakly continuous, we have

$$w - \lim_{h \downarrow 0} \frac{T(t + h)x - T(t)x}{h} = w - \lim_{h \downarrow 0} T(t)\left( \frac{T(h)x - x}{h} \right)$$

$$= T(t)\left( w - \lim_{h \downarrow 0} \frac{T(h)x - x}{h} \right) = T(t)\tilde{A}x.$$

$$(1.2)$$

Therefore, if $x \in D(\tilde{A})$ and $x^* \in X^*$ then

$$D^+\langle x^*, T(t)x \rangle = \langle x^*, T(t)\tilde{A}x \rangle, \tag{1.3}$$

i.e., the right derivative of $\langle x^*, T(t)x \rangle$ exists on $[0, \infty[$ and equals $\langle x^*, T(t)\tilde{A}x \rangle$. But $t \to \langle x^*, T(t)\tilde{A}x \rangle$ is continuous in $t$ and so by Corollary 1.2 $\langle x^*, T(t)x \rangle$ is continuously differentiable on $[0, \infty[$ and its derivative is $\langle x^*, T(t)\tilde{A}x \rangle$. Furthermore,

$$\langle x^*, T(t)x - x \rangle = \langle x^*, T(t)x \rangle - \langle x^*, x \rangle = \int_0^t \langle x^*, T(s)\tilde{A}x \rangle\, ds$$

$$= \left\langle x^*, \int_0^t T(s)\tilde{A}x\, ds \right\rangle. \tag{1.4}$$

Since (1.4) holds for every $x^* \in X^*$, it follows from the Hahn-Banach theorem that

$$T(t)x - x = \int_0^t T(s)\tilde{A}x\, ds. \tag{1.5}$$

Dividing (1.5) by $t > 0$ and letting $t \downarrow 0$, we obtain

$$\lim_{t \downarrow 0} \frac{T(t)x - x}{t} = \tilde{A}x. \tag{1.6}$$

Therefore $x \in D(A)$ and $Ax = \tilde{A}x$. This implies $A \supset \tilde{A}$ and thus $A = \tilde{A}$. $\square$

Another result in which weak implies strong is the next theorem which we state here without proof.

**Theorem 1.4.** *If $T(t)$ is a semigroup of bounded linear operators on a Banach space $X$ (Definition 1.1.1) satisfying*

$$w - \lim_{t \downarrow 0} T(t)x = x \qquad \text{for every} \quad x \in X \tag{1.7}$$

*then $T(t)$ is a $C_0$ semigroup of bounded linear operators.*

## 2.2. Spectral Mapping Theorems

Let $T(t)$ be a $C_0$ semigroup on a Banach space $X$ and let $A$ be its infinitesimal generator. In this section we will be interested in the relations between the spectrum of $A$ and the spectrum of each one of the operators $T(t)$, $t \geq 0$. From a purely formal point of view one would expect the relation $\sigma(T(t)) = \exp\{t\sigma(A)\}$. This, however, is not true in general as is shown by the following example.

EXAMPLE 2.1. Let $X$ be the Banach space of continuous functions on $[0, 1]$ which are equal to zero at $x = 1$ with the supremum norm. Define

$$(T(t)f)(x) = \begin{cases} f(x + t) & \text{if} \quad x + t \leq 1 \\ 0 & \text{if} \quad x + t > 1 \end{cases}$$

$T(t)$ is obviously a $C_0$ semigroup of contractions on $X$. Its infinitesimal generator $A$ is given by

$$D(A) = \{f : f \in C^1([0, 1]) \cap X, f' \in X\}$$

and

$$Af = f' \qquad \text{for} \quad f \in D(A).$$

One checks easily that for every $\lambda \in \mathbf{C}$ and $g \in X$ the equation $\lambda f - f' = g$ has a unique solution $f \in X$ given by

$$f(t) = \int_t^1 e^{\lambda(t-s)} g(s) \, ds.$$

Therefore $\sigma(A) = \phi$. On the other hand, since for every $t \geq 0$, $T(t)$ is a bounded linear operator, $\sigma(T(t)) \neq \phi$ for all $t \geq 0$ and the relation $\sigma(T(t)) = \exp\{t\sigma(A)\}$ does not hold for any $t \geq 0$.

**Lemma 2.2.** *Let $T(t)$ be a $C_0$ semigroup and let $A$ be its infinitesimal generator. If*

$$B_\lambda(t)x = \int_0^t e^{\lambda(t-s)}T(s)x\,ds \qquad (2.1)$$

*then*

$$(\lambda I - A)B_\lambda(t)x = e^{\lambda t}x - T(t)x \qquad \text{for every} \quad x \in X \qquad (2.2)$$

*and*

$$B_\lambda(t)(\lambda I - A)x = e^{\lambda t}x - T(t)x \qquad \text{for every} \quad x \in D(A). \qquad (2.3)$$

PROOF. For every fixed $\lambda$ and $t$, $B_\lambda(t)$ defined by (2.1) is a bounded linear operator on $X$. Moreover, for every $x \in X$ we have

$$\frac{T(h) - I}{h}B_\lambda(t)x = \frac{e^{\lambda h} - 1}{h}\int_h^t e^{\lambda(t-s)}T(s)x\,ds + \frac{e^{\lambda h}}{h}\int_t^{t+h}e^{\lambda(t-s)}T(s)x\,ds$$

$$- \frac{1}{h}\int_0^h e^{\lambda(t-s)}T(s)x\,ds. \qquad (2.4)$$

As $h \downarrow 0$ the right-hand side of (2.4) converges to $\lambda B_\lambda(t)x + T(t)x - e^{\lambda t}x$ and consequently $B_\lambda(t)x \in D(A)$ and

$$AB_\lambda(t)x = \lambda B_\lambda(t)x + T(t)x - e^{\lambda t}x \qquad (2.5)$$

which implies (2.2). From the definition of $B_\lambda(t)$ it is clear that for $x \in D(A)$, $AB_\lambda(t)x = B_\lambda(t)Ax$ and (2.3) follows. $\qquad \square$

**Theorem 2.3.** *Let $T(t)$ be a $C_0$ semigroup and let $A$ be its infinitesimal generator. Then,*

$$\sigma(T(t)) \supset e^{t\sigma(A)} \qquad \text{for} \quad t \geq 0. \qquad (2.6)$$

PROOF. Let $e^{\lambda t} \in \rho(T(t))$ and let $Q = (e^{\lambda t}I - T(t))^{-1}$. The operators $B_\lambda(t)$, defined by (2.1), and $Q$ clearly commute. From (2.2) and (2.3) we deduce

$$(\lambda I - A)B_\lambda(t)Qx = x \qquad \text{for every} \quad x \in X \qquad (2.7)$$

and

$$QB_\lambda(t)(\lambda I - A)x = x \qquad \text{for every} \quad x \in D(A). \qquad (2.8)$$

Since $B_\lambda(t)$ and $Q$ commute we also have

$$B_\lambda(t)Q(\lambda I - A)x = x \qquad \text{for every} \quad x \in D(A). \qquad (2.9)$$

Therefore, $\lambda \in \rho(A)$, $B_\lambda(t)Q = (\lambda I - A)^{-1} = R(\lambda : A)$ and $\rho(T(t)) \subset \exp\{t\rho(A)\}$ which implies (2.6). $\qquad \square$

We recall that the spectrum of $A$ consists of three mutually exclusive parts; the point spectrum $\sigma_p(A)$ the continuous spectrum $\sigma_c(A)$ and the residual spectrum $\sigma_r(A)$. These are defined as follows: $\lambda \in \sigma_p(A)$ if $\lambda I - A$

is not one-to-one, $\lambda \in \sigma_c(A)$ if $\lambda I - A$ is one-to-one, $\lambda I - A$ is not onto but its range is dense in $X$ and finally $\lambda \in \sigma_r(A)$ if $\lambda I - A$ is one-to-one and its range is not dense in $X$. From these definitions it is clear that $\sigma_p(A)$, $\sigma_c(A)$ and $\sigma_r(A)$ are mutually exclusive and that their union is $\sigma(A)$. In the rest of this section we will study the relations between each part of the spectrum of $A$ and the corresponding part of the spectrum of $T(t)$. We start with the point spectrum.

**Theorem 2.4.** *Let $T(t)$ be a $C_0$ semigroup and let $A$ be its infinitesimal generator. Then*

$$e^{t\sigma_p(A)} \subset \sigma_p(T(t)) \subset e^{t\sigma_p(A)} \cup \{0\}. \tag{2.10}$$

*More precisely if $\lambda \in \sigma_p(A)$ then $e^{\lambda t} \in \sigma_p(T(t))$ and if $e^{\lambda t} \in \sigma_p(T(t))$ there exists a $k, k \in \mathbb{Z}$ such that $\lambda_k = \lambda + 2\pi i k/t \in \sigma_p(A)$.*

PROOF. If $\lambda \in \sigma_p(A)$ then there is an $x_0 \in D(A)$, $x_0 \neq 0$, such that $(\lambda I - A)x_0 = 0$. From (2.3) it then follows that $(e^{\lambda t}I - T(t))x_0 = 0$ and therefore $e^{\lambda t} \in \sigma_p(T(t))$ which proves the first inclusion. To prove the second inclusion let $e^{\lambda t} \in \sigma_p(T(t))$ and let $x_0 \neq 0$ satisfy $(e^{\lambda t}I - T(t))x_0 = 0$. This implies that the continuous function $s \to e^{-\lambda s}T(s)x_0$ is periodic with period $t$ and since it does not vanish identically one of its Fourier coefficients must be different from zero. Therefore there is a $k, k \in \mathbb{Z}$ such that

$$x_k = \frac{1}{t} \int_0^t e^{-(2\pi i k/t)s}\left(e^{-\lambda s}T(s)x_0\right) ds \neq 0. \tag{2.11}$$

We will show that $\lambda_k = \lambda + 2\pi i k/t$ is an eigenvalue of $A$. Let $\|T(t)\| \leq Me^{\omega t}$. For $\operatorname{Re}\mu > \omega$ we have

$$R(\mu : A)x_0 = \int_0^\infty e^{-\mu s}T(s)x_0 \, ds = \sum_{n=0}^\infty \int_{nt}^{(n+1)t} e^{-\mu s}T(s)x_0 \, ds$$

$$= \sum_{n=0}^\infty e^{n(\lambda-\mu)t} \int_0^t e^{-\mu s}T(s)x_0 \, ds$$

$$= \left(1 - e^{(\lambda-\mu)t}\right)^{-1} \int_0^t e^{-\mu s}T(s)x_0 \, ds \tag{2.12}$$

where we used the periodicity of $e^{-\lambda s}T(s)x_0$. The integral on the right-hand side of (2.12) is clearly an entire function and therefore $R(\mu : A)x_0$ can be extended by (2.12) to a meromorphic function with possible poles at $\lambda_n = \lambda + 2\pi i n/t$, $n \in \mathbb{Z}$. Using (2.12) it is easy to show that

$$\lim_{\mu \to \lambda_k} (\mu - \lambda_k)R(\mu : A)x_0 = x_k \tag{2.13}$$

and

$$\lim_{\mu \to \lambda_k} (\lambda_k I - A)[(\mu - \lambda_k)R(\mu : A)x_0] = 0. \tag{2.14}$$

From the closedness of $A$ and (2.13), (2.14) it follows that $x_k \in D(A)$ and that $(\lambda_k I - A)x_k = 0$, i.e., $\lambda_k \in \sigma_p(A)$.                     $\square$

We turn now to the residual spectrum of $A$.

**Theorem 2.5.** *Let $T(t)$ be a $C_0$ semigroup and let $A$ be its infinitesimal generator. Then,*

(i) *If $\lambda \in \sigma_r(A)$ and none of the $\lambda_n = \lambda + 2\pi in/t$, $n \in \mathbb{Z}$ is in $\sigma_p(A)$ then $e^{\lambda t} \in \sigma_r(T(t))$.*

(ii) *If $e^{\lambda t} \in \sigma_r(T(t))$ then none of the $\lambda_n = \lambda + 2\pi in/t$, $n \in \mathbb{Z}$ is in $\sigma_p(A)$ and there exists a $k$, $k \in \mathbb{Z}$ such that $\lambda_k \in \sigma_r(A)$.*

PROOF. If $\lambda \in \sigma_r(A)$ then there is an $x^* \in X^*$, $x^* \neq 0$, such that $\langle x^*, (\lambda I - A)x \rangle = 0$ for all $x \in D(A)$. From (2.2) it then follows that $\langle x^*, (e^{\lambda t}I - T(t))x \rangle = 0$ for all $x \in X$ and therefore the range of $e^{\lambda t}I - T(t)$ is not dense in $X$. If $e^{\lambda t}I - T(t)$ is not one-to-one then by Theorem 2.4 there is a $k \in \mathbb{Z}$ such that $\lambda_k \in \sigma_p(A)$ contradicting our assumption that $\lambda_n \notin \sigma_p(A)$. Therefore $e^{\lambda t}I - T(t)$ is one-to-one and $e^{\lambda t} \in \sigma_r(T(t))$ which concludes the proof of (i).

To prove (ii) we note first that if for some $k$, $\lambda_k = \lambda + 2\pi ik/t \in \sigma_p(A)$ then by Theorem 2.4 $e^{\lambda t} \in \sigma_p(T(t))$ contradicting the assumption that $e^{\lambda t} \in \sigma_r(T(t))$. It suffices therefore to show that for some $k \in \mathbb{Z}$, $\lambda_k \in \sigma_r(A)$. This follows at once if we show that $\{\lambda_n\} \subset \rho(A) \cup \sigma_c(A)$ is impossible. From (2.3) we have

$$(e^{\lambda_n t}I - T(t))x = B_{\lambda_n}(t)(\lambda_n I - A)x \qquad \text{for} \quad x \in D(A) \qquad n \in \mathbb{Z}.$$
(2.15)

Since by our assumption $e^{\lambda t} = e^{\lambda_n t} \in \sigma_r(T(t))$ the left hand side of (2.15) belongs to a fixed nondense linear subspace $Y$ of $X$. On the other hand if $\lambda_n \in \rho(A) \cup \sigma_c(A)$ then the range of $\lambda_n I - A$ is dense in $X$ which implies by (2.15) that the range of $B_{\lambda_n}(t)$ belongs to $Y$ for every $n \in \mathbb{Z}$. Writing the Fourier series of the continuous function $e^{-\lambda s}T(s)x$ we have

$$e^{-\lambda s}T(s)x \sim \frac{e^{-\lambda t}}{t} \sum_{n=-\infty}^{\infty} e^{(2\pi in/t)s}B_{\lambda_n}(t)x \qquad (2.16)$$

and each term on the right-hand side of (2.16) belongs to $Y$. As in the classical numerical case the series (2.16) is (C.1) summable to $e^{-\lambda s}T(s)x$ for $0 < s < t$ and therefore for $0 < s < t$, $e^{-\lambda s}T(s)x \in Y$. Letting $s \downarrow 0$ it follows that every $x \in D(A)$ satisfies $x \in \overline{Y}$ which is impossible since $\overline{Y}$ is a proper closed subspace of $X$ and $D(A)$ is dense in $X$.                     $\square$

**Theorem 2.6.** *Let $T(t)$ be a $C_0$ semigroup and let $A$ be its infinitesimal generator. If $\lambda \in \sigma_c(A)$ and if none of the $\lambda_n = \lambda + 2\pi in/t$ is in $\sigma_p(A) \cup \sigma_r(A)$ then $e^{\lambda t} \in \sigma_c(T(t))$.*

PROOF. From Theorem 2.3 it follows that if $\lambda \in \sigma_c(A)$ then $e^{\lambda t} \in \sigma(T(t))$. If $e^{\lambda t} \in \sigma_p(T(t))$ then by Theorem 2.4 some $\lambda_k \in \sigma_p(A)$ and therefore $e^{\lambda t} \notin \sigma_p(T(t))$. Similarly if $e^{\lambda t} \in \sigma_r(T(t))$ then some $\lambda_k \in \sigma_r(A)$ and again $e^{\lambda t} \notin \sigma_r(T(t))$.                                                                □

**Remark.** The converse of Theorem 2.6 does not hold. It is possible that $e^{\lambda t} \in \sigma_c(T(t))$ while all $\lambda_n = \lambda + 2\pi i n/t$ are in $\rho(A)$.

## 2.3. Semigroups of Compact Operators

**Definition 3.1.** A $C_0$ semigroup $T(t)$ is called *compact* for $t > t_0$ if for every $t > t_0$, $T(t)$ is a compact operator. $T(t)$ is called *compact* if it is compact for $t > 0$.

Note that if $T(t)$ is compact for $t \geq 0$, then in particular the identity is compact and $X$ is necessarily finite dimensional. Note also that if for some $t_0 > 0$, $T(t_0)$ is compact, then so is $T(t)$ for every $t \geq t_0$ since $T(t) = T(t - t_0)T(t_0)$ and $T(t - t_0)$ is bounded.

**Theorem 3.2.** *Let $T(t)$ be a $C_0$ semigroup. If $T(t)$ is compact for $t > t_0$, then $T(t)$ is continuous in the uniform operator topology for $t > t_0$.*

PROOF. Let $\|T(s)\| \leq M$ for $0 \leq s \leq 1$ and let $\varepsilon > 0$ be given. If $t > t_0$ then the set $U_t = \{T(t)x : \|x\| \leq 1\}$ is compact and therefore, there exist $x_1, x_2, \ldots, x_N$ such that the open balls with radius $\varepsilon/2(M + 1)$ centered at $T(t)x_j$, $1 \leq j \leq N$ cover $U_t$. From the strong continuity of $T(t)$ it is clear that there exists an $0 < h_0 \leq 1$ such that

$$\left\|T(t + h)x_j - T(t)x_j\right\| < \varepsilon/2 \quad \text{for} \quad 0 \leq h \leq h_0 \quad \text{and} \quad 1 \leq j \leq N. \tag{3.1}$$

Let $x \in X$, $\|x\| \leq 1$, then there is an index $j$, $1 \leq j \leq N$ ($j$ depending on $x$) such that

$$\|T(t)x - T(t)x_j\| < \varepsilon/2(M + 1). \tag{3.2}$$

Thus, for $0 \leq h \leq h_0$ and $\|x\| \leq 1$, we have

$$\|T(t + h)x - T(t)x\| \leq \|T(h)\| \|T(t)x - T(t)x_j\|$$
$$+ \|T(t + h)x_j - T(t)x_j\| + \|T(t)x_j - T(t)x\| < \varepsilon \quad (3.3)$$

which proves the continuity of $T(t)$ in the uniform operator topology for $t > t_0$.                                                                          □

**Theorem 3.3.** *Let $T(t)$ be a $C_0$ semigroup and let $A$ be its infinitesimal generator. $T(t)$ is a compact semigroup if and only if $T(t)$ is continuous in the uniform operator topology for $t > 0$ and $R(\lambda : A)$ is compact for $\lambda \in \rho(A)$.*

PROOF. Let $\|T(t)\| \leq Me^{\omega t}$. If $T(t)$ is compact for $t > 0$, then by Theorem 3.2, $T(t)$ is continuous in the uniform operator topology for $t > 0$. Therefore,

$$R(\lambda : A) = \int_0^\infty e^{-\lambda s} T(s)\, ds \qquad \text{for} \quad \operatorname{Re} \lambda > \omega \qquad (3.4)$$

and the integral exists in the uniform operator topology. Let $\varepsilon > 0$, $\operatorname{Re} \lambda > \omega$ and

$$R_\varepsilon(\lambda) = \int_\varepsilon^\infty e^{-\lambda s} T(s)\, ds. \qquad (3.5)$$

Since $T(s)$ is compact for every $s > 0$, $R_\varepsilon(\lambda)$ is compact. But

$$\|R(\lambda : A) - R_\varepsilon(\lambda)\| \leq \left\| \int_0^\varepsilon e^{-\lambda s} T(s)\, ds \right\| \leq \varepsilon M e^{\omega \varepsilon} \to 0 \qquad \text{as} \quad \varepsilon \downarrow 0$$

and therefore $R(\lambda : A)$ is compact as a uniform limit of compact operators. From the resolvent identity

$$R(\lambda : A) - R(\mu : A) = (\mu - \lambda) R(\lambda : A) R(\mu : A) \qquad \lambda, \mu \in \rho(A)$$

it follows that if $R(\mu : A)$ is compact for some $\mu \in \rho(A)$, $R(\lambda : A)$ is compact for every $\lambda \in \rho(A)$. The conditions of the theorem are therefore necessary.

Assume now that $R(\lambda : A)$ is compact for $\lambda \in \rho(A)$ and that $T(t)$ is continuous in the uniform operator topology for $t > 0$. It follows that (3.4) holds and that

$$\lambda R(\lambda : A) T(t) - T(t) = \lambda \int_0^\infty e^{-\lambda s} (T(t + s) - T(t))\, ds. \qquad (3.6)$$

If $\lambda$ is real, $\lambda > \omega$, then for every $\delta > 0$ we have

$$\|\lambda R(\lambda : A) T(t) - T(t)\| \leq \int_0^\delta \lambda e^{-\lambda s} \|T(t + s) - T(t)\| ds$$

$$+ \int_\delta^\infty \lambda e^{-\lambda s} \|T(t + s) - T(t)\| ds$$

$$\leq \sup_{0 \leq s \leq \delta} \|T(t + s) - T(t)\|$$

$$+ 2\lambda (\lambda - \omega)^{-1} M e^{\omega(t + \delta)} e^{-\lambda \delta}$$

which implies

$$\limsup_{\lambda \to \infty} \|\lambda R(\lambda : A) T(t) - T(t)\| \leq \sup_{0 \leq s \leq \delta} \|T(t + s) - T(t)\|$$

$$\text{for every} \quad \delta > 0 \quad (3.7)$$

Since $\delta > 0$ is arbitrary we have

$$\lim_{\lambda \to \infty} \|\lambda R(\lambda : A) T(t) - T(t)\| = 0.$$

But $\lambda R(\lambda : A) T(t)$ is compact for every $\lambda > \omega$ and therefore $T(t)$ is compact. $\qquad \square$

**Corollary 3.4.** *Let $T(t)$ be a $C_0$ semigroup and let $A$ be its infinitesimal generator. If $R(\lambda : A)$ is compact for some $\lambda \in \rho(A)$ and $T(t)$ is continuous in the uniform operator topology for $t > t_0$, then $T(t)$ is compact for $t > t_0$.*

PROOF. From our assumptions it follows that $R(\lambda : A)$ is compact for every $\lambda \in \rho(A)$ and that (3.6) holds for every $t > t_0$. The rest of the proof is identical to the end of the proof of Theorem 3.3.                                                          □

**Corollary 3.5.** *Let $T(t)$ be a uniformly continuous semigroup (Definition 1.1.1). $T(t)$ is a compact semigroup if and only if $R(\lambda : A)$ is compact for every $\lambda \in \rho(A)$.*

The characterization of compact semigroups in Theorem 3.3 is not completely satisfactory since it does not characterize the compact semigroup $T(t)$ solely in terms of properties of its infinitesimal generator $A$. The reason for this is that so far, there are no known necessary and sufficient conditions, in terms of $A$ or the resolvent $R(\lambda : A)$, which assure the continuity for $t > 0$ of $T(t)$ in the uniform operator topology. A necessary condition for $T(t)$ to be continuous, in the uniform operator topology, for $t > 0$ is given next.

**Theorem 3.6.** *Let $T(t)$ be a $C_0$ semigroup and let $A$ be its infinitesimal generator. If $T(t)$ is continuous in the uniform operator topology for $t > 0$, then there exists a function $\psi : [0, \infty[ \to [0, \infty[$ such that*

$$\rho(A) \supset \{\lambda : \lambda = \sigma + i\tau, |\tau| \geq \psi(|\sigma|)\}, \tag{3.8}$$

*and*

$$\lim_{|\tau| \to \infty} \|R(\sigma + i\tau : A)\| = 0 \qquad \text{for every real } \sigma. \tag{3.9}$$

PROOF. We will assume without loss of generality that $\rho(A) \supset \{\lambda : \operatorname{Re} \lambda > 0\}$ and that $\|T(t)\| \leq M$. Otherwise, we consider $S(t) = e^{-\omega t}T(t)$ with $\omega$ chosen so that these conditions are satisfied. Obviously $T(t)$ is continuous in the uniform operator topology for $t > 0$ if and only if $S(t)$ has this property.

If $\sigma > 0$ then by our assumption $\lambda = \sigma + i\tau \in \rho(A)$. Substituting $x = R(\lambda : A)y$ in (2.3), we obtain

$$e^{\lambda t}R(\lambda : A)y - T(t)R(\lambda : A)y = B_\lambda(t)y \qquad \text{for } y \in X \tag{3.10}$$

which implies

$$(e^{\sigma t} - M)\|R(\lambda : A)\| \leq e^{\sigma t}\left\|\int_0^t e^{-i\tau s}e^{-\sigma s}T(s)\,ds\right\|$$

Choosing $t > \sigma^{-1} \log M$ yields

$$\|R(\sigma + i\tau : A)\| \leq C\left\|\int_0^t e^{-i\tau s}e^{-\sigma s}T(s)\,ds\right\| \tag{3.11}$$

for some constant $C$ independent of $\tau$. The right-hand side of (3.11) tends to zero as $|\tau| \to \infty$ by the lemma of Riemann-Lebesgue. For $\sigma \leq 0$ we write

$$R(\lambda : A) = \sum_{k=0}^{\infty} R(1 + i\tau : A)^{k+1}(1 + i\tau - \lambda)^k \qquad (3.12)$$

and set

$$\varphi(|\tau|) = \max_{|t| \geq |\tau|} \|R(1 + it : A)\|.$$

By what we have already proved above, $\varphi(|\tau|) \to 0$ as $|\tau| \to \infty$. The series (3.12) clearly converges (in the uniform operator topology) for $|1 - \sigma| \leq 1/2\varphi(|\tau|)$, which implies (3.8). Moreover, for any fixed $\sigma$ satisfying $|1 - \sigma| \leq 1/2\varphi(|\tau|)$ we have

$$\|R(\sigma + i\tau : A)\| \leq 2\|R(1 + i\tau : A)\| \leq 2\varphi(|\tau|)$$

and therefore (3.9) holds and the proof is complete.                         □

**Corollary 3.7.** *Let $T(t)$ be a compact $C_0$ semigroup and let $A$ be its infinitesimal generator. For every $-\infty < \alpha \leq \beta < \infty$, the intersection of the strip $\alpha \leq \operatorname{Re} \lambda \leq \beta$ with $\sigma(A)$ contains at most a finite number of eigenvalues of $A$.*

PROOF. The compactness of the semigroup $T(t)$ implies the compactness of $R(\lambda : A)$ for $\lambda \in \rho(A)$ (Theorem 3.3). Therefore the spectrum of $R(\lambda : A)$ consists of zero and a sequence, which may be finite or even empty, of eigenvalues converging to zero if the sequence is infinite. This implies that $\sigma(A)$ consists of a sequence of eigenvalues with $\infty$ as the only possible limit point. From Theorem 3.6 it follows that the intersection of $\sigma(A)$ and the strip $\alpha \leq \operatorname{Re} \lambda \leq \beta$ is compact and hence can contain only a finite number of eigenvalues of $A$.                                              □

Note that in Corollary 3.7 we have proved that if $T(t)$ is a compact $C_0$ semigroup the spectrum $\sigma(A)$ of its infinitesimal generator consists solely of eigenvalues.

## 2.4. Differentiability

**Definition 4.1.** Let $T(t)$ be a $C_0$ semigroup on a Banach space $X$. The semigroup $T(t)$ is called *differentiable* for $t > t_0$ if for every $x \in X$, $t \to T(t)x$ is differentiable for $t > t_0$. $T(t)$ is called *differentiable* if it is differentiable for $t > 0$.

We have seen in Theorem 1.2.4 (c) that if $T(t)$ is a $C_0$ semigroup with infinitesimal generator $A$ and $x \in D(A)$ then $t \to T(t)x$ is differentiable for $t \geq 0$. If $T(t)$ is moreover differentiable then for every $x \in X$, $t \to T(t)x$ is differentiable for $t > 0$. Note that if $t \rightarrow T(t)x$ is differentiable for every $x \in X$ and $t \geq 0$ then $D(A) = X$ and since $A$ is closed it is necessarily bounded.

Example 2.1 provides a simple example of a $C_0$ semigroup which is differentiable for $t > 1$.

**Lemma 4.2.** *Let $T(t)$ be a $C_0$ semigroup which is differentiable for $t > t_0$ and let $A$ be its infinitesimal generator, then*

(a) *For $t > nt_0$, $n = 1, 2, \ldots$, $T(t) : X \rightarrow D(A^n)$ and $T^{(n)}(t) = A^n T(t)$ is a bounded linear operator.*

(b) *For $t > nt_0$, $n = 1, 2, \ldots$, $T^{(n-1)}(t)$ is continuous in the uniform operator topology.*

PROOF. We start with $n = 1$. By our assumption $t \rightarrow T(t)x$ is differentiable for $t > t_0$ and all $x \in X$. Therefore $T(t)x \in D(A)$ and $T'(t)x = AT(t)x$ for every $x \in X$ and $t > t_0$. Moreover, since $A$ is closed and $T(t)$ is bounded, $AT(t)$ is closed. For $t > t_0$, $AT(t)$ is defined on all of $X$ and therefore, by the closed graph theorem, it is a bounded linear operator. This concludes the proof of (a) for $n = 1$. To prove (b) let $\|T(t)\| \leq M_1$ for $0 \leq t \leq 1$ and let $t_0 < t_1 \leq t_2 \leq t_1 + 1$ then,

$$T(t_2)x - T(t_1)x = \int_{t_1}^{t_2} AT(s)x\, ds = \int_{t_1}^{t_2} T(s - t_1)AT(t_1)x\, ds \quad (4.1)$$

and therefore

$$\|T(t_2)x - T(t_1)x\| \leq (t_2 - t_1)M_1\|AT(t_1)\|\, \|x\|$$

which implies the continuity of $T(t)$ for $t > t_0$ in the uniform operator topology.

We now proceed by induction on $n$. Assume that (a) and (b) are true for $n$ and let $t > (n + 1)t_0$. Choose $s > nt_0$ such that $t - s > t_0$. Then

$$T^{(n)}(t)x = T(t - s)A^n T(s)x \qquad \text{for every } x \in X. \quad (4.2)$$

The right-hand side of (4.2) is differentiable since $t - s > t_0$ and therefore $T(t)x$ is $(n + 1)$-times differentiable and $T^{(n+1)}(t)x = A^{n+1}T(t)x$ for every $x \in X$ and $t > (n + 1)t_0$. This implies like in the case $n = 1$ that $T(t) : X \rightarrow D(A^{n+1})$ and that $A^{n+1}T(t)$ is a bounded linear operator for $t > (n + 1)t_0$. This concludes the proof of (a). The continuity of $T^{(n)}(t)$ for $t > (n + 1)t_0$ in the uniform operator topology is proved exactly as for the case $n = 1$, using the fact that $A^n T(t)$ is bounded for $t > (n + 1)t_0$.  $\square$

**Corollary 4.3.** *Let $T(t)$ be a $C_0$ semigroup which is differentiable for $t > t_0$. If $t > (n + 1)t_0$ then $T(t)$ is $n$-times differentiable in the uniform operator topology.*

PROOF. From part (b) of Lemma 4.2 it follows that for $t > (n + 1)t_0$, $A^k T(t)$, $1 \leq k \leq n$ is continuous in the uniform operator topology. Therefore if $t > (n + 1)t_0$, we have

$$T^{(k-1)}(t + h) - T^{(k-1)}(t) = \int_t^{t+h} A^k T(s)\, ds \qquad \text{for } 1 \le k \le n,$$

which implies the differentiability of $T^{(k-1)}(t)$ in the uniform operator topology for $1 \le k \le n$ and $t > (n + 1)t_0$ and thus $T(t)$ is $n$-times differentiable in the uniform operator topology.                                    $\square$

**Corollary 4.4.** *If $T(t)$ is a differentiable $C_0$ semigroup, then $T(t)$ is differentiable infinitely many times in the uniform operator topology for $t > 0$.*

**Lemma 4.5.** *Let $T(t)$ be a differentiable $C_0$ semigroup and let $A$ be its infinitesimal generator. Then*

$$T^{(n)}(t) = \left( AT\left(\frac{t}{n}\right) \right)^n = \left( T'\left(\frac{t}{n}\right) \right)^n \qquad n = 1, 2, \dots. \qquad (4.3)$$

PROOF. The lemma is proved by induction on $n$. For $n = 1$ the result has been proved in Lemma 4.2. If (4.3) holds for $n$ and $t \ge s$ then

$$T^{(n)}(t) = \left( AT\left(\frac{t}{n}\right) \right)^n = T(t - s)\left( AT\left(\frac{s}{n}\right) \right)^n. \qquad (4.4)$$

Differentiating (4.4) with respect to $t$ we find

$$T^{(n+1)}(t) = AT(t - s)\left( AT\left(\frac{s}{n}\right) \right)^n. \qquad (4.5)$$

Substituting $s = nt/n + 1$ in (4.5) yields the result for $n + 1$.                $\square$

We turn now to the characterization of the infinitesimal generator of a $C_0$ semigroup which is differentiable for $t > t_0$. Before turning to the main result we will need one more preliminary.

**Lemma 4.6.** *Let $T(t)$ be a $C_0$ semigroup and let $A$ be its infinitesimal generator. If $T(t)$ is differentiable for $t > t_0$ and $\lambda \in \sigma(A)$, $t > t_0$ then $\lambda e^{\lambda t} \in \sigma(AT(t))$.*

PROOF. We define

$$B_\lambda(t)x = \int_0^t e^{\lambda(t-s)} T(s)x\, ds.$$

$B_\lambda(t)x$ is clearly differentiable in $t$ and differentiating it we find

$$B_\lambda'(t)x = T(t)x + \lambda B_\lambda(t)x.$$

$B_\lambda'(t)$ is a bounded linear operator in $X$. Assuming now that $t > t_0$ and differentiating (2.2) with respect to $t$, we obtain

$$\lambda e^{\lambda t} x - AT(t)x = (\lambda I - A) B_\lambda'(t)x \qquad \text{for every } x \in X. \qquad (4.6)$$

Let

$$C(t)x = \lambda e^{\lambda t}x - AT(t)x.$$

For $t > t_0$, $C(t)$ is a bounded linear operator. It is easy to check that $B'_\lambda(t)$ and $C(t)$ commute and that for $x \in D(A)$, $AB'_\lambda(t)x = B'_\lambda(t)Ax$. If $\lambda e^{\lambda t} \in \rho(AT(t))$ then $C(t)$ is invertible and from (4.6) it follows that

$$x = (\lambda I - A)B'_\lambda(t)C(t)^{-1}x \qquad \text{for every} \quad x \in X,$$

i.e., $B'_\lambda(t)C(t)^{-1}$ is a right inverse of $\lambda I - A$. Multiplying (4.6) from the left by $C(t)^{-1}$ we have

$$x = C(t)^{-1}(\lambda I - A)B'_\lambda(t)x.$$

Choosing $x \in D(A)$ we can commute $B'_\lambda(t)$ and $\lambda I - A$. Then using the commutativity of $B'_\lambda(t)$ and $C(t)$ and therefore also of $B'_\lambda(t)$ and $C(t)^{-1}$, we obtain

$$x = B'_\lambda(t)C(t)^{-1}(\lambda I - A)x \qquad \text{for every} \quad x \in D(A).$$

Therefore, $B'_\lambda(t)C(t)^{-1}$ is the inverse of $\lambda I - A$, $\lambda \in \rho(A)$ and the result follows.                                                                                                    $\square$

**Theorem 4.7.** *Let $T(t)$ be a $C_0$ semigroup and let $A$ be its infinitesimal generator. If $\|T(t)\| \le Me^{\omega t}$ then the following two assertions are equivalent:*

(i) *There exists a $t_0 > 0$ such that $T(t)$ is differentiable for $t > t_0$.*
(ii) *There exist real constants $a$, $b$ and $C$ such that $b > 0$, $C > 0$,*

$$\rho(A) \supset \Sigma = \{\lambda : \operatorname{Re}\lambda \ge a - b\log|\operatorname{Im}\lambda|\} \qquad (4.7)$$

*and*

$$\|R(\lambda : A)\| \le C|\operatorname{Im}\lambda| \qquad \text{for} \quad \lambda \in \Sigma, \operatorname{Re}\lambda \le \omega. \qquad (4.8)$$

PROOF. We may assume without loss of generality that $\omega < 0$. Otherwise we consider the semigroup $T_1(t) = e^{-(\omega+\varepsilon)t}T(t)$ satisfying $\|T_1(t)\| \le Me^{-\varepsilon t}$ and for which (i) or (ii) hold if and only if they hold for $T(t)$. We will therefore assume that $\omega < 0$.

We start by showing that (ii) implies (i). Let $\Gamma$ be a path in $\Sigma$ composed of three parts; $\Gamma_1$ given by $\operatorname{Re}\lambda = 2a - b\log(-\operatorname{Im}\lambda)$ for $-\infty < \operatorname{Im}\lambda \le -L = -e^{2a/b}$, $\Gamma_2$ is given by $\operatorname{Re}\lambda = 0$ for $-L \le \operatorname{Im}\lambda \le L$ and $\Gamma_3$ is given by $\operatorname{Re}\lambda = 2a - b\log(\operatorname{Im}\lambda)$ for $L \le \operatorname{Im}\lambda < \infty$. $\Gamma$ is oriented so that $\operatorname{Im}\lambda$ increases along $\Gamma$. By changing the constant $C$ in (4.8) we can assume that (4.8) holds for $\lambda \in \Gamma_j$, $j = 1, 3$. Let $\Gamma_n = \Gamma \cap \{\lambda : |\operatorname{Im}\lambda| < n\}$. Since $\lambda \to e^{\lambda t}R(\lambda : A)$ is a continuous function from $\rho(A) \subset \mathbb{C}$ into $B(X)$ (the space of all bounded linear operators on $X$) the integrals

$$S_n(t) = \frac{1}{2\pi i}\int_{\Gamma_n} e^{\lambda t}R(\lambda : A)\,d\lambda$$

are well defined. If $S_n(t)$ converge in $B(X)$ as $n \to \infty$ we define the limit to be the improper integral

$$S(t) = \frac{1}{2\pi i} \int_\Gamma e^{\lambda t} R(\lambda : A)\, d\lambda \qquad (4.9)$$

and say that the integral (4.9) converges in $B(X)$, i.e., in the uniform operator topology. Moreover, it is easy to see that $S_n(t)$ are differentiable in $B(X)$ and their derivatives $S_n'(t)$ are given by

$$S_n'(t) = \frac{1}{2\pi i} \int_{\Gamma_n} \lambda e^{\lambda t} R(\lambda : A)\, d\lambda. \qquad (4.10)$$

If $S_n(t)$ and $S_n'(t)$ converge in $B(X)$ uniformly, say for $t \geq t_1$, as $n \to \infty$ then, for $t \geq t_1$ the limit $S'(t)$ of $S_n'(t)$ is obviously the derivative of $S(t)$ in $B(X)$. We will show that (4.9) converges in $B(X)$ for $t > 2/b$ and that

$$S'(t) = \frac{1}{2\pi i} \int_\Gamma \lambda e^{\lambda t} R(\lambda : A)\, d\lambda \qquad (4.11)$$

converges in $B(X)$ for $t > 3/b$. Moreover (4.9) and (4.11) converge uniformly in $t$ for $t \geq 2/b + \delta$ and $t \geq 3/b + \delta$ respectively, for every $\delta > 0$. To prove these claims we set $\Gamma_{j,n} = \Gamma_j \cap \{\lambda : |\mathrm{Im}\ \lambda| < n\}$ $j = 1, 2, 3$,

$$S_{j,n}(t) = \frac{1}{2\pi i} \int_{\Gamma_{j,n}} e^{\lambda t} R(\lambda : A)\, d\lambda \qquad j = 1, 2, 3, \qquad (4.12)$$

and

$$S_j(t) = \frac{1}{2\pi i} \int_{\Gamma_j} e^{\lambda t} R(\lambda : A)\, d\lambda \qquad j = 1, 2, 3. \qquad (4.13)$$

Taking $n > L$ it is clear that $\Gamma_{2,n} = \Gamma_2$ and $S_{2,n}(t) = S_2(t)$ and thus $S_2(t)$ is well defined for every $t \geq 0$. To prove the convergence of the integrals $S_{j,n}(t)$, $j = 1, 3$, we estimate their integrands on the respective paths of integration and find for $\lambda = \sigma + i\tau \in \Gamma_j, j = 1, 3$

$$\|e^{\lambda t} R(\lambda : A)\| = |e^{\lambda t}|\ \|R(\lambda : A)\| \leq e^{2at} |\tau|^{-bt} C|\tau| = C e^{2at} |\tau|^{1-bt}. \qquad (4.14)$$

Therefore, for $n > m \geq L$ we have,

$$\left\| S_{j,n}(t) - S_{j,m}(t) \right\| = \frac{1}{2\pi} \left\| \int_{\Gamma \cap \{m < |\mathrm{Im}\ \lambda| < n\}} e^{\lambda t} R(\lambda : A)\, d\lambda \right\|$$

$$\leq C_1 e^{2at} \int_m^n |\tau|^{1-bt}\, d|\tau| \qquad (4.15)$$

where $C_1$ is a constant independent of $t$. Thus for $t > 2/b$, $S_{j,n}(t) j = 1, 3$ converge in $B(X)$ and the convergence is uniform in $t$ on every compact subinterval of $(2/b, \infty)$. This concludes the proof of the convergence of (4.9). To prove the convergence of (4.11) we proceed similarly. First, we note that $S_2'(t)$ exists for $t \geq 0$. Then we estimate the integrands of $S_{j,n}'(t)$, $j = 1, 3$ on their respective paths of integration.

$$\|\lambda e^{\lambda t} R(\lambda : A)\| \leq |\lambda| C_1 e^{2at} |\tau|^{1-bt} \leq C_2 e^{2at} |\tau|^{2-bt} \qquad (4.16)$$

where $C_2$ is a constant independent of $t$. The convergence of $S_{j,n}'(t), j = 1, 3$ for $t > 3/b$ now follows exactly as the convergence of $S_{j,n}(t) \, j = 1, 3$ for $t > 2/b$. Thus $S(t)$ exists for $t > 2/b$ and is differentiable for $t > 3/b$. To conclude that $T(t)$ is differentiable for $t > 3/b$ we will now show that for $t > 2/b$, $S(t) = T(t)$.

Let $x \in D(A^2)$. From Corollary 1.7.5 it follows that

$$T(t)x = \lim_{|\tau| \to \infty} \frac{1}{2\pi i} \int_{\gamma - i\tau}^{\gamma + i\tau} e^{\lambda t} R(\lambda : A) x \, d\lambda \qquad \text{for every} \quad \gamma > 0.$$
$$(4.17)$$

But for $x \in D(A^2)$ we have

$$R(\lambda : A)x = \frac{x}{\lambda} + \frac{Ax}{\lambda^2} + \frac{R(\lambda : A)A^2 x}{\lambda^2}. \qquad (4.18)$$

From (4.9) it follows that for every $x \in X$ and $t > 2/b$

$$S(t)x = \frac{1}{2\pi i} \int_\Gamma e^{\lambda t} R(\lambda : A) x \, d\lambda. \qquad (4.19)$$

Taking $x \in D(A^2)$ in (4.19), using the estimates (4.18) and (4.8), we observe that one can shift the path of integration in (4.19) from $\Gamma$ to the line $\gamma + i\tau$, $-\infty < \tau < \infty$. Therefore, for $t > 2/b$ and $x \in D(A^2)$, $T(t)x = S(t)x$. Since for $t > 2/b$ both $S(t)$ and $T(t)$ are bounded operators and since $D(A^2)$ is dense in $X$ it follows that $S(t) = T(t)$ for $t > 2/b$ and consequently $T(t)$ is differentiable for $t > 3/b$ even in the uniform operator topology. Thus (ii) implies (i).

Next we show that (i) implies (ii). If $t_1 > t_0$ then $AT(t_1)$ is a bounded linear operator. Set $\|AT(t_1)\| = M(t_1)$. From Lemma 4.6 it follows that

$$\sigma(A) \subset \{\lambda : \lambda e^{\lambda t_1} \in \sigma(AT(t_1))\} \subset \{\lambda : |\lambda e^{\lambda t_1}| \leq M(t_1)\}. \quad (4.20)$$

Consequently

$$\rho(A) \supset \{\lambda : \operatorname{Re} \lambda > t_1^{-1} \log M(t_1) - t_1^{-1} \log |\operatorname{Im} \lambda|\}.$$

Set

$$\Sigma = \{\lambda : \operatorname{Re} \lambda > t_1^{-1} \log (1 + \delta) M(t_1) - t_1^{-1} \log |\operatorname{Im} \lambda|\}$$
$$\text{for some} \quad \delta > 0. \quad (4.21)$$

Obviously $\Sigma \subset \rho(A)$, which proves (4.7). To prove (4.8) substitute $R(\lambda : A)x$ for $x$ in (4.6). The result is,

$$\lambda e^{\lambda t} R(\lambda : A)x = AT(t)R(\lambda : A)x + T(t)x + \lambda \int_0^t e^{\lambda(t-s)} T(s)x \, ds.$$
$$(4.22)$$

Estimating (4.22) with $t = t_1$ and $\lambda = \sigma + i\tau \in \Sigma$ we find

$$\|R(\lambda : A)x\| \le |\tau|^{-1}e^{-\sigma t_1}\big(\|AT(t_1)\|\ \|R(\lambda : A)\|$$
$$+ \|T(t_1)\|\big)\|x\| + \left\|\int_0^{t_1} e^{-\lambda s}T(s)x\, ds\right\|.$$

But for $\lambda \in \Sigma$, $|\tau|^{-1}e^{-\sigma t_1}\|AT(t_1)\| \le (1 + \delta)^{-1}$. Choosing $|\tau| \ge 1$ and $\sigma \le \omega < 0$ we find

$$\|R(\lambda : A)x\| \le \frac{1 + \delta}{\delta}\left[|\tau|^{-1}e^{(\omega-\sigma)t_1}M\|x\| + \left\|\int_0^{t_1} e^{-\lambda s}T(s)x\, ds\right\|\right]$$
$$\le \left(\frac{1 + \delta}{\delta}\right)Me^{(\omega-\sigma)t_1}\big(|\tau|^{-1} + t_1\big)\|x\|$$
$$\le \frac{M(1 + t_1)e^{\omega t_1}}{\delta\|AT(t_1)\|}|\tau|\ \|x\| = C|\tau|\ \|x\|.$$

Thus, for $\lambda \in \Sigma$, $\operatorname{Re}\lambda \le \omega$, $\|R(\lambda : A)\| \le C|\operatorname{Im}\lambda|$ and the proof is complete. □

From the proof of Theorem 4.7 it follows that if $T(t)$ is a $C_0$ semigroup satisfying (4.7) and (4.8) then $T(t)$ is differentiable for $t > t_0 = 3/b$, and if $T(t)$ is differentiable for $t > t_0$ then for every $t_1 > t_0$ the constant $b$ in (4.7) can be taken as $b = 1/t_1$. These remarks enable us to give the following characterization of the infinitesimal generator of a differentiable semigroup.

**Theorem 4.8.** *Let $T(t)$ be a $C_0$ semigroup satisfying $\|T(t)\| \le Me^{\omega t}$ and let $A$ be its infinitesimal generator. $T(t)$ is a differentiable semigroup if and only if for every $b > 0$ there are constants $a_b$ real and $C_b$ positive such that*

$$\rho(A) \supset \Sigma_b = \{\lambda : \operatorname{Re}\lambda > a_b - b\log|\operatorname{Im}\lambda|\}, \tag{4.23}$$

*and*

$$\|R(\lambda : A)\| \le C_b|\operatorname{Im}\lambda| \quad \text{for} \quad \lambda \in \Sigma_b, \operatorname{Re}\lambda \le \omega. \tag{4.24}$$

Our next theorem is a simple consequence of Theorem 4.7.

**Theorem 4.9.** *Let $A$ be the infinitesimal generator of a $C_0$ semigroup $T(t)$ satisfying $\|T(t)\| \le Me^{\omega t}$. If for some $\mu \ge \omega$*

$$\limsup_{|\tau| \to \infty} \log|\tau|\ \|R(\mu + i\tau : A)\| = C < \infty \tag{4.25}$$

*then $T(t)$ is differentiable for $t > 3C$.*

PROOF. We will show that (4.25) implies condition (ii) of Theorem 4.7. Developing the resolvent $R(\lambda : A)$ into a Taylor series around the point

$\mu + i\tau$ we obtain

$$R(\lambda : A) = \sum_{k=0}^{\infty} R(\mu + i\tau : A)^{k+1}(\mu + i\tau - \lambda)^k. \qquad (4.26)$$

This series converges in the uniform operator topology as long as $\|R(\mu + i\tau : A)\|\, |\mu + i\tau - \lambda| < 1$. Let $\varepsilon > 0$ be fixed and let $\tau_0$ be such that for $|\tau| > \tau_0$,

$$\|R(\mu + i\tau : A)\| \leq \frac{C + \varepsilon/2}{\log |\tau|}$$

holds. Choosing $\lambda = \sigma + i\tau$ we see that (4.26) converges in the region $|\tau| > \tau_0$, $|\sigma - \mu| < (C + \varepsilon)^{-1} \log |\tau|$, i.e., the resolvent exists for

$$\sigma > C_0 - (C + \varepsilon)^{-1} \log |\tau|, \; |\tau| > \tau_0, \qquad (4.27)$$

where $C_0 = \max(\mu, \omega + (\varepsilon + C^{-1}) \log \tau_0)$. Moreover, in this region

$$\|R(\lambda : A)\| \leq \frac{C_1}{\log |\mathrm{Im}\, \lambda|} \leq C_2.$$

From the remarks following Theorem 4.7 we have that $T(t)$ is differentiable for $t > 3(C + \varepsilon)$ and since $\varepsilon > 0$ was arbitrary, $T(t)$ is differentiable for $t > 3C$. $\qquad \square$

**Corollary 4.10.** *Let $A$ be the infinitesimal generator of a $C_0$ semigroup $T(t)$ satisfying $\|T(t)\| \leq Me^{\omega t}$. If for some $\mu \geq \omega$*

$$\limsup_{|\tau| \to \infty} \log |\tau| \, \|R(\mu + i\tau : A)\| = 0, \qquad (4.28)$$

*then $T(t)$ is a differentiable semigroup.*

We conclude this section with some results that give the connection between the differentiability of a $C_0$ semigroup for $t > t_0$ and the behavior of $\|T(t) - I\|$ as $t \to 0$. We already know (see Theorem 1.1.2) that if $\|T(t) - I\| \to 0$ as $t \downarrow 0$ then $T(t)$ is a differentiable semigroup. In this case $T(t)$ is differentiable in the uniform operator topology for $t \geq 0$ and its generator $A$ is a bounded linear operator. Our next theorem is a considerable generalization of this result.

**Theorem 4.11.** *Let $T(t)$ be a $C_0$ semigroup satisfying $\|T(t)\| \leq Me^{\omega t}$. If there are constants $C > 0$ and $\delta_C > 0$ such that*

$$\|T(t) - I\| \leq 2 - Ct \log(1/t) \qquad for \quad 0 < t < \delta_C, \qquad (4.29)$$

*then $T(t)$ is differentiable for $t > 3M/C$.*

PROOF. We first prove the result for uniformly bounded semigroups, that is, $\|T(t)\| \leq M$. If $\alpha$ is real and $x \in D(A)$ then by (2.3) we have

$$T(t)x - e^{i\alpha t}x = \int_0^t e^{i\alpha(t-s)}T(s)(A - i\alpha I)x \, ds. \qquad (4.30)$$

This implies

$$\|T(t)x - e^{i\alpha t}x\| \leq tM\|(A - i\alpha I)x\|.$$

Substituting $\alpha = \pm\pi/t$ we obtain

$$\|T(t)x + x\| \leq tM\left\|\left(A \pm i\frac{\pi}{t}I\right)x\right\|.$$

From (4.29) it follows that

$$\|(I + T(t)x\| \geq 2\|x\| - \|(I - T(t))x\| \geq (Ct\log(1/t))\|x\|$$

$$\text{for} \quad 0 < t < \delta_C,$$

and therefore,

$$\|(A - i\tau I)x\| \geq \left(\frac{C}{M}\log\frac{|\tau|}{\pi}\right)\|x\| \tag{4.31}$$

where $\tau = \pm\pi/t$.

Thus, for $|\tau|$ sufficiently large $A - i\tau I$ is one-to-one and has closed range. We will show that the range of $A - i\tau I$ is all of $X$. Since $A$ is the infinitesimal generator of a uniformly bounded $C_0$ semigroup $T(t)$, it follows from Remark 1.5.4, that for every $\rho > 0$, $(A - (\rho + i\tau)I)^{-1}$ is a bounded linear operator in $X$ and its norm is bounded by $M\rho^{-1}$. Let $f \in X$ and set

$$(A - (\rho + i\tau)I)x_\rho = f.$$

Then $\|x_\rho\| \leq M\rho^{-1}\|f\|$ and therefore,

$$\|(A - i\tau I)x_\rho\| \leq \rho\|x_\rho\| + \|f\| \leq (M + 1)\|f\|. \tag{4.32}$$

From (4.31) and (4.32) it follows that $\|x_\rho\|$ is bounded as $\rho \to 0$. Therefore,

$$\|(A - i\tau I)x_\rho - f\| \leq \rho\|x_\rho\| \to 0 \quad \text{as} \quad \rho \to 0. \tag{4.33}$$

Using (4.31) again we deduce from (4.33) that $x_\rho$ converges to some $x$ as $\rho \to 0$. Since $A$ is closed it follows that $x \in D(A)$ and $(A - i\tau I)x = f$ and therefore $A - i\tau I$ is onto and from (4.31) we have

$$\|(A - i\tau I)^{-1}\| \leq \frac{M}{C}\left(\log\frac{|\tau|}{\pi}\right)^{-1},$$

which implies

$$\limsup_{|\tau| \to \infty} \log|\tau| \ \|(A - i\tau I)^{-1}\| \leq \frac{M}{C}$$

and the result follows from Theorem 4.9. This concludes the proof for the case $\|T(t)\| \leq M$. If $\|T(t)\| \leq Me^{\omega t}$, $\omega > 0$, we consider $S(t) = e^{-\omega t}T(t)$. Then $\|S(t)\| \leq M$ and

$$\|S(t) - I\| \leq e^{-\omega t}\|T(t) - I\| + |e^{-\omega t} - 1|$$

$$\leq 2 - Ct\log\frac{1}{t} + |e^{-\omega t} - 1| \leq 2 - C_1 t\log\frac{1}{t},$$

for every $C > C_1 > 0$ and $0 < t < \delta_C$. Therefore, $S(t)$ is differentiable for $t > 3M/C_1$ by the first part of the proof. Since $T(t) = e^{\omega t}S(t)$, $T(t)$ is differentiable for $t > 3M/C_1$ and since $C_1 < C$ is arbitrary, $T(t)$ is differentiable for $t > 3M/C$. □

**Corollary 4.12.** *Let $T(t)$ be a $C_0$ semigroup satisfying $\|T(t) - I\| \le 2 - Ct \log 1/t$ for $0 < t < \delta_C$. If $T(t)$ can be extended to a group, then its infinitesimal generator is necessarily bounded.*

PROOF. From Theorem 4.11 it follows that for $t$ large enough, $T(t)$ is differentiable and therefore, by Lemma 4.2, $AT(t)$ is bounded. Since $A = T(-t)AT(t)$ it follows that $A$ is bounded as a product of two bounded operators. □

**Corollary 4.13.** *Let $T(t)$ be a $C_0$ group and let $A$ be its infinitesimal generator. If $A$ is unbounded, then*

$$\limsup_{t \downarrow 0} \|I - T(t)\| \ge 2. \tag{4.34}$$

PROOF. From Corollary 4.12 it follows that for every $C > 0$ and $\alpha > 0$

$$\sup_{0 \le t \le \alpha} \left( \|T(t) - I\| + Ct \log \frac{1}{t} \right) \ge 2. \tag{4.35}$$

Letting $\alpha \downarrow 0$ in (4.35), (4.34) follows. □

# 2.5. Analytic Semigroups

Up to this point we dealt with semigroups whose domain was the real nonnegative axis. We will now consider the possibility of extending the domain of the parameter to regions in the complex plane that include the nonnegative real axis. It is clear that in order to preserve the semigroup structure, the domain in which the complex parameter should vary must be an additive semigroup of complex numbers. In this section however, we will restrict ourselves to very special complex domains, namely, angles around the positive real axis.

**Definition 5.1.** Let $\Delta = \{z : \varphi_1 < \arg z < \varphi_2, \ \varphi_1 < 0 < \varphi_2\}$ and for $z \in \Delta$ let $T(z)$ be a bounded linear operator. The family $T(z)$, $z \in \Delta$ is an *analytic semigroup* in $\Delta$ if

(i) $z \to T(z)$ is analytic in $\Delta$.
(ii) $T(0) = I$ and $\lim_{\substack{z \to 0 \\ z \in \Delta}} T(z)x = x$ for every $x \in X$. *similar to def g semigroup except No. (duction)*
(iii) $T(z_1 + z_2) = T(z_1)T(z_2)$ for $z_1, z_2 \in \Delta$.

A semigroup $T(t)$ will be called *analytic* if it is analytic in some sector $\Delta$ containing the nonnegative real axis.

Clearly, the restriction of an analytic semigroup to the real axis is a $C_0$ semigroup. We will be interested below in the possibility of extending a given $C_0$ semigroup to an analytic semigroup in some sector $\Delta$ around the nonnegative real axis.

Since multiplication of a $C_0$ semigroup $T(t)$ by $e^{\omega t}$ does not effect the possibility or impossibility of extending it to an analytic semigroup in some sector $\Delta$, we will restrict ourselves in many of the results of this section to the case of uniformly bounded $C_0$ semigroups. The results for general $C_0$ semigroups follow from the corresponding results for uniformly bounded $C_0$ semigroups in an obvious way. For convenience we will also often assume that $0 \in \rho(A)$ where $A$ is the infinitesimal generator of the semigroup $T(t)$. This again can be always achieved by multiplying the uniformly bounded semigroup $T(t)$ by $e^{-\varepsilon t}$ for $\varepsilon > 0$.

We start the discussion by recalling Theorem 1.7.7 which claims that a densely defined operator $A$ in $X$ satisfying

$$\rho(A) \supset \Sigma = \left\{ \lambda : |\arg \lambda| < \frac{\pi}{2} + \delta \right\} \cup \{0\} \qquad \text{for some} \quad 0 < \delta < \frac{\pi}{2}, \tag{5.1}$$

and

$$\|R(\lambda : A)\| \leq M / |\lambda| \qquad \text{for} \quad \lambda \in \Sigma, \lambda \neq 0 \tag{5.2}$$

is the infinitesimal generator of a uniformly bounded $C_0$ semigroup $T(t)$. More is actually true. The semigroup $T(t)$ generated by a densely defined $A$ satisfying (5.1) and (5.2), can be extended to an analytic semigroup in the sector $\Delta_\delta = \{z : |\arg z| < \delta\}$ and in every closed subsector $\bar{\Delta}_{\delta'} = \{z : |\arg z| \leq \delta' < \delta\}$, $\|T(z)\|$ is uniformly bounded. This and much more follow from our next theorem.

**Theorem 5.2.** *Let $T(t)$ be a uniformly bounded $C_0$ semigroup. Let $A$ be the infinitesimal generator of $T(t)$ and assume $0 \in \rho(A)$. The following statements are equivalent:*

(a) *$T(t)$ can be extended to an analytic semigroup in a sector $\Delta_\delta = \{z : |\arg z| < \delta\}$ and $\|T(z)\|$ is uniformly bounded in every closed subsector $\bar{\Delta}_{\delta'}$, $\delta' < \delta$, of $\Delta_\delta$.*

(b) *There exists a constant $C$ such that for every $\sigma > 0$, $\tau \neq 0$*

$$\|R(\sigma + i\tau : A)\| \leq \frac{C}{|\tau|}. \tag{5.3}$$

(c) *There exist $0 < \delta < \pi/2$ and $M > 0$ such that*

$$\rho(A) \supset \Sigma = \left\{ \lambda : |\arg \lambda| < \frac{\pi}{2} + \delta \right\} \cup \{0\} \tag{5.4}$$

*and*

$$\|R(\lambda : A)\| \leq \frac{M}{|\lambda|} \quad \text{for} \quad \lambda \in \Sigma, \lambda \neq 0. \tag{5.5}$$

(d) $T(t)$ *is differentiable for* $t > 0$ *and there is a constant* $C$ *such that*

$$\|AT(t)\| \leq \frac{C}{t} \quad \text{for} \quad t > 0. \tag{5.6}$$

**PROOF.** (a) $\Rightarrow$ (b). Let $0 < \delta' < \delta$ be such that $\|T(z)\| \leq C_1$ for $z \in \overline{\Delta}_{\delta'} = \{z : |\arg z| \leq \delta'\}$. For $x \in X$ and $\sigma > 0$ we have

$$R(\sigma + i\tau : A)x = \int_0^\infty e^{-(\sigma + i\tau)t} T(t)x \, dt. \tag{5.7}$$

From the analyticity and the uniform boundedness of $T(z)$ in $\overline{\Delta}_{\delta'}$ it follows that we can shift the path of integration in (5.7) from the positive real axis to any ray $\rho e^{i\vartheta}$, $0 < \rho < \infty$ and $|\vartheta| \leq \delta'$. For $\tau > 0$, shifting the path of integration to the ray $\rho e^{i\delta'}$ and estimating the resulting integral we find

$$\|R(\sigma + i\tau : A)x\| \leq \int_0^\infty e^{-\rho(\sigma \cos \delta' + \tau \sin \delta')} C_1 \|x\| \, d\rho$$

$$\leq \frac{C_1 \|x\|}{\sigma \cos \delta' + \tau \sin \delta'} \leq \frac{C}{\tau} \|x\|.$$

Similarly for $\tau < 0$ we shift the path of integration to the ray $\rho e^{-i\delta'}$ and obtain $\|R(\sigma + i\tau : A)\| \leq -C/\tau$ and thus (5.3) holds.

(b) $\Rightarrow$ (c). Since $A$ is by assumption the infinitesimal generator of a $C_0$ semigroup we have $\|R(\lambda : A)\| \leq M_1/\text{Re } \lambda$ for $\text{Re } \lambda > 0$. From (b) we have for $\text{Re } \lambda > 0$, $\|R(\lambda : A)\| \leq C/|\text{Im } \lambda|$ and therefore, $\|R(\lambda : A)\| \leq C_1/|\lambda|$ for $\text{Re } \lambda > 0$. Let $\sigma > 0$ and write the Taylor expansion for $R(\lambda : A)$ around $\lambda = \sigma + i\tau$

$$R(\lambda : A) = \sum_{n=0}^\infty R(\sigma + i\tau : A)^{n+1} (\sigma + i\tau - \lambda)^n. \tag{5.8}$$

This series converges in $B(X)$ for $\|R(\sigma + i\tau : A)\| \, |\sigma + i\tau - \lambda| \leq k < 1$. Choosing $\lambda = \text{Re } \lambda + i\tau$ in (5.8) and using (5.3) we see that the series converges uniformly in $B(X)$ for $|\sigma - \text{Re } \lambda| \leq k|\tau|/C$ since both $\sigma > 0$ and $k < 1$ are arbitrary it follows that $\rho(A)$ contains the set of all $\lambda$ with $\text{Re } \lambda \leq 0$ satisfying $|\text{Re } \lambda|/|\text{Im } \lambda| < 1/C$ and in particular

$$\rho(A) \supset \left\{ \lambda : |\arg \lambda| \leq \frac{\pi}{2} + \delta \right\} \tag{5.9}$$

where $\delta = k \arctan 1/C$, $0 < k < 1$. Moreover, in this region

$$\|R(\lambda : A)\| \leq \frac{C}{1 - k} \cdot \frac{1}{|\tau|} \leq \frac{\sqrt{C^2 + 1}}{(1 - k)} \frac{1}{|\lambda|} = \frac{M}{|\lambda|}. \tag{5.10}$$

Since by assumption $0 \in \rho(A)$, $A$ satisfies (c).

(c) $\Rightarrow$ (d). If $A$ satisfies (c) it follows from Theorem 1.7.7 that

$$T(t) = \frac{1}{2\pi i} \int_\Gamma e^{\lambda t} R(\lambda : A)\, d\lambda \qquad (5.11)$$

where $\Gamma$ is the path composed from the two rays $\rho e^{i\vartheta}$ and $\rho e^{-i\vartheta}$, $0 < \rho < \infty$ and $\pi/2 < \vartheta < \pi/2 + \delta$. $\Gamma$ is oriented so that Im $\lambda$ increases along $\Gamma$. The integral (5.11) converges in $B(X)$ for $t > 0$. Differentiating (5.11) with respect to $t$ (first just formally) yields

$$T'(t) = \frac{1}{2\pi i} \int_\Gamma \lambda e^{\lambda t} R(\lambda : A)\, d\lambda. \qquad (5.12)$$

But, the integral (5.12) converges in $B(X)$ for every $t > 0$ since

$$\|T'(t)\| \le \frac{1}{\pi} \int_0^\infty M e^{-\rho \cos \vartheta t}\, d\rho = \left(\frac{M}{\pi \cos \vartheta}\right) \frac{1}{t}. \qquad (5.13)$$

Therefore the formal differentiation of $T(t)$ is justified, $T(t)$ is differentiable for $t > 0$ and

$$\|AT(t)\| = \|T'(t)\| \le C/t \qquad \text{for} \quad t > 0. \qquad (5.14)$$

(d) $\Rightarrow$ (a). Since $T(t)$ is differentiable for $t > 0$ it follows from Lemma 4.5 that $\|T^{(n)}(t)\| = \|T'(t/n)^n\| \le \|T'(t/n)\|^n$. Using this fact together with (5.14) and $n!e^n \ge n^n$ we have

$$\frac{1}{n!} \|T^{(n)}(t)\| \le \left(\frac{Ce}{t}\right)^n. \qquad (5.15)$$

We consider now the power series

$$T(z) = T(t) + \sum_{n=1}^\infty \frac{T^{(n)}(t)}{n!} (z - t)^n. \qquad (5.16)$$

This series converges uniformly in $B(X)$ for $|z - t| \le k(t/eC)$ for every $k < 1$. Therefore $T(z)$ is analytic in $\Delta = \{z : |\arg z| < \arctan 1/Ce\}$. Since obviously for real values of $z$, $T(z) = T(t)$, $T(z)$ extends $T(t)$ to the sector $\Delta$. By the analyticity of $T(z)$ it follows that $T(z)$ satisfies the semigroup property and from (5.16) one sees that $T(z)x \to x$ as $z \to 0$ in $\Delta$. Finally, reducing the sector $\Delta$ to every closed subsector $\overline{\Delta}_\varepsilon = \{z : |\arg z| \le \arctan(1/Ce) - \varepsilon\}$ we see that $\|T(z)\|$ is uniformly bounded in $\overline{\Delta}_\varepsilon$ and the proof is complete.                                                   $\square$

There are several relations between the different constants that appear in the statement of Theorem 5.2. These relations can be discovered by checking carefully the details of the proof. In particular, as we have mentioned before the statement of the theorem, the $\delta$ in (5.4) implies the same $\delta$ in part (a) of the theorem. This follows easily by checking the regions of convergence of the integrals (5.11) and (5.12).

In the part (d) $\Rightarrow$ (a) of the theorem we saw that if $\|AT(t)\| \le C/t$ then $T(t)$ can be extended to an analytic semigroup in a sector around the positive real axis. If the constant $C$ is small enough the opening angle of the sector becomes greater than $2\pi$ and $T(t)$ is analytic in the whole plane which implies in particular that $A$ is bounded. More precisely we have

**Theorem 5.3.** *Let $T(t)$ be a $C_0$ semigroup which is differentiable for $t > 0$. Let $A$ be the infinitesimal generator of $T(t)$. If*

$$\limsup_{t \to 0} t \|AT(t)\| < \frac{1}{e} \tag{5.17}$$

*then $A$ is a bounded operator and $T(t)$ can be extended analytically to the whole complex plane.*

PROOF. From (5.17) it follows that

$$\limsup_{n \to \infty} \frac{t}{n} \left\| T'\left(\frac{t}{n}\right) \right\| < 1/e$$

and therefore the series

$$T(z) = \sum_{n=0}^{\infty} \frac{(z - t)^n}{n!} T^{(n)}(t) = \sum_{n=0}^{\infty} \frac{(z - t)^n}{t^n} \frac{n^n}{n!} \left(\frac{t}{n} T'\left(\frac{t}{n}\right)\right)^n$$

converges in the uniform operator topology for $|z - t|/t < 1 + \delta$ for some $\delta > 0$. But this domain contains the origin as an interior point. Therefore $\lim_{t \to 0} \|T(t) - I\| = 0$ which by Theorem 1.1.2 implies that $A$ is bounded. But a bounded linear operator clearly generates a semigroup $T(t)$ which can be extended analytically to the whole plane.                                          □

EXAMPLE 5.4. Let $X = l_2$ and for every $a = \{a_n\} \in l_2$ let

$$T(t)a = \{e^{-nt}a_n\}. \tag{5.18}$$

It is clear that $T(t)$ defined by (5.18) is a $C_0$ semigroup on $X$. Its infinitesimal generator $A$ is defined on $D(A) = \{\{a_n\} : \{na_n\} \in l_2\}$ and for $a \in D(A)$,

$$A\{a_n\} = \{-na_n\}.$$

Also,

$$\|AT(t)\| = \frac{1}{t} \sup_n \left(nte^{-nt}\right) = \frac{1}{te}$$

and therefore

$$\limsup_{t \to 0} t \|AT(t)\| = \frac{1}{e}.$$

Since $A$ is unbounded, this example shows that the constant $1/e$ in Theorem 5.3 is the best possible one.

The characterization of the infinitesimal generator $A$ of a $C_0$ semigroup involves usually only estimates of $R(\lambda : A)$ for real values of $\lambda$ (see e.g.,

Theorems 1.3.1 and 1.5.2) while in the characterization of the infinitesimal generator of an analytic semigroup we used, in Theorem 5.2 estimates of $R(\lambda : A)$ for complex values of $\lambda$. Our next theorem gives a characterization of the infinitesimal generator of an analytic semigroup in terms of estimates of $R(\lambda : A)$ for only real values of $\lambda$.

**Theorem 5.5.** *Let $A$ be the infinitesimal generator of a $C_0$ semigroup $T(t)$ satisfying $\|T(t)\| \leq Me^{\omega t}$. Then $T(t)$ is analytic if and only if there are constants $C > 0$ and $\Lambda \geq 0$ such that*

$$\|AR(\lambda : A)^{n+1}\| \leq \frac{C}{n\lambda^n} \qquad \text{for} \quad \lambda > n\Lambda \qquad n = 1, 2, \ldots . \quad (5.19)$$

PROOF. We note first that from Theorem 5.2 it follows easily that $T(t)$ is analytic if and only if it is differentiable for $t > 0$ and there are constants $C_1 > 0$ and $\omega_1 > 0$ such that

$$\|AT(t)\| \leq \frac{C_1}{t}e^{\omega_1 t} \qquad \text{for} \quad t > 0. \qquad (5.20)$$

If $A$ satisfies (5.19) then for $\lambda > n\Lambda$ and $x \in D(A)$ we have

$$\|AR(\lambda : A)^{n+1}x\| = \|R(\lambda : A)^{n+1}Ax\| \leq \frac{C}{n\lambda^n}\|x\|. \qquad (5.21)$$

Choosing $t < 1/\Lambda$ and substituting $\lambda = n/t$ in (5.21) we find

$$\left\|A\left(\frac{n}{t}R\left(\frac{n}{t} : A\right)\right)^{n+1}x\right\| = \left\|\left(\frac{n}{t}R\left(\frac{n}{t} : A\right)\right)^{n+1}Ax\right\| \leq \frac{C}{t}\|x\|$$

$$\text{for} \quad x \in D(A).$$

Letting $n \to \infty$ it follows from Theorem 1.8.3 and the closedness of $A$ that

$$\|AT(t)x\| \leq \frac{C}{t}\|x\| \qquad \text{for} \quad x \in D(A), \qquad 0 < t < 1/\Lambda. \quad (5.22)$$

Since $D(A)$ is dense in $X$ and $AT(t)$ is closed, it follows that (5.22) holds for every $x \in X$. Therefore there are constants $C_1 > 0$ and $\omega_1 > 0$ such that (5.20) holds and $T(t)$ is analytic.

For the converse, we differentiate the formula

$$R(\lambda : A)x = \int_0^\infty e^{-\lambda t}T(t)x\, dt$$

$n$ times with respect to $\lambda$ and find

$$R(\lambda : A)^{(n)}x = (-1)^n n! R(\lambda : A)^{n+1}x = (-1)^n \int_0^\infty t^n e^{-\lambda t}T(t)x\, dt.$$

$$(5.23)$$

Operating with $A$ on both sides of (5.23) and estimating the right-hand side using (5.20) yields

$$n!\|AR(\lambda : A)^{n+1}x\| \leq C_1\left(\int_0^\infty t^{n-1}e^{-(\lambda-\omega_1)t}\, dt\right)\|x\| = \frac{C_1}{(\lambda-\omega_1)^n}(n-1)!\|x\|$$

and therefore, for $\lambda > n\Lambda$

$$\|AR(\lambda : A)^{n+1}\| \leq \frac{C_1}{n\lambda^n}\left(\frac{1}{1 - \dfrac{\omega_1}{\Lambda n}}\right)^n \leq \frac{C_2}{n\lambda^n}. \qquad \square$$

The characterizations of analytic semigroups given so far in this section are based on conditions on the infinitesimal generator $A$ of the semigroup or on conditions on the resolvent $R(\lambda : A)$ of $A$. A different type of characterization of analytic semigroups based on the behavior of $T(t)$ near its spectral radius is the subject of our next theorem.

**Theorem 5.6.** *For a uniformly bounded $C_0$ semigroup $T(t)$ the following conditions are equivalent:*

(a) *$T(t)$ is analytic in a sector around the nonnegative real axis.*
(b) *For every complex $\zeta$, $\zeta \neq 1$, $|\zeta| \geq 1$ there exist positive constants $\delta$ and $K$ such that $\zeta \in \rho(T(t))$ and*

$$\|(\zeta I - T(t))^{-1}\| \leq K \qquad \text{for} \quad 0 < t < \delta. \qquad (5.23)$$

(c) *There exist a complex number $\zeta$, $|\zeta| = 1$, and positive constants $K$ and $\delta$ such that*

$$\|(\zeta I - T(t))x\| \geq \frac{1}{K}\|x\| \qquad \text{for every} \quad x \in X, 0 < t < \delta. \qquad (5.24)$$

PROOF. (a) $\Rightarrow$ (b). Let $T(t)$ be analytic in a sector around the nonnegative real axis. From Theorems 5.2 and 1.7.7, it follows that for $t > 0$

$$T(t) = \frac{1}{2\pi i}\int_{\Gamma'} e^{\lambda t}R(\lambda : A)\, d\lambda$$

where $\Gamma'$ is a path composed from two rays $\rho e^{i\vartheta_1}$, $\pi/2 < \vartheta_1 < \pi$, $\rho \geq 1$ and $\rho e^{i\vartheta_2}$, $-\pi < \vartheta_2 < -\pi/2$, $\rho \geq 1$ and a curve $\rho = \rho(\vartheta)$ joining $e^{i\vartheta_1}$ to $e^{i\vartheta_2}$ inside the resolvent set. $\Gamma'$ is oriented in a way that Im $\lambda$ increases along $\Gamma'$.
Changing variables to $z = \lambda t$ we obtain

$$T(t) = \frac{1}{2\pi i}\int_{\Gamma} e^{z}(zI - tA)^{-1}\, dz. \qquad (5.25)$$

Given $\zeta \neq 1$, $|\zeta| \geq 1$, the path of integration $\Gamma$ can be chosen to be independent of $t$, for $0 < t < \delta$ and such that $e^z \neq \zeta$ for all $z$ on and to the left of $\Gamma$. Having chosen such $\delta$ and $\Gamma$, we define

$$B(t) = \frac{1}{2\pi i}\int_{\Gamma} e^{z}(e^{z} - \zeta)^{-1}(zI - tA)^{-1}\, dz \qquad 0 < t < \delta. \qquad (5.26)$$

This integral converges in the uniform operator topology and thus defines a bounded linear operator $B(t)$. Since on $\Gamma$ we have $\|(zI - tA)^{-1}\| \leq$

$M_1|z|^{-1}$, it follows that

$$\|B(t)\| \leq \frac{M'}{2\pi} \int_\Gamma \left|z^{-1}e^z(e^z - \zeta)^{-1}\right| |dz| \leq M''. \tag{5.27}$$

Since $e^z$ and $e^z(e^z - \zeta)^{-1}$ are analytic on and to the left of $\Gamma$ and tend rapidly to zero as $\mathrm{Re}\, z \to -\infty$ and since $e^z \cdot e^z(e^z - \zeta)^{-1} = e^z + \zeta e^z(e^z - \zeta)^{-1}$, we obtain from the Dunford-Taylor operator calculus that

$$T(t)B(t) = T(t) + \zeta B(t),$$

which implies

$$(I - B(t))(\zeta I - T(t)) = (\zeta I - T(t))(I - B(t)) = \zeta I. \tag{5.28}$$

Therefore,

$$(\zeta I - T(t))^{-1} = \zeta^{-1}(I - B(t))$$

and

$$\left\|(\zeta I - T(t))^{-1}\right\| \leq |\zeta|^{-1}(1 + M''). \tag{5.29}$$

(b) $\Rightarrow$ (c) is trivial.

(c) $\Rightarrow$ (a). Substituting $\lambda = -i\alpha$ with $\alpha$ real into (2.3) we have

$$e^{-it\alpha}T(t)x - x = \int_0^t e^{-is\alpha}T(s)(A - i\alpha I)x\, ds \qquad \text{for} \quad x \in D(A). \tag{5.30}$$

Since $\|T(t)\| \leq M$, (5.30) implies

$$\left\|(T(t) - e^{it\alpha}I)x\right\| \leq Mt\|(A - i\alpha I)x\|. \tag{5.31}$$

If $e^{\pm it\alpha} = e^{i\vartheta} = \zeta, \vartheta > 0$, then by assumption (c) we have

$$K^{-1}\|x\| \leq \left\|(T(t) - e^{\pm i\alpha t}I)x\right\| \leq Mt\|(A \pm i\alpha I)x\|$$

$$\leq M\vartheta|\alpha|^{-1}\|(A \pm i\alpha I)x\|, \tag{5.32}$$

which implies

$$\|(A \pm i\alpha I)x\| \geq |\alpha|(KM\vartheta)^{-1}\|x\|, \tag{5.33}$$

and therefore, $A \pm i\alpha I$ is one to one and has closed range. We will now show that it is onto. Since $A$ is the infinitesimal generator of a uniformly bounded $C_0$ semigroup, $(A - (\varepsilon \pm i\alpha)I)^{-1}$ exists and $\|(A - (\varepsilon \pm i\alpha)I)^{-1}\| \leq M\varepsilon^{-1}$. For every $f \in X$ let

$$(A - (\varepsilon \pm i\alpha)I)x_\varepsilon = f. \tag{5.34}$$

Then $\|x_\varepsilon\| \leq M\varepsilon^{-1}\|f\|$ and $\varepsilon\|x_\varepsilon\| \leq M\|f\|$. This together with (5.34) imply $\|(A \pm i\alpha I)x_\varepsilon\| \leq (M + 1)\|f\|$. From (5.33) we then deduce that $\|x_\varepsilon\| \leq C$ and therefore,

$$\|(A \pm i\alpha I)x_\varepsilon - f\| \leq \varepsilon\|x_\varepsilon\| \to 0 \qquad \text{as} \quad \varepsilon \downarrow 0. \tag{5.35}$$

Combining (5.35) with (5.33), we see that $x_\varepsilon$ is a Cauchy net as $\varepsilon \to 0$ and therefore $x_\varepsilon \to x$. From the closedness of $A$ it then follows that $x \in D(A)$ and $(A \pm i\alpha I)x = f$. Thus the range of $(A \pm i\alpha I)$ is all of $X$ and (5.33) implies

$$\|(A \pm i\alpha I)^{-1}\| \leq \frac{MK\vartheta}{|\alpha|} \tag{5.36}$$

which implies that $T(t)$ can be extended analytically to a sector $\Delta$ around the nonnegative real axis by Theorem 5.2 (b).                                         $\square$

**Corollary 5.7.** *Let $T(t)$ be a $C_0$ semigroup. If*

$$\limsup_{t \downarrow 0} \|I - T(t)\| < 2 \tag{5.37}$$

*then $T(t)$ is analytic in a sector around the nonnegative real axis.*

PROOF. From (5.37) it follows that there exist $\delta > 0$ and $\varepsilon > 0$ such that

$$\|I - T(t)\| \leq 2 - \varepsilon \qquad \text{for} \quad 0 < t < \delta. \tag{5.38}$$

But then

$$\|(-I - T(t))x\| \geq 2\|x\| - \|(I - T(t))x\| \geq \varepsilon\|x\| \qquad \text{for} \quad 0 < t < \delta.$$

This implies by Theorem 5.6 (c) with $\zeta = -1$ that $T(t)$ is analytic.                    $\square$

Corollary 5.7 shows that a certain behavior of $\|I - T(t)\|$ at $t = 0$ can be translated into the analyticity of $T(t)$ in an angle around the nonnegative real axis. It is natural to ask whether the analyticity of $T(t)$ implies (5.37). In general the answer to this is negative. Under certain restricting assumptions however, it is positive. A simple result in this direction is,

**Corollary 5.8.** *If $T(t)$ is an analytic semigroup of contractions on a uniformly convex Banach space $X$, then*

$$\limsup_{t \downarrow 0} \|I - T(t)\| < 2. \tag{5.39}$$

PROOF. Since $T(t)$ is a semigroup of contractions $\|I - T(t)\| \leq 2$. If

$$\limsup_{t \downarrow 0} \|I - T(t)\| = 2$$

then there exist sequences $x_n$ and $t_n$ such that $\|x_n\| = 1$, $t_n \to 0$ and

$$\|(I - T(t_n))x_n\| \geq 2 - 1/n. \tag{5.40}$$

Since $\|T(t_n)x_n\| \leq 1$ it follows from (5.40) and the uniform convexity of $X$ that

$$\|(I + T(t_n))x_n\| \to 0 \qquad \text{as} \quad n \to \infty. \tag{5.41}$$

But this contradicts Theorem 5.6 (b) and therefore (5.39) must hold.            $\square$

## 2.6. Fractional Powers of Closed Operators

In this section we define fractional powers of certain unbounded linear operators and study some of their properties. We concentrate mainly on fractional powers of operators $A$ for which $-A$ is the infinitesimal generator of an analytic semigroup. The results of this section will be used in the study of solutions of semilinear initial value problems.

For our definition we will make the following assumption.

**Assumption 6.1.** *Let $A$ be a densely defined closed linear operator for which*

$$\rho(A) \supset \Sigma^+ = \{\lambda : 0 < \omega < |\arg \lambda| \le \pi\} \cup V \tag{6.1}$$

*where $V$ is a neighborhood of zero, and*

$$\|R(\lambda : A)\| \le \frac{M}{1 + |\lambda|} \qquad for \quad \lambda \in \Sigma^+. \tag{6.2}$$

If $M = 1$ and $\omega = \pi/2$ then $-A$ is the infinitesimal generator of a $C_0$ semigroup. If $\omega < \pi/2$ then, by Theorem 5.2, $-A$ is the infinitesimal generator of an analytic semigroup. The assumption that $0 \in \rho(A)$ and therefore a whole neighborhood $V$ of zero is in $\rho(A)$ was made mainly for convenience. Most of the results on fractional powers that we will obtain in this section remain true even if $0 \notin \rho(A)$.

For an operator $A$ satisfying Assumption 6.1 and $\alpha > 0$ we define

$$A^{-\alpha} = \frac{1}{2\pi i} \int_C z^{-\alpha}(A - zI)^{-1} \, dz \tag{6.3}$$

where the path $C$ runs in the resolvent set of $A$ from $\infty e^{-i\vartheta}$ to $\infty e^{i\vartheta}$, $\omega < \vartheta < \pi$, avoiding the negative real axis and the origin and $z^{-\alpha}$ is taken to be positive for real positive values of $z$. The integral (6.3) converges in the uniform operator topology for every $\alpha > 0$ and thus defines a bounded linear operator $A^{-\alpha}$. If $\alpha = n$ the integrand is analytic in $\Sigma^+$ and it is easy to check that the path of integration $C$ can be transformed to a small circle around the origin. Then using the residue theorem it follows that the integral equals $A^{-n}$ and thus for positive integral values of $\alpha$ the definition (6.3) coincides with the classical definition of $(A^{-1})^n$.

For $0 < \alpha < 1$ we can deform the path of integration $C$ into the upper and lower sides of the negative real axis and obtain

$$A^{-\alpha} = \frac{\sin \pi \alpha}{\pi} \int_0^\infty t^{-\alpha}(tI + A)^{-1} \, dt \qquad 0 < \alpha < 1. \tag{6.4}$$

If $\omega < \pi/2$, i.e., if $-A$ is the infinitesimal generator of an analytic semigroup $T(t)$ we obtain still another representation of $A^{-\alpha}$. This representation turns out to be very useful and therefore in the rest of this section we will assume, unless we state explicitly otherwise, that $\omega < \pi/2$. In this case since by Assumption 6.1 $0 \in \rho(A)$ there exists a constant $\delta > 0$ such that $-A + \delta$ is still an infinitesimal generator of an analytic semigroup. This implies the following estimates;

$$\|T(t)\| \le Me^{-\delta t} \tag{6.5}$$

$$\|AT(t)\| \le M_1 t^{-1} e^{-\delta t} \tag{6.6}$$

$$\|A^m T(t)\| \le M_m t^{-m} e^{-\delta t}. \tag{6.7}$$

The two first estimates are simple consequences of the results of Section 5 while (6.7) follows from (6.6) since

$$\|A^m T(t)\| = \left\|\left(AT\left(\frac{t}{m}\right)\right)^m\right\| \le \left\|AT\left(\frac{t}{m}\right)\right\|^m$$

$$\le \left(M_1 t^{-1} e^{-\delta t/m}\right)^m = M_m t^{-m} e^{-\delta t}.$$

Furthermore, we know that

$$(tI + A)^{-1} = \int_0^\infty e^{-st} T(s)\, ds \tag{6.8}$$

converges uniformly for $t \ge 0$ in the uniform operator topology, by (6.5). Substituting (6.8) into (6.4) and using Fubini's theorem, we have

$$A^{-\alpha} = \frac{\sin \pi\alpha}{\pi} \int_0^\infty t^{-\alpha} \left(\int_0^\infty e^{-st} T(s)\, ds\right) dt$$

$$= \frac{\sin \pi\alpha}{\pi} \int_0^\infty T(s) \left(\int_0^\infty t^{-\alpha} e^{-st}\, dt\right) ds$$

$$= \frac{\sin \pi\alpha}{\pi} \left(\int_0^\infty u^{-\alpha} e^{-u}\, du\right) \int_0^\infty s^{\alpha-1} T(s)\, ds.$$

Since

$$\int_0^\infty u^{-\alpha} e^{-u}\, du = \frac{\pi}{\sin \pi\alpha} \frac{1}{\Gamma(\alpha)}$$

we finally obtain

$$A^{-\alpha} = \frac{1}{\Gamma(\alpha)} \int_0^\infty t^{\alpha-1} T(t)\, dt \tag{6.9}$$

where the integral converges in the uniform operator topology for every $\alpha > 0$. In the case where $\omega < \pi/2$, i.e., $-A$ is the infinitesimal generator of an analytic semigroup $T(t)$ we will use (6.9) as the definition of $A^{-\alpha}$ for $\alpha > 0$ and we define $A^{-0} = I$.

**Lemma 6.2.** *For $\alpha, \beta \ge 0$*

$$A^{-(\alpha+\beta)} = A^{-\alpha} \cdot A^{-\beta}. \tag{6.10}$$

PROOF.

$$A^{-\alpha}A^{-\beta} = \frac{1}{\Gamma(\alpha)\Gamma(\beta)} \int_0^\infty \int_0^\infty t^{\alpha-1}s^{\beta-1}T(t)T(s)\,dt\,ds$$

$$= \frac{1}{\Gamma(\alpha)\Gamma(\beta)} \int_0^\infty t^{\alpha-1} \int_t^\infty (u-t)^{\beta-1}T(u)\,du\,dt$$

$$= \frac{1}{\Gamma(\alpha)\Gamma(\beta)} \int_0^\infty \left( \int_0^u t^{\alpha-1}(u-t)^{\beta-1}\,dt \right)T(u)\,du$$

$$= \frac{1}{\Gamma(\alpha)\Gamma(\beta)} \int_0^1 v^{\alpha-1}(1-v)^{\beta-1}\,dv \cdot \int_0^\infty u^{\alpha+\beta-1}T(u)\,du$$

$$= \frac{1}{\Gamma(\alpha+\beta)} \int_0^\infty u^{\alpha+\beta-1}T(u)\,du = A^{-(\alpha+\beta)}. \qquad \square$$

**Lemma 6.3.** *There exists a constant $C$ such that*

$$\|A^{-\alpha}\| \leq C \qquad for \ \ 0 \leq \alpha \leq 1. \tag{6.11}$$

PROOF. For $\alpha = 0$ and $\alpha = 1$, (6.11) is obvious. For $0 < \alpha < 1$ we use (6.4) to obtain

$$\|A^{-\alpha}\| \leq \left\| \frac{\sin\pi\alpha}{\pi} \int_0^1 t^{-\alpha}(tI+A)^{-1}\,dt \right\| + \left\| \frac{\sin\pi\alpha}{\pi} \int_1^\infty t^{-1-\alpha}t(tI+A)^{-1}\,dt \right\|.$$

Let $\|(tI+A)^{-1}\| \leq C_0$ for $0 \leq t \leq 1$. From Assumption 6.1 we have $\|t(tI+A)^{-1}\| \leq C_1$ for $t \geq 1$ and therefore

$$\|A^{-\alpha}\| \leq \left| \frac{\sin\pi(1-\alpha)}{\pi(1-\alpha)} \right| C_0 + \left| \frac{\sin\pi\alpha}{\pi\alpha} \right| C_1 \leq C. \qquad \square$$

**Lemma 6.4.** *For every $x \in X$ we have*

$$\lim_{\alpha \to 0} A^{-\alpha}x = x. \tag{6.12}$$

PROOF. Assume first that $x \in D(A)$. Since $0 \in \rho(A)$ we have $x = A^{-1}y$ for some $y \in X$. Therefore,

$$A^{-\alpha}x - x = A^{-(1+\alpha)}y - A^{-1}y = \int_0^\infty \left( \frac{t^\alpha}{\Gamma(1+\alpha)} - 1 \right)T(t)y\,dt.$$

Using (6.5) we therefore have

$$\|A^{-\alpha}x - x\| \leq C_1 \int_0^k \left| \frac{t^\alpha}{\Gamma(1+\alpha)} - 1 \right| e^{-\delta t}\,dt + C_2 \int_k^\infty te^{-\delta t}\,dt \tag{6.13}$$

for every $k > 0$ and $0 \leq \alpha \leq 1$. Given $\varepsilon > 0$ we first choose $k$ so large that the second term on the right of (6.13) is less than $\varepsilon/2$ and then choose $\alpha$ so small that the first term on the right of (6.13) is less than $\varepsilon/2$. Thus for $x \in D(A)$ we have $A^{-\alpha}x \to x$ as $\alpha \to 0$. Since $D(A)$ is dense in $X$ and by Lemma 6.3, $A^{-\alpha}$ are uniformly bounded (6.12) follows for every $x \in X$. $\quad\square$

Combining the previous results we have

**Corollary 6.5.** *If $A$ satisfies Assumption 6.1 with $\omega < \pi/2$ then $A^{-t}$ is a $C_0$ semigroup of bounded linear operators.*

**Lemma 6.6.** $A^{-\alpha}$ *defined by* (6.9) *is one-to-one.*

PROOF. It is clear that $A^{-1}$ is one-to-one. Therefore, for every integer $n \geq 1$, $A^{-n}$ is one-to-one. Let $A^{-\alpha}x = 0$. Take $n \geq \alpha$ then $A^{-n}x = A^{-n+\alpha}A^{-\alpha}x = 0$. This implies $x = 0$ and thus $A^{-\alpha}$ is one-to-one.                    □

**Definition 6.7.** Let $A$ satisfy Assumption 6.1 with $\omega < \pi/2$. For every $\alpha > 0$ we define

$$A^{\alpha} = \left(A^{-\alpha}\right)^{-1}. \tag{6.14}$$

For $\alpha = 0$, $A^{\alpha} = I$.

In the rest of this section we assume that $A$ satisfies Assumption 6.1 with $\omega < \pi/2$ and collect some simple properties of $A^{\alpha}$ in our next theorem.

**Theorem 6.8.** *Let $A^{\alpha}$ be defined by Definition 6.7 then,*

(a) $A^{\alpha}$ *is a closed operator with domain* $D(A^{\alpha}) = R(A^{-\alpha}) =$ *the range of* $A^{-\alpha}$.
(b) $\alpha \geq \beta > 0$ *implies* $D(A^{\alpha}) \subset D(A^{\beta})$.
(c) $\overline{D(A^{\alpha})} = X$ *for every* $\alpha \geq 0$.
(d) *If $\alpha$, $\beta$ are real then*

$$A^{\alpha+\beta}x = A^{\alpha} \cdot A^{\beta}x \tag{6.15}$$

*for every $x \in D(A^{\gamma})$ where $\gamma = \max(\alpha, \beta, \alpha + \beta)$.*

PROOF. For $\alpha \leq 0$, $A^{\alpha}$ is bounded and (a) is clear. If $\alpha > 0$, $A^{\alpha}$ is invertible and therefore $0 \in \rho(A^{\alpha})$. This implies that $A^{\alpha}$ is closed. For $\alpha \geq \beta$ we have by Lemma 6.2, $A^{-\alpha} = A^{-\beta} \cdot A^{-(\alpha-\beta)}$ and therefore $R(A^{-\alpha}) \subset R(A^{-\beta})$ which implies (b). Since by Theorem 1.2.7, $\overline{D(A^n)} = X$ for every $n = 1, 2, \ldots$ and for $\alpha \leq n$ $D(A^{\alpha}) \supset D(A^n)$ by (b), we have (c). Finally, (d) is again a simple consequence of the definition of $A^{\alpha}$ and Lemma 6.2, for example if $\alpha > 0$ and $\beta > 0$ and $x \in D(A^{\alpha}A^{\beta})$ then $x \in D(A^{\beta})$ and $A^{\beta}x \in D(A^{\alpha})$. Let $y = A^{\alpha}A^{\beta}x$ then $A^{\beta}x = A^{-\alpha}y$ and $x = A^{-\beta}A^{-\alpha}y = A^{-(\alpha+\beta)}y$. Therefore $x \in D(A^{\alpha+\beta})$ and $A^{\alpha+\beta}x = y = A^{\alpha}A^{\beta}x$. Similarly if $x \in D(A^{\alpha+\beta})$ we have $x \in D(A^{\alpha}A^{\beta})$ and $A^{\alpha+\beta} = A^{\alpha} \cdot A^{\beta}$.                    □

In Definition 6.7 we define $A^{\alpha}$ in an indirect way. For $x \in D(A) \subset D(A^{\alpha})$, $0 < \alpha < 1$, we have the following explicit formula for $A^{\alpha}x$.

**Theorem 6.9.** *Let $0 < \alpha < 1$. If $x \in D(A)$ then*

$$A^{\alpha}x = \frac{\sin \pi\alpha}{\pi} \int_0^{\infty} t^{\alpha-1}A(tI + A)^{-1}x \, dt. \tag{6.16}$$

PROOF. We have $0 < 1 - \alpha < 1$. Therefore by (6.4) we have

$$A^{\alpha-1}x = \frac{\sin \pi\alpha}{\pi} \int_0^\infty t^{\alpha-1}(tI + A)^{-1}x \, dt. \tag{6.17}$$

The integrand on the right-hand side of (6.17) is in $D(A)$ for every $t > 0$ and $t^{\alpha-1}A(tI + A)^{-1}x$ is integrable on $[0, \infty[$ since near $t = 0$ $\|A(tI + A^{-1})\|$ is uniformly bounded and near $t = \infty$ $\|t^{\alpha-1}A(tI + A)^{-1}x\| \le t^{\alpha-2}M\|Ax\|$. Finally if $x \in D(A^\alpha)$, $A^{\alpha-1}x \in D(A)$ and from the closedness of $A$ we deduce

$$A^\alpha x = A(A^{\alpha-1}x) = \frac{\sin \pi\alpha}{\pi} \int_0^\infty t^{\alpha-1}A(tI + A)^{-1}x \, dt. \qquad \square$$

**Remark.** From the proof of Theorem 6.9 it is clear that (6.16) holds for every $x \in D(A^\gamma)$ with $\gamma > \alpha$.

**Theorem 6.10.** *Let $0 < \alpha < 1$. There exists a constant $C_0 > 0$ such that for every $x \in D(A)$ and $\rho > 0$, we have*

$$\|A^\alpha x\| \le C_0(\rho^\alpha\|x\| + \rho^{\alpha-1}\|Ax\|) \tag{6.18}$$

*and*

$$\|A^\alpha x\| \le 2C_0\|x\|^{1-\alpha}\|Ax\|^\alpha. \tag{6.19}$$

PROOF. By our assumptions on $A$ there exists a constant $M$ satisfying $\|(tI + A)^{-1}\| \le M/t$ for every $t > 0$. If $x \in D(A)$ then by Theorem 6.9,

$$\|A^\alpha x\| \le \left| \frac{\sin \pi\alpha}{\pi} \right| \left| \int_0^\rho t^{\alpha-1}\|A(tI + A)^{-1}\| \, \|x\| \, dt \right.$$

$$+ \left| \frac{\sin \pi\alpha}{\pi} \right| \left| \int_\rho^\infty t^{\alpha-1}\|(tI + A)^{-1}\| \, \|Ax\| \, dt \right.$$

$$\le \left| \frac{\sin \pi\alpha}{\pi\alpha} \right| (1 + M)\rho^\alpha\|x\| + \left| \frac{\sin \pi(1 - \alpha)}{\pi(1 - \alpha)} \right| M\rho^{\alpha-1}\|Ax\|$$

$$\le C_0(\rho^\alpha\|x\| + \rho^{\alpha-1}\|Ax\|).$$

For $x = 0$, (6.19) is obvious. For $x \neq 0$, (6.19) follows from (6.18) taking $\rho = \|Ax\|/\|x\|$. $\qquad \square$

**Corollary 6.11.** *Let $B$ be a closed linear operator satisfying $D(B) \supset D(A^\alpha)$, $0 < \alpha \le 1$ then*

$$\|Bx\| \le C\|A^\alpha x\| \qquad \text{for every} \quad x \in D(A^\alpha) \tag{6.20}$$

*and there is a constant $C_1$ such that for every $\rho > 0$ and $x \in D(A)$*

$$\|Bx\| \le C_1(\rho^\alpha\|x\| + \rho^{\alpha-1}\|Ax\|). \tag{6.21}$$

PROOF. Consider the closed operator $BA^{-\alpha}$. Since $D(B) \supset D(A^\alpha)$, $BA^{-\alpha}$ is defined on all of $X$ and by the closed graph theorem it is bounded. This proves (6.20). For $x \in D(A)$, (6.21) is a direct consequence of (6.20) and (6.18).                                                                                           □

A sufficient condition for $D(B) \supset D(A^{\alpha})$ is given in our next theorem.

**Theorem 6.12.** *Let $B$ be a closed linear operator satisfying $D(B) \supset D(A)$. If for some $\gamma$, $0 < \gamma < 1$, and every $\rho \geq \rho_0 > 0$ we have*

$$\|Bx\| \leq C(\rho^{\gamma}\|x\| + \rho^{\gamma-1}\|Ax\|) \qquad for \quad x \in D(A) \qquad (6.22)$$

*then*

$$D(B) \supset D(A^{\alpha}) \qquad for \; every \quad \gamma < \alpha \leq 1. \qquad (6.23)$$

PROOF. Let $x \in D(A^{1-\alpha})$ then $A^{-\alpha}x \in D(A) \subset D(B)$. Since $B$ is closed

$$BA^{-\alpha}x = \frac{1}{\Gamma(\alpha)} \int_0^{\infty} t^{\alpha-1}BT(t)x\,dt$$

provided that the integral is convergent. But

$$\|BA^{-\alpha}x\| \leq \frac{1}{\Gamma(\alpha)}\left(\int_0^{\delta} t^{\alpha-1}\|BT(t)x\|dt + \int_{\delta}^{\infty} t^{\alpha-1}\|BT(t)x\|dt\right).$$

$$(6.24)$$

Since $T(t)$ is an analytic semigroup $T(t)x \in D(A)$ for every $t > 0$. Choosing $\delta = \rho_0^{-1}$ and using (6.22) with $\rho = t^{-1}$ in the first integral on the right-hand side of (6.24) and with $\rho = \rho_0$ in the second integral and making use of (6.5) and (6.6) we find $\|BA^{-\alpha}x\| \leq C\|x\|$. This is true for all $x \in D(A^{1-\alpha})$. Since $BA^{-\alpha}$ is closed and $D(A^{1-\alpha})$ is dense in $X$, $\|BA^{-\alpha}x\| \leq C\|x\|$ for every $x \in X$ and therefore $D(B) \supset D(A^{\alpha})$.                □

It can be shown that if $A$ satisfies the Assumption 6.1 without any restriction on $\omega$, then $-A^{\alpha}$ with $\alpha \leq 1/2$ is the infinitesimal generator of a $C_0$ semigroup of bounded linear operators. If $\omega < \pi/2$ as we assume, $-A^{\alpha}$ is the infinitesimal generator of an analytic semigroup for all $0 < \alpha \leq 1$.

We conclude this section with some results relating $A^{\alpha}$ and the analytic semigroup $T(t)$ generated by $-A$.

**Theorem 6.13.** *Let $-A$ be the infinitesimal generator of an analytic semigroup $T(t)$. If $0 \in \rho(A)$ then,*

(a)  $T(t): X \to D(A^{\alpha})$ *for every $t > 0$ and $\alpha \geq 0$.*
(b)  *For every $x \in D(A^{\alpha})$ we have $T(t)A^{\alpha}x = A^{\alpha}T(t)x$.*
(c)  *For every $t > 0$ the operator $A^{\alpha}T(t)$ is bounded and*

$$\|A^{\alpha}T(t)\| \leq M_{\alpha}t^{-\alpha}e^{-\delta t}. \qquad (6.25)$$

(d)  *Let $0 < \alpha \leq 1$ and $x \in D(A^{\alpha})$ then*

$$\|T(t)x - x\| \leq C_{\alpha}t^{\alpha}\|A^{\alpha}x\|. \qquad (6.26)$$

PROOF. Our assumptions on $A$ imply that it satisfies Assumption 6.1 with $\omega < \pi/2$ and therefore we have the existence of $A^\alpha$ for $\alpha \geq 0$. Since $T(t)$ is analytic we have $T(t): X \to \cap_{n=0}^\infty D(A^n) \subset D(A^\alpha)$ for every $\alpha \geq 0$ which proves (a).

Let $x \in D(A^\alpha)$ then $x = A^{-\alpha}y$ for some $y \in X$ and

$$T(t)x = T(t)A^{-\alpha}y = \frac{1}{\Gamma(\alpha)} \int_0^\infty s^{\alpha-1}T(s)T(t)y\,ds$$

$$= A^{-\alpha}T(t)y = A^{-\alpha}T(t)A^\alpha x$$

and (b) follows.

Since $A^\alpha$ is closed so is $A^\alpha T(t)$. By part (a) $A^\alpha T(t)$ is everywhere defined and therefore by the closed graph theorem $A^\alpha T(t)$ is bounded. Let $n-1 < \alpha \leq n$ then using (6.7) we have

$$\|A^\alpha T(t)\| = \|A^{\alpha-n}A^n T(t)\| \leq \frac{1}{\Gamma(n-\alpha)} \int_0^\infty s^{n-\alpha-1}\|A^n T(t+s)\|\,ds$$

$$\leq \frac{M_n}{\Gamma(n-\alpha)} \int_0^\infty s^{n-\alpha-1}(t+s)^{-n}e^{-\delta(t+s)}\,ds$$

$$\leq \frac{M_n e^{-\delta t}}{\Gamma(n-\alpha)t^\alpha} \int_0^\infty u^{n-\alpha-1}(1+u)^{-n}\,du = \frac{M_\alpha}{t^\alpha}e^{-\delta t}.$$

Finally,

$$\|T(t)x - x\| = \left\| \int_0^t AT(s)x\,ds \right\| = \left\| \int_0^t A^{1-\alpha}T(s)A^\alpha x\,ds \right\|$$

$$\leq C \int_0^t s^{\alpha-1}\|A^\alpha x\|\,ds = C_\alpha t^\alpha\|A^\alpha x\|. \qquad \square$$

# CHAPTER 3

# Perturbations and Approximations

## 3.1. Perturbations by Bounded Linear Operators

**Theorem 1.1.** *Let $X$ be a Banach space and let $A$ be the infinitesimal generator of a $C_0$ semigroup $T(t)$ on $X$, satisfying $\|T(t)\| \leq Me^{\omega t}$. If $B$ is a bounded linear operator on $X$ then $A + B$ is the infinitesimal generator of a $C_0$ semigroup $S(t)$ on $X$, satisfying $\|S(t)\| \leq Me^{(\omega + M\|B\|)t}$.*

PROOF. From Lemma 1.5.1 and Theorem 1.5.3 it follows that there exists a norm $|\cdot|$ on $X$ such that $\|x\| \leq |x| \leq M\|x\|$ for every $x \in X$, $|T(t)| \leq e^{\omega t}$ and $|R(\lambda : A)| \leq (\lambda - \omega)^{-1}$ for real $\lambda$ satisfying $\lambda > \omega$. Thus, for $\lambda > \omega + |B|$ the bounded operator $BR(\lambda : A)$ satisfies $|BR(\lambda : A)| < 1$ and therefore $I - BR(\lambda : A)$ is invertible for $\lambda > \omega + |B|$. Set

$$R = R(\lambda : A)(I - BR(\lambda : A))^{-1} = \sum_{k=0}^{\infty} R(\lambda : A)[BR(\lambda : A)]^k \quad (1.1)$$

then

$$(\lambda I - A - B)R = (I - BR(\lambda : A))^{-1}$$
$$- BR(\lambda : A)(I - BR(\lambda : A))^{-1} = I$$

and

$$R \cdot (\lambda I - A - B)x = R(\lambda : A)(\lambda I - A - B)x$$
$$+ \sum_{k=1}^{\infty} R(\lambda : A)[BR(\lambda : A)]^k(\lambda I - A - B)x$$
$$= x - R(\lambda : A)Bx + \sum_{k=1}^{\infty} [R(\lambda : A)B]^k x$$
$$- \sum_{k=2}^{\infty} [R(\lambda : A)B]^k x = x$$

for every $x \in D(A)$. Therefore, the resolvent of $A + B$ exists for $\lambda > \omega + |B|$ and it is given by the operator $R$. Moreover,

$$|(\lambda I - A - B)^{-1}| = \left| \sum_{k=0}^{\infty} R(\lambda : A)[BR(\lambda : A)]^k \right|$$

$$\leq (\lambda - \omega)^{-1}(1 - |BR(\lambda : A)|)^{-1} \leq (\lambda - \omega - |B|)^{-1}.$$

From Corollary 1.3.8 it follows that $A + B$ is the infinitesimal generator of a $C_0$ semigroup $S(t)$, satisfying $|S(t)| \leq e^{(\omega + |B|)t}$. Returning to the original norm $\| \ \|$ on $X$ we have,

$$\|S(t)\| \leq M e^{(\omega + M\|B\|)t}. \qquad \square$$

We are now interested in the relations between the semigroup $T(t)$ generated by $A$ and the semigroup $S(t)$ generated by $A + B$. To this end we consider the operator $H(s) = T(t - s)S(s)$. For $x \in D(A) = D(A + B)$, $s \to H(s)x$ is differentiable and $H'(s)x = T(t - s)BS(s)x$. Integrating $H'(s)x$ from 0 to $t$ yields

$$S(t)x = T(t)x + \int_0^t T(t - s)BS(s)x \, ds \qquad \text{for} \quad x \in D(A). \quad (1.2)$$

Since the operators on both sides of (1.2) are bounded, (1.2) holds for every $x \in X$. The semigroup $S(t)$ is therefore a solution of the integral equation (1.2). For such integral equations we have:

**Proposition 1.2.** *Let $T(t)$ be a $C_0$ semigroup satisfying $\|T(t)\| \leq M e^{\omega t}$. Let $B$ be a bounded operator on $X$. Then there exists a unique family $V(t), t \geq 0$ of bounded operators on $X$ such that $t \to V(t)x$ is continuous on $[0, \infty[$ for every $x \in X$ and*

$$V(t)x = T(t)x + \int_0^t T(t - s)BV(s)x \, ds \qquad \text{for} \quad x \in X. \quad (1.3)$$

PROOF. Set

$$V_0(t) = T(t) \qquad (1.4)$$

and define $V_n(t)$ inductively by

$$V_{n+1}(t)x = \int_0^t T(t - s)BV_n(s)x \, ds \qquad \text{for} \quad x \in X, n \geq 0. \quad (1.5)$$

From this definition it is obvious that $t \to V_n(t)x$ is continuous for $x \in X$, $t \geq 0$ and every $n \geq 0$. Next we prove by induction that,

$$\|V_n(t)\| \leq M e^{\omega t} \frac{M^n \|B\|^n t^n}{n!}. \qquad (1.6)$$

Indeed, for $n = 0$, (1.6) holds by our assumptions on $T(t)$ and the definition of $V_0(t)$. Assume (1.6) holds for $n$ then by (1.5) we have

$$\|V_{n+1}(t)x\| \leq \int_0^t Me^{\omega(t-s)}\|B\|Me^{\omega s}\frac{M^n\|B\|^n s^n}{n!}\|x\|\,ds$$

$$= Me^{\omega t}\frac{M^{n+1}\|B\|^{n+1}t^{n+1}}{(n+1)!}\|x\|$$

and thus (1.6) holds for $n > 0$. Defining

$$V(t) = \sum_{n=0}^{\infty} V_n(t), \tag{1.7}$$

it follows from (1.6) that the series (1.7) converges uniformly in the uniform operator topology on bounded intervals. Therefore $t \rightarrow V(t)x$ is continuous for every $x \in X$ and moreover by (1.4) and (1.5) it follows that for every $x \in X$, $V(t)x$ satisfies the equation (1.3). This concludes the proof of the existence statement. To prove the uniqueness let $U(t)$, $t \geq 0$ be a family of bounded operators for which $t \rightarrow U(t)x$ is continuous for every $x \in X$ and

$$U(t)x = T(t)x + \int_0^t T(t-s)BU(s)x\,ds \qquad \text{for} \quad x \in X. \tag{1.8}$$

Subtracting (1.8) from (1.2) and estimating the difference yields

$$\|(V(t) - U(t))x\| \leq \int_0^t Me^{\omega(t-s)}\|B\|\|(V(s) - U(s))x\|\,ds. \tag{1.9}$$

But (1.9) implies, for example by Gronwall's inequality, that $\|(V(t) - U(t))x\| = 0$ for every $t \geq 0$ and therefore $V(t) = U(t)$. □

From Proposition 1.2 and the fact that the semigroup $S(t)$ generated by $A + B$ satisfies the integral equation (1.2) we immediately obtain the following explicit representation of $S(t)$ in terms of $T(t)$:

$$S(t) = \sum_{n=0}^{\infty} S_n(t) \tag{1.10}$$

where $S_0(t) = T(t)$,

$$S_{n+1}(t)x = \int_0^t T(t-s)BS_n(s)x\,ds \qquad x \in X \tag{1.11}$$

and the convergence in (1.10) is in the uniform operator topology.
For the difference between $T(t)$ and $S(t)$ we have:

**Corollary 1.3.** *Let $A$ be the infinitesimal generator of a $C_0$ semigroup $T(t)$ satisfying $\|T(t)\| \leq Me^{\omega t}$. Let $B$ be a bounded operator and let $S(t)$ be the $C_0$ semigroup generated by $A + B$. Then*

$$\|S(t) - T(t)\| \leq Me^{\omega t}(e^{M\|B\|t} - 1). \tag{1.12}$$

PROOF. From the integral equation (1.2) and Theorem 1.1 we have

$$\|S(t)x - T(t)x\| \leq \int_0^t \|T(t-s)\| \, \|B\| \, \|S(s)\| \, \|x\| \, ds$$

$$\leq M^2 e^{\omega t} \|B\| \int_0^t e^{M\|B\|s} \|x\| \, ds$$

$$= M e^{\omega t} (e^{M\|B\|t} - 1) \|x\|. \qquad \square$$

The main result of this section, Theorem 1.1, shows that the addition of a bounded linear operator $B$ to an infinitesimal generator $A$ of a $C_0$ semigroup, does not destroy this property of $A$. It is natural to ask which special properties of the semigroup $T(t)$ generated by $A$ are preserved when $A$ is perturbed by a bounded operator $B$. It is not difficult to show that if $A$ is the infinitesimal generator of a compact or analytic semigroup so is $A + B$. We will prove here the statement about compact semigroups. The statement about analytic semigroups will follow from the results of the next section (see Corollary 2.2).

**Proposition 1.4.** *Let $A$ be the infinitesimal generator of a compact $C_0$ semigroup $T(t)$. Let $B$ be a bounded operator, then $A + B$ is the infinitesimal generator of a compact $C_0$ semigroup $S(t)$.*

PROOF. From Theorem 2.3.3 it follows that $T(t)$ is continuous in the uniform operator topology for $t > 0$ and that $R(\lambda : A)$ is compact for $\lambda \in \rho(A)$. Since

$$R(\lambda : A + B) = \sum_{k=0}^{\infty} R(\lambda : A)[BR(\lambda : A)]^k \qquad (1.13)$$

and $\|R(\lambda : A)\| \leq M(\lambda - \omega)^{-1}$ for $\lambda > \omega$, it follows that for $\lambda > \omega + M\|B\| + 1$, (1.13) converges in $B(X)$ and since each one of the terms on the right-hand side of (1.13) is compact so is $R(\lambda : A + B)$ for $\lambda > \omega + M\|B\| + 1$. From the resolvent identity it follows that $R(\lambda : A + B)$ is compact for every $\lambda \in \rho(A + B)$. To show that $S(t)$ is a compact semigroup it is therefore sufficient, by Theorem 2.3.3., to show that $S(t)$ is continuous in the uniform operator topology for $t > 0$. To show this we note that if $T(t)$ is continuous in the uniform operator topology for $t > 0$ then each one of the operators $S_n(t)$ defined by (1.11) is continuous in the uniform operator topology for $t > 0$. Since $S(t)$ is the uniform limit (on bounded $t$-sets) in the uniform operator topology of $\sum_{j=0}^n S_j(t)$ it follows that $S(t)$ is continuous in the uniform operator topology for $t > 0$. $\qquad \square$

Not all the properties of the semigroup $T(t)$ are preserved by a bounded perturbation of its infinitesimal generator. For example it is known that if $A$ is the infinitesimal generator of a semigroup $T(t)$ which is continuous in the uniform operator topology for $t \geq t_0 > 0$, or is differentiable for $t \geq t_0 > 0$

or is compact for $t \geq t_0 > 0$ then $S(t)$, the semigroup generated by $A + B$ where $B$ is a bounded operator need not have the corresponding property.

## 3.2. Perturbations of Infinitesimal Generators of Analytic Semigroups

**Theorem 2.1.** *Let $A$ be the infinitesimal generator of an analytic semigroup. Let $B$ be a closed linear operator satisfying $D(B) \supset D(A)$ and*

$$\|Bx\| \leq a\|Ax\| + b\|x\| \qquad for \quad x \in D(A). \tag{2.1}$$

*There exists a positive number $\delta$ such that if $0 \leq a \leq \delta$ then $A + B$ is the infinitesimal generator of an analytic semigroup.*

PROOF. Assume first that the semigroup $T(t)$ generated by $A$ is uniformly bounded. Then $\rho(A) \supset \Sigma = \{\lambda : |\arg \lambda| \leq \pi/2 + \omega\}$ for some $\omega > 0$ and in $\Sigma$, $\|R(\lambda : A)\| \leq M|\lambda|^{-1}$. Consider the bounded operator $BR(\lambda : A)$. From (2.1) it follows that for every $x \in X$

$$\|BR(\lambda : A)x\| \leq a\|AR(\lambda : A)x\| + b\|R(\lambda : A)x\|$$

$$\leq a(M + 1)\|x\| + \frac{bM}{|\lambda|}\|x\|. \tag{2.2}$$

Choosing $\delta = \frac{1}{2}(1 + M)^{-1}$ and $|\lambda| > 2bM$ we have $\|BR(\lambda : A)\| < 1$ and therefore the operator $I - BR(\lambda : A)$ is invertible. A simple computation shows that

$$(\lambda I - (A + B))^{-1} = R(\lambda : A)(I - BR(\lambda : A))^{-1}. \tag{2.3}$$

Thus for $|\lambda| > 2bM$ and $|\arg \lambda| \leq \pi/2 + \omega$ we obtain from (2.3) that

$$\|R(\lambda : A + B)\| \leq M'|\lambda|^{-1} \tag{2.4}$$

which implies that $A + B$ is the infinitesimal generator of an analytic semigroup.

If $T(t)$ is not uniformly bounded, let $\|T(t)\| \leq Me^{\omega t}$. Consider the semigroup $e^{-\omega t}T(t)$ generated by $A_0 = A - \omega I$. From (2.1) we have

$$\|Bx\| \leq a\|A_0 x\| + (a\omega + b)\|x\| \qquad for \quad x \in D(A).$$

Therefore, by the first part of the proof if $0 \leq a \leq \delta$, $A_0 + B = A + B - \omega I$ is the infinitesimal generator of an analytic semigroup which implies that $A + B$ is also the infinitesimal generator of an analytic semigroup. $\qquad \square$

**Remark.** In Theorem 2.1, the semigroup $S(t)$ generated by $A + B$ satisfies $\|S(t)\| \leq Me^{(\omega + \Lambda(b))t}$ where $\lim_{b \to 0} \Lambda(b) = 0$.

From the case $a = 0$ in Theorem 2.1 we obtain,

**Corollary 2.2.** *Let $A$ be the infinitesimal generator of an analytic semigroup. If $B$ is a bounded linear operator then $A + B$ is the infinitesimal generator of an analytic semigroup.*

From the proof of Theorem 2.1 one deduces easily the following corollary.

**Corollary 2.3.** *Let $A$ be the infinitesimal generator of a uniformly bounded analytic semigroup. Let $B$ be a closed operator satisfying $D(B) \supset D(A)$ and*

$$\|Bx\| \le a\|Ax\| \qquad for \quad x \in D(A). \tag{2.5}$$

*Then there exists a positive constant $\delta$ such that for $0 \le a \le \delta$, $A + B$ is the infinitesimal generator of a uniformly bounded analytic semigroup.*

**Corollary 2.4.** *Let $A$ be the infinitesimal generator of an analytic semigroup. Let $B$ be closed and suppose that for some $0 < \alpha < 1$, $D(B) \supset D(A^\alpha)$ then $A + B$ is the infinitesimal generator of an analytic semigroup.*

PROOF. Since $D(B) \supset D(A^\alpha)$ we obviously have $D(B) \supset D(A)$. From Corollary 2.6.11 it follows that

$$\|Bx\| \le C\big(\rho^\alpha\|x\| + \rho^{\alpha-1}\|Ax\|\big) \qquad for \quad x \in D(A) \text{ and } \rho > 0. \tag{2.6}$$

Choosing $\rho > 0$ so large that $C\rho^{\alpha-1} < \delta$ where $\delta$ is the constant given in the statement of Theorem 2.1, the result follows readily from Theorem 2.1.  $\square$

## 3.3. Perturbations of Infinitesimal Generators of Contraction Semigroups

We start with a definition.

**Definition 3.1.** A dissipative operator $A$ for which $R(I - A) = X$, is called *m-dissipative.*

If $A$ is dissipative so is $\mu A$ for all $\mu > 0$ and therefore if $A$ is *m*-dissipative then $R(\lambda I - A) = X$ for every $\lambda > 0$. In terms of *m*-dissipative operators the Lumer-Phillips theorem can be restated as: A densely defined operator $A$ is the infinitesimal generator of a $C_0$ semigroup of contractions if and only if it is *m*-dissipative.

The main result of this section is the following perturbation theorem for *m*-dissipative operators.

**Theorem 3.2.** *Let $A$ and $B$ be linear operators in $X$ such that $D(B) \supset D(A)$ and $A + tB$ is dissipative for $0 \le t \le 1$. If*

$$\|Bx\| \le \alpha\|Ax\| + \beta\|x\| \qquad for \quad x \in D(A) \tag{3.1}$$

*where* $0 \leq \alpha < 1$, $\beta \geq 0$ *and for some* $t_0 \in [0, 1]$, $A + t_0 B$ *is* m-*dissipative then* $A + tB$ *is* m-*dissipative for all* $t \in [0, 1]$.

PROOF. We will show that there is a $\delta > 0$ such that if $A + t_0 B$ is m-dissipative, $A + tB$ is m-dissipative for all $t \in [0, 1]$ satisfying $|t - t_0| \leq \delta$. Since every point in $[0, 1]$ can be reached from every other point by a finite number of steps of length $\delta$ or less this implies the result.

Assume that for some $t_0 \in [0, 1]$ $A + t_0 B$ is m-dissipative. Then $I - (A + t_0 B)$ is invertible. Denoting $(I - (A + t_0 B))^{-1}$ by $R(t_0)$ we have $\|R(t_0)\| \leq 1$. We show now that the operator $BR(t_0)$ is a bounded linear operator. From (3.1) and the triangle inequality we have for $x \in D(A)$

$$\|Bx\| \leq \alpha \|(A + t_0 B)x\| + \alpha t_0 \|Bx\| + \beta \|x\|$$

$$\leq \alpha \|(A + t_0 B)x\| + \alpha \|Bx\| + \beta \|x\|$$

and therefore

$$\|Bx\| \leq \frac{\alpha}{1 - \alpha} \|(A + t_0 B)x\| + \frac{\beta}{1 - \alpha} \|x\|. \tag{3.2}$$

Since $R(t_0): X \to D(A)$ and $(A + t_0 B)R(t_0) = R(t_0) - I$ it follows from (3.2) that

$$\|BR(t_0)x\| \leq \frac{\alpha}{1 - \alpha} \|(R(t_0) - I)x\| + \frac{\beta}{1 - \alpha} \|R(t_0)x\|$$

$$\leq \frac{2\alpha + \beta}{1 - \alpha} \|x\| \qquad \text{for all} \quad x \in X \tag{3.3}$$

and so $BR(t_0)$ is bounded. To show that $A + tB$ is m-dissipative we will show that $I - (A + tB)$ is invertible and thus its range is all of $X$. We have

$$I - (A + tB) = I - (A + t_0 B) + (t_0 - t)B$$

$$= (I + (t_0 - t)BR(t_0))(I - (A + t_0 B)). \tag{3.4}$$

Therefore $I - (A + tB)$ is invertible if and only if $I + (t_0 - t)BR(t_0)$ is invertible. But $I + (t_0 - t)BR(t_0)$ is invertible for all $t$ satisfying $|t - t_0| < (1 - \alpha)(2\alpha + \beta)^{-1} \leq \|BR(t_0)\|^{-1}$ and we can therefore choose $\delta = (1 - \alpha)(4\alpha + 2\beta)^{-1}$ to conclude the proof. $\qquad \square$

Theorem 3.2 is usually used through the following simple corollary.

**Corollary 3.3.** *Let* $A$ *be the infinitesimal generator of a* $C_0$ *semigroup of contractions. Let* $B$ *be dissipative and satisfy* $D(B) \supset D(A)$ *and*

$$\|Bx\| \leq \alpha \|Ax\| + \beta \|x\| \qquad \text{for} \quad x \in D(A) \tag{3.5}$$

*where* $0 \leq \alpha < 1$ *and* $\beta \geq 0$. *Then* $A + B$ *is the infinitesimal generator of a* $C_0$ *semigroup of contractions.*

PROOF. By Lumer-Phillips' theorem (Theorem 1.4.3), $\overline{D(A)} = X$ and $A$ is m-dissipative. Therefore $A + tB$ is dissipative for every $0 \leq t \leq 1$. This follows from the fact that if $A$ is $m$-dissipative $\operatorname{Re}\langle Ax, x^*\rangle \leq 0$ for every $x^* \in F(x)$. Indeed, if $B$ is dissipative with $D(B) \supset D(A)$, then for every $x \in D(A)$ there is an $x^* \in F(x)$ such that $\operatorname{Re}\langle Bx, x^*\rangle \leq 0$ and for this same $x^*$, $\operatorname{Re}\langle Ax + tBx, x^*\rangle \leq 0$. From Theorem 3.2 it follows that $A + tB$ is $m$-dissipative for all $t \in [0, 1]$ and in particular $A + B$ is $m$-dissipative. Since $D(A + B) = D(A)$ is dense in $X$, $A + B$ is the infinitesimal generator of a $C_0$ semigroup by Lumer-Phillips' theorem. □

Note that Corollary 3.3 can be stated in a slightly more symmetric form as follows: If $A + tB$ is dissipative for $t \in [0, 1]$, $D(B) \supset D(A)$, $\overline{D(A)} = X$ and (3.5) holds then either both $A$ and $A + B$ are $m$-dissipative or neither $A$ nor $A + B$ are $m$-dissipative.

Theorem 3.2 and Corollary 3.3 do not hold in general if $\alpha < 1$ in (3.1) is replaced by $\alpha = 1$. One of the reasons for this is that in this case it is no more true that $A + B$ is necessarily closed. If $A + B$ is not closed it cannot be the infinitesimal generator of a $C_0$ semigroup. A simple example of this kind of situation is provided by a self adjoint operator $iA$ in a Hilbert space. If $iA$ is self adjoint both $A$ and $-A$ are infinitesimal generators of $C_0$ semigroups of contractions (see Theorem 1.10.8). Taking $B = -A$ in Theorem 3.2 we have the estimate (3.1) with $\alpha = 1$ and $\beta = 0$, but $A + B$ restricted to $D(A)$ is not closed. In this simple example however, the closure of $A + B$, i.e., the zero operator on the whole space, is the infinitesimal generator of a $C_0$ semigroup of contractions. Our next theorem shows that under a certain additional assumption this is always the case.

**Theorem 3.4.** *Let $A$ be the infinitesimal generator of a $C_0$ semigroup of contractions. Let $B$ be dissipative such that $D(B) \supset D(A)$ and*

$$\|Bx\| \leq \|Ax\| + \beta\|x\| \qquad for \quad x \in D(A) \tag{3.6}$$

*where $\beta \geq 0$ is a constant. If $B^*$, the adjoint of $B$, is densely defined then the closure $\overline{A + B}$ of $A + B$ is the infinitesimal generator of a $C_0$ semigroup of contractions.*

PROOF. $A + B$ is dissipative and densely defined since $A$ is $m$-dissipative and $B$ is dissipative with $D(B) \supset D(A)$. Therefore, by Theorem 1.4.5, $A + B$ is closable and its closure $\overline{A + B}$ is dissipative. To prove that $\overline{A + B}$ is the infinitesimal generator of a $C_0$ semigroup of contractions it is therefore sufficient to show that $R(I - (\overline{A + B})) = X$. Since $\overline{A + B}$ is dissipative and closed, it follows from Theorem 1.4.2. that $R(I - (\overline{A + B}))$ is closed and therefore it suffices to show that $R(I - (\overline{A + B}))$ is dense in $X$.

Let $y^* \in X^*$ be "orthogonal" to the range of $I - (\overline{A + B})$, that is, $\langle y^*, z\rangle = 0$ for every $z \in R(I - (\overline{A + B}))$. Let $y \in X$ be such that $\|y^*\| \leq \langle y^*, y\rangle$. From Corollary 3.3 it follows that $A + tB$ is $m$-dissipative for $0 \leq t < 1$ and therefore the equation

$$x_t - Ax_t - tBx_t = y \tag{3.7}$$

has a unique solution $x_t$ for every $0 \leq t < 1$. Moreover, since $A + tB$ is dissipative $\|x_t\| \leq \|y\|$. From (3.6) it follows that

$$\|Bx_t\| \leq \|Ax_t\| + \beta\|x_t\| \leq \|(A + tB)x_t\| + t\|Bx_t\| + \beta\|x_t\|$$
$$\leq \|y - x_t\| + t\|Bx_t\| + \beta\|x_t\|$$

and therefore,

$$(1 - t)\|Bx_t\| \leq \|y - x_t\| + \beta\|x_t\| \leq (2 + \beta)\|y\|. \qquad (3.8)$$

Let $z^* \in D(B^*)$ then

$$|\langle z^*, (1 - t)Bx_t \rangle| = (1 - t)|\langle B^*z^*, x_t \rangle|$$
$$\leq (1 - t)\|B^*z^*\|\,\|y\| \to 0 \quad \text{as} \quad t \to 1. \qquad (3.9)$$

Since $D(B^*)$ is dense in $X^*$ and since by (3.8) $(1 - t)Bx_t$ is uniformly bounded it follows from (3.9) that $(1 - t)Bx_t$ tends weakly to zero as $t \to 1$. In particular by the choice of $y^*$ we have

$$\|y^*\| \leq \langle y^*, y \rangle = \langle y^*, x_t - Ax_t - tBx_t \rangle$$
$$= \langle y^*, (1 - t)Bx_t \rangle \to 0 \qquad \text{as} \quad t \to 1$$

which implies $y^* = 0$ and the range of $I - \overline{(A + B)}$ is dense in $X$. $\qquad \square$

Let $X$ be a reflexive Banach space and let $T$ be a closable densely defined operator in $X$. Then it is well known that $T^*$ is closed and $D(T^*)$ is dense in $X^*$ (see Lemma 1.10.5). Therefore, for reflexive Banach spaces we have:

**Corollary 3.5.** *Let $X$ be a reflexive Banach space and let $A$ be the infinitesimal generator of a $C_0$ semigroup of contractions in $X$. Let $B$ be dissipative such that $D(B) \supset D(A)$ and*

$$\|Bx\| \leq \|Ax\| + \beta\|x\| \qquad \textit{for} \quad x \in D(A)$$

*where $\beta \geq 0$. Then $\overline{A + B}$, the closure of $A + B$, is the infinitesimal generator of a $C_0$ semigroup of contractions in $X$.*

## 3.4. The Trotter Approximation Theorem

In this section we study, roughly speaking, the continuous dependence of a semigroup $T(t)$ on its infinitesimal generator $A$ and the continuous dependence of $A$ on $T(t)$. We show that the convergence (in an appropriate sense) of a sequence of infinitesimal generators is equivalent to the convergence of the corresponding semigroups. We start with a lemma.

**Lemma 4.1.** *Let A and B be the infinitesimal generators of $C_0$ semigroups $T(t)$ and $S(t)$ respectively. For every $x \in X$ and $\lambda \in \rho(A) \cap \rho(B)$ we have*

$$R(\lambda:B)[T(t) - S(t)]R(\lambda:A)x$$

$$= \int_0^t S(t-s)[R(\lambda:A) - R(\lambda:B)]T(s)x\,ds. \quad (4.1)$$

PROOF. For every $x \in X$ and $\lambda \in \rho(A) \cap \rho(B)$ the $X$ valued function $s \to S(t-s)R(\lambda:B)T(s)R(\lambda:A)x$ is differentiable. A simple computation yields

$$\frac{d}{ds}[S(t-s)R(\lambda:B)T(s)R(\lambda:A)x]$$

$$= S(t-s)[-BR(\lambda:B)T(s) + R(\lambda:B)T(s)A]R(\lambda:A)x$$

$$= S(t-s)[R(\lambda:A) - R(\lambda:B)]T(s)x$$

where we have used the fact that $R(\lambda:A)T(s)x = T(s)R(\lambda:A)x$. Integrating the last equation from 0 to $t$ yields (4.1).                    □

In the sequel we will use the notation $A \in G(M, \omega)$ for an operator $A$ which is the infinitesimal generator of a $C_0$ semigroup $T(t)$ satisfying $\|T(t)\| \leq Me^{\omega t}$.

**Theorem 4.2.** *Let $A$, $A_n \in G(M, \omega)$ and let $T(t)$ and $T_n(t)$ be the semigroups generated by $A$ and $A_n$ respectively then the following are equivalent:*

(a) *For every $x \in X$ and $\lambda$ with $\mathrm{Re}\,\lambda > \omega$. $R(\lambda:A_n)x \to R(\lambda:A)x$ as $n \to \infty$.*
(b) *For every $x \in X$ and $t \geq 0$, $T_n(t)x \to T(t)x$ as $n \to \infty$.*

*Moreover, the convergence in part* (b) *is uniform on bounded t-intervals.*

PROOF. We start by showing that (a) ⇒ (b). Fix $x \in X$ and an interval $0 \leq t \leq T$ and consider

$$\|(T_n(t) - T(t))R(\lambda:A)x\| \leq \|T_n(t)(R(\lambda:A) - R(\lambda:A_n))x\|$$

$$+ \|R(\lambda:A_n)(T_n(t) - T(t))x\|$$

$$+ \|(R(\lambda:A_n) - R(\lambda:A))T(t)x\|$$

$$= D_1 + D_2 + D_3 \quad (4.2)$$

Since $\|T_n(t)\| \leq Me^{\omega T}$ for $0 \leq t \leq T$ it follows from (a) that $D_1 \to 0$ as $n \to \infty$ uniformly on $[0, T]$. Also, since $t \to T(t)x$ is continuous the set $\{T(t)x : 0 \leq t \leq T\}$ is compact in $X$ and therefore $D_3 \to 0$ as $n \to \infty$

uniformly on $[0, T]$. Finally, using Lemma 4.1 with $B = A_n$ we have

$$\|R(\lambda : A_n)(T_n(t) - T(t))R(\lambda : A)x\|$$

$$\leq \int_0^t \|T_n(t - s)\| \, \|(R(\lambda : A) - R(\lambda : A_n))T(s)x\| \, ds$$

$$\leq \int_0^t \|T_n(t - s)\| \, \|(R(\lambda : A) - R(\lambda : A_n))T(s)x\| \, dx. \qquad (4.3)$$

The integrand on the right hand side of (4.3) is bounded by $2M^3 e^{2\omega t}(\mathrm{Re}\, \lambda - \omega)^{-1}\|x\|$ and it tends to zero as $n \to \infty$. By Lebesgue's bounded convergence theorem the right hand side of (4.3) tends to zero and therefore

$$\lim_{n \to \infty} \|R(\lambda : A_n)(T_n(t) - T(t))R(\lambda : A)x\| = 0 \qquad (4.4)$$

and the limit in (4.4) is uniform on $[0, T]$. Since every $x \in D(A)$ can be written as $x = R(\lambda : A)z$ for some $z \in X$ it follows from (4.4) that for $x \in D(A)$, $D_2 \to 0$ as $n \to \infty$ uniformly on $[0, T]$. From (4.2) it then follows that for $x \in D(A^2)$

$$\lim_{n \to \infty} \|(T_n(t) - T(t))x\| = 0 \qquad (4.5)$$

and the limit in (4.5) is uniform on $[0, T]$. Since $\|T_n(t) - T(t)\|$ are uniformly bounded on $[0, T]$ and since $D(A^2)$ is dense in $X$ (see Theorem 1.2.7) it follows that (4.5) holds for every $x \in X$ uniformly on $[0, T]$ and (a) $\Rightarrow$ (b).

Assume now that (b) holds and $\mathrm{Re}\, \lambda > \omega$ then

$$\|R(\lambda : A_n)x - R(\lambda : A)x\| \leq \int_0^\infty e^{-\mathrm{Re}\, \lambda t}\|(T_n(t) - T(t))x\| \, dt. \quad (4.6)$$

The right-hand side of (4.6) tends to zero as $n \to \infty$ by (b) and Lebesgue's dominated convergence theorem and therefore (b) $\Rightarrow$ (a).                                        $\square$

**Remark.** From the proof of Theorem 4.2 it is clear that a weaker version of (a) namely, for all $x \in X$ and some $\lambda_0$ with $\mathrm{Re}\, \lambda_0 > \omega$, $R(\lambda_0 : A_n)x \to R(\lambda_0 : A)x$ as $n \to \infty$, still implies (b).

We say that a sequence of operators $A_n$, *r-converges* to an operator $A$ if for some complex $\lambda$, $R(\lambda : A_n)x \to R(\lambda : A)x$ for all $x \in X$. In Theorem 4.2 we assumed the existence of the *r*-limit $A$ of the sequence $A_n$ and furthermore assumed that $A \in G(M, \omega)$. It turns out that these assumptions are unnecessary. This will follow from our next theorem.

**Theorem 4.3.** *Let $A_n \in G(M, \omega)$. If there exists a $\lambda_0$ with $\mathrm{Re}\, \lambda_0 > \omega$ such that*

(a) *for every $x \in X$, $R(\lambda_0 : A_n)x \to R(\lambda_0)x$ as $n \to \infty$ and*
(b) *the range of $R(\lambda_0)$ is dense in $X$,*

*then there exists a unique operator $A \in G(M, \omega)$ such that $R(\lambda_0) = R(\lambda_0 : A)$.*

PROOF. We will assume without loss of generality that $\omega = 0$ and start by proving that $R(\lambda : A_n)x$ converges as $n \to \infty$ for every $\lambda$ with Re $\lambda > 0$. Indeed, let $S = \{\lambda : \text{Re } \lambda > 0, R(\lambda : A_n)x \text{ converges as } n \to \infty\}$. $S$ is open. To see this expand $R(\lambda : A_n)$ in a Taylor series around a point $\mu$ at which $R(\mu : A_n)x$ converges as $n \to \infty$. Then

$$R(\lambda : A_n) = \sum_{k=0}^{\infty} (\mu - \lambda)^k R(\mu : A_n)^{k+1}. \qquad (4.7)$$

Since by Remark 1.5.4 $\|R(\mu : A_n)^k\| \leq M(\text{Re } \mu)^{-k}$, the series (4.7) converges in the uniform operator topology for all $\lambda$ satisfying $|\mu - \lambda|(\text{Re } \mu)^{-1} < 1$. The convergence is uniform in $\lambda$ for $\lambda$ satisfying $|\mu - \lambda|(\text{Re } \mu)^{-1} \leq \vartheta < 1$ and the series of constants $\sum_{k=0}^{\infty} M \vartheta^{k+1}$ is majorant to the series $\sum_{k=0}^{\infty} |\mu - \lambda|^k \|R(\mu : A_n)^{k+1}\|$. This implies the convergence of $R(\lambda : A_n)x$ as $n \to \infty$ for all $\lambda$ satisfying $|\mu - \lambda|(\text{Re } \mu)^{-1} \leq \vartheta < 1$, and the set $S$ is open as claimed. Let $\lambda$ be a cluster point of $S$ with Re $\lambda > 0$. Given $0 < \vartheta < 1$ there exists a point $\mu \in S$ such that $|\mu - \lambda|(\text{Re } \mu)^{-1} \leq \vartheta < 1$ and therefore by the first part of the proof $R(\lambda : A_n)x$ converges as $n \to \infty$, i.e., $\lambda \in S$. Thus $S$ is relatively closed in Re $\lambda > 0$. Since by assumption $\lambda_0 \in S$ we conclude that $S = \{\lambda : \text{Re } \lambda > 0\}$.

For every $\lambda$ with Re $\lambda > 0$ we define a linear operator $R(\lambda)$ by

$$R(\lambda)x = \lim_{n \to \infty} R(\lambda : A_n)x. \qquad (4.8)$$

Clearly,

$$R(\lambda) - R(\mu) = (\mu - \lambda)R(\lambda)R(\mu) \qquad \text{for} \quad \text{Re } \lambda > 0 \text{ and Re } \mu > 0 \qquad (4.9)$$

and therefore $R(\lambda)$ is a pseudo resolvent on Re $\lambda > 0$ (see Definition 1.9.1). Since for a pseudo resolvent the range of $R(\lambda)$ is independent of $\lambda$ (see Lemma 1.9.2.) we have by (b) that the range of $R(\lambda)$ is dense in $X$. Also, from the definition of $R(\lambda)$ it is clear that

$$\|R(\lambda)^k\| \leq M(\text{Re } \lambda)^{-k} \qquad \text{for} \quad \text{Re } \lambda > 0, k = 1, 2, \ldots . \qquad (4.10)$$

In particular for real $\lambda$, $\lambda > 0$

$$\|\lambda R(\lambda)\| \leq M \qquad \text{for all} \quad \lambda > 0. \qquad (4.11)$$

It follows from Theorem 1.9.4 that there exists a unique closed densely defined linear operator $A$ for which $R(\lambda) = R(\lambda : A)$. Finally, from (4.10) and Theorem 1.5.2 it follows that $A \in G(M, 0)$ and the proof is complete. $\square$

A direct consequence of Theorems 4.2 and 4.3 is the following theorem.

**Theorem 4.4** (Trotter-Kato). *Let $A_n \in G(M, \omega)$ and let $T_n(t)$ be the semigroup whose infinitesimal generator is $A_n$. If for some $\lambda_0$ with Re $\lambda_0 > \omega$ we have:*

(a) *As $n \to \infty$, $R(\lambda_0 : A_n)x \to R(\lambda_0)x$ for all $x \in X$ and*
(b) *the range of $R(\lambda_0)$ is dense in $X$,*

*then there exists a unique operator $A \in G(M, \omega)$ such that $R(\lambda_0) = R(\lambda_0 : A)$. If $T(t)$ is the $C_0$ semigroup generated by $A$ then as $n \to \infty$, $T_n(t)x \to T(t)x$ for all $t \geq 0$ and $x \in X$. The limit is uniform in $t$ for $t$ in bounded intervals.*

A somewhat different consequence of the previous results is the following theorem.

**Theorem 4.5.** *Let $A_n \in G(M, \omega)$ and assume*

(a) *As $n \to \infty$, $A_n x \to Ax$ for every $x \in D$ where $D$ is a dense subset of $X$.*
(b) *There exists a $\lambda_0$ with $\operatorname{Re} \lambda_0 > \omega$ for which $(\lambda_0 I - A)D$ is dense in $X$, then the closure $\bar{A}$ of $A$ is in $G(M, \omega)$. If $T_n(t)$ and $T(t)$ are the $C_0$ semigroups generated by $A_n$ and $\bar{A}$ respectively then*

$$\lim_{n \to \infty} T_n(t)x = T(t)x \qquad \text{for all} \quad t \geq 0, x \in X \qquad (4.12)$$

*and the limit in (4.12) is uniform in $t$ for $t$ in bounded intervals.*

PROOF. Let $y \in D$, $x = (\lambda_0 I - A)y$ and $x_n = (\lambda_0 I - A_n)y$. Since $A_n y \to Ay$, $x_n \to x$ as $n \to \infty$. Also since $\|R(\lambda_0 : A_n)\| \leq M(\operatorname{Re} \lambda_0 - \omega)^{-1}$ it follows that

$$\lim_{n \to \infty} R(\lambda_0 : A_n)x = \lim_{n \to \infty} \left( R(\lambda_0 : A_n)(x - x_n) + y \right) = y \quad (4.13)$$

i.e., $R(\lambda_0 : A_n)$ converges on the range of $\lambda_0 I - A$. But by (b) this range is dense in $X$ and by our assumptions $\|R(\lambda_0 : A_n)\|$ are uniformly bounded. Therefore $R(\lambda_0 : A_n)x$ converges for every $x \in X$. Let

$$\lim_{n \to \infty} R(\lambda_0 : A_n)x = R(\lambda_0)x. \qquad (4.14)$$

From (4.13) it follows that the range of $R(\lambda_0)$ contains $D$ and is therefore dense in $X$. Theorem 4.3 implies the existence of an operator $A' \in G(M, \omega)$ satisfying $R(\lambda_0) = R(\lambda_0 : A')$. To conclude the proof we show that $\bar{A} = A'$. Let $x \in D$ then

$$\lim_{n \to \infty} R(\lambda_0 : A_n)(\lambda_0 I - A)x = R(\lambda_0 : A')(\lambda_0 I - A)x. \qquad (4.15)$$

On the other hand as $n \to \infty$

$$R(\lambda_0 : A_n)(\lambda_0 I - A)x = R(\lambda_0 : A_n)(\lambda_0 I - A_n)x$$
$$+ R(\lambda_0 : A_n)(A_n - A)x$$
$$= x + R(\lambda_0 : A_n)(A_n - A)x \to x,$$

since $\|R(\lambda_0 : A_n)\|$ are uniformly bounded and for $x \in D$, $A_n x \to Ax$. Therefore

$$R(\lambda_0 : A')(\lambda_0 I - A)x = x \qquad \text{for} \quad x \in D. \qquad (4.16)$$

But (4.16) implies $A'x = Ax$ for $x \in D$ and therefore $A' \supset A$. Since $A'$ is closed, $A$ is closable. Next we show that $\bar{A} \supset A'$. Let $y' = A'x'$. Since $(\lambda_0 I - A)D$ is dense in $X$ there exists a sequence $x_n \in D$ such that

$$y_n = (\lambda_0 I - A')x_n = (\lambda_0 I - A)x_n \to \lambda_0 x' - y'$$

$$= (\lambda_0 I - A')x' \qquad \text{as} \quad n \to \infty. \tag{4.17}$$

Therefore,

$$x_n = R(\lambda_0 : A')y_n \to R(\lambda_0 : A')(\lambda_0 I - A')x' = x' \qquad \text{as} \quad n \to \infty \tag{4.18}$$

and

$$Ax_n = \lambda_0 x_n - y_n \to y' \qquad \text{as} \quad n \to \infty. \tag{4.19}$$

From (4.18) and (4.19) it follows that $y' = \bar{A}x'$ and $\bar{A} \supset A'$. Thus $\bar{A} = A'$. The rest of the assertions of the theorem follow now directly from Theorem 4.4.                                                                        □

## 3.5. A General Representation Theorem

Using the results of the previous section we will obtain in this section a representation theorem which generalizes considerably the results of Section 1.8. We start with a preliminary estimate.

**Lemma 5.1.** *Let $T$ be a bounded linear operator satisfying*

$$\|T^k\| \le MN^k, \qquad k = 1, 2, \ldots \tag{5.1}$$

*where $N \ge 1$. Then for every $n \ge 0$ we have*

$$\|e^{(T-I)n}x - T^n x\| \le MN^{n-1}e^{(N-1)n}\left[n^2(N-1)^2 + nN\right]^{1/2}\|x - Tx\|. \tag{5.2}$$

PROOF. Let $k, n \ge 0$ be integers. If $k \ge n$ then

$$\|T^k x - T^n x\| = \left\| \sum_{j=n}^{k-1} (T^{j+1}x - T^j x) \right\|$$

$$\le M\|x - Tx\| \sum_{j=n}^{k-1} N^j \le (k-n)MN^{k-1}\|x - Tx\|$$

$$\le |k - n|MN^{n+k-1}\|x - Tx\|. \tag{5.3}$$

From the symmetry of the estimate (5.3) with respect to $k$ and $n$ it is clear that (5.3) holds also for $n > k$. For $k = n$ we have equality and therefore

(5.3) is valid for all integers $k, n \geq 0$. Now

$$\left\| e^{t(T-I)}x - T^n x \right\| = \left\| e^{-t} \sum_{k=0}^{\infty} \frac{t^k}{k!} (T^k x - T^n x) \right\| \leq e^{-t} \sum_{k=0}^{\infty} \frac{t^k}{k!} \| T^k x - T^n x \|$$

$$\leq M N^{n-1} \| x - Tx \| e^{-t} \sum_{k=0}^{\infty} \frac{(tN)^k}{k!} |k - n|. \tag{5.4}$$

Using the Cauchy-Schwartz inequality we have

$$\sum_{k=0}^{\infty} \frac{(tN)^k}{k!} |n - k| \leq \left( \sum_{k=0}^{\infty} \frac{(tN)^k}{k!} \right)^{1/2} \left( \sum_{k=0}^{\infty} \frac{(tN)^k}{k!} (n-k)^2 \right)^{1/2}$$

$$= e^{tN} \left( (n - Nt)^2 + Nt \right)^{1/2}. \tag{5.5}$$

Combining (5.4) and (5.5) we obtain

$$\| e^{t(T-I)}x - T^n x \| \leq M N^{n-1} e^{(N-1)t} \left[ (n - Nt)^2 + Nt \right]^{1/2} \| x - Tx \|. \tag{5.6}$$

Substituting $t = n$ in (5.6) we get (5.2).                                                          $\square$

**Remark.** The function $e^{t(T-I)}x$ is the solution of the differential equation $du/dt = (T - I)u$ with $u(0) = x$. The elements $T^n x$ are the polygonal approximations with steps of length 1 of the solution of this equation, i.e., the solution of the difference equations

$$u(j + 1) - u(j) = (T - I)u(j), \qquad u(0) = x.$$

**Corollary 5.2.** *If $T$ is nonexpansive on $X$, i.e., $\|T\| \leq 1$, then for every $n \geq 0$ we have*

$$\| e^{(T-I)n}x - T^n x \| \leq \sqrt{n} \, \| x - Tx \|. \tag{5.7}$$

We now turn to the representation theorem.

**Theorem 5.3.** *Let $F(\rho)$, $\rho \geq 0$ be a family of bounded linear operators satisfying*

$$\| F(\rho)^k \| \leq M e^{\omega \rho k} \qquad k = 1, 2, \ldots \tag{5.8}$$

*for some constants $\omega \geq 0$ and $M \geq 1$. Let $D$ be a dense subset of $X$ and let*

$$\lim_{\rho \to 0} \rho^{-1}(F(\rho)x - x) = Ax \qquad \text{for} \quad x \in D. \tag{5.9}$$

*If for some $\lambda_0$ with $\mathrm{Re}\, \lambda_0 > \omega$, $(\lambda_0 I - A)D$ is dense in $X$ then $A$ is closable and $\bar{A}$, the closure of $A$, satisfies $\bar{A} \in G(M, \omega)$. Moreover, if $T(t)$ is the $C_0$ semigroup generated by $\bar{A}$ then for every sequence of positive integers $k_n \to \infty$ satisfying $k_n \rho_n \to t$ we have*

$$\lim_{n \to \infty} F(\rho_n)^{k_n} x = T(t)x \qquad \text{for} \quad x \in X. \tag{5.10}$$

*Choosing $\rho_n k_n = t$ for every n, the limit in (5.10) is uniform on bounded t intervals.*

PROOF. For $\rho > 0$ consider the bounded operators $A_\rho = \rho^{-1}(F(\rho) - I)$. These operators are the infinitesimal generators of uniformly continuous semigroups $S_\rho(t)$ satisfying:

$$\|S_\rho(t)\| \leq e^{-t/\rho} \sum_{k=0}^{\infty} \left(\frac{t}{\rho}\right)^k \frac{\|F(\rho)^k\|}{k!} \leq M \exp\left\{\frac{t}{\rho}(e^{\omega\rho} - 1)\right\}.$$

Let $\varepsilon > 0$ be such that $\operatorname{Re} \lambda_0 > \omega + \varepsilon$ and let $\rho_0 > 0$ be such that for $0 < \rho \leq \rho_0$, $(e^{\omega\rho} - 1)\rho^{-1} < \omega + \varepsilon$. Then

$$\|S_\rho(t)\| \leq M e^{(\omega+\varepsilon)t} \qquad \text{for} \quad 0 < \rho \leq \rho_0.$$

From Theorem 4.5 it follows that $A$ is closable and that $\bar{A} \in G(M, \omega + \varepsilon)$. If $T(t)$ is the semigroup generated by $\bar{A}$ then Theorem 4.5 implies further that

$$\|S_\rho(t)x - T(t)x\| \to 0 \qquad \text{as} \quad \rho \to 0 \tag{5.11}$$

uniformly on bounded $t$ intervals.

On the other hand, it follows from Lemma 5.1 that

$$\|S_{\rho_n}(\rho_n k_n)x - F(\rho_n)^{k_n}x\| \leq M \exp\{\omega\rho_n(k_n - 1) + (e^{\omega\rho_n} - 1)k_n\}$$
$$\cdot \left[k_n^2(e^{\omega\rho_n} - 1)^2 + k_n e^{\omega\rho_n}\right]^{1/2} \cdot \rho_n \left\|\frac{F(\rho_n)x - x}{\rho_n}\right\|.$$

Choosing $x \in D$, $\rho_n \to 0$, $k_n \to \infty$ such that $\rho_n k_n \to t$ it is obvious that $\rho_n k_n$, $(e^{\omega\rho_n} - 1)k_n$ and $\rho_n^{-1}\|F(\rho_n)x - x\|$ stay bounded as $n \to \infty$. Therefore we have

$$\|S_{\rho_n}(\rho_n k_n)x - F(\rho_n)^{k_n}x\| \leq C\rho_n^{1/2} \to 0 \qquad \text{as} \quad n \to \infty. \tag{5.12}$$

If $\rho_n = t/k_n$ one can choose the constant $C$ independent of $t$ for $0 \leq t \leq T$, which implies, in this case, uniform convergence on bounded intervals in (5.12). For $x \in D$ we have

$$\|T(t)x - F(\rho_n)^{k_n}x\| \leq \|T(t)x - S_{\rho_n}(t)x\| + \|S_{\rho_n}(t)x - S_{\rho_n}(k_n\rho_n)x\|$$
$$+ \|S_{\rho_n}(k_n\rho_n)x - F(\rho_n)^{k_n}x\| = I_1 + I_2 + I_3.$$

From (5.11) and (5.12) it follows that $I_1 \to 0$ and $I_3 \to 0$ as $n \to \infty$. To show that $I_2 \to 0$ as $n \to \infty$ we observe that for $x \in D$, $0 \leq t \leq T$ we have for large values of $n$

$$\|S_{\rho_n}(t)x - S_{\rho_n}(\rho_n k_n)x\| \leq M e^{(\omega+\varepsilon)T}|t - \rho_n k_n| \left\|\frac{F(\rho_n)x - x}{\rho_n}\right\| \to 0$$

$$\text{as} \quad n \to \infty.$$

If $\rho_n = t/k_n$ then $I_2 \equiv 0$. This concludes the proof of (5.10) for $x \in D$.

Since $D$ is dense in $X$ and $\|T(t) - F(\rho_n)^{k_n}\|$ are uniformly bounded (5.10) holds for every $x \in X$.

Finally, the semigroup $T(t)$ generated by $\bar{A}$ satisfies $\|T(t)\| \le Me^{(\omega+\varepsilon)t}$ for every small enough $\varepsilon > 0$ and therefore it also satisfies $\|T(t)\| \le Me^{\omega t}$ and $\bar{A} \in G(M, \omega)$.                                                                    $\square$

**Corollary 5.4.** *Let* $F(\rho)$, $\rho \ge 0$ *be a family of bounded linear operators satisfying*

$$\|F(\rho)^k\| \le Me^{\omega \rho k} \qquad k = 1, 2, \ldots, \tag{5.13}$$

*for some constants* $\omega \ge 0$ *and* $M \ge 1$. *Let* $A$ *be the infinitesimal generator of a* $C_0$ *semigroup* $T(t)$. *If*

$$\rho^{-1}(F(\rho)x - x) \to Ax \qquad as \quad \rho \to 0 \text{ for every } x \in D(A) \tag{5.14}$$

*then,*

$$T(t)x = \lim_{n \to \infty} F\left(\frac{t}{n}\right)^n x \qquad for \quad x \in X \tag{5.15}$$

*and the limit is uniform on bounded t-intervals.*

PROOF. Since $A$ is the infinitesimal generator of a $C_0$ semigroup it is closed and for every real $\lambda$ large enough the range of $\lambda I - A$ is all of $X$. Therefore, our result follows readily from Theorem 5.3.                                           $\square$

As a simple consequence of Corollary 5.4, we can prove the exponential formula

$$T(t)x = \lim_{n \to \infty} \left(I - \frac{t}{n}A\right)^{-n} x \qquad for \quad x \in X \tag{5.16}$$

where $T(t)$ is a $C_0$ semigroup and $A$ is its infinitesimal generator. This formula has already been proved in Theorem 1.8.3 by a different method.

To prove (5.16) assume that $A \in G(M, \omega)$ and set $F(\rho) = (I - \rho A)^{-1} = (1/\rho)R(1/\rho : A)$, for $0 < \rho < 1/\omega$. From Theorem 1.5.3. it follows that

$$\|F(\rho)^n\| \le M(1 - \rho\omega)^{-n} \le Me^{2\omega\rho n}$$

for $\rho$ small enough. Also from Lemma 1.3.2 it follows that if $x \in D(A)$ then

$$\frac{1}{\rho}(F(\rho) - I)x = A\left(\frac{1}{\rho}R\left(\frac{1}{\rho} : A\right)x\right) \to Ax.$$

Therefore $F(\rho) = (I - \rho A)^{-1}$ satisfies the conditions of Corollary 5.4 and (5.16) is then a direct consequence of this corollary.

**Corollary 5.5.** *Let* $A_j \in G(M_j, \omega_j)$, $j = 1, 2, \ldots, k$ *and let* $S_j(t)$ *be the semigroup generated by* $A_j$. *Let* $\cap_{j=1}^k D(A_j)$ *be dense in* $X$ *and*

$$\left\|(S_1(t)S_2(t) \cdots S_k(t))^n\right\| \le Me^{\omega nt} \qquad n = 1, 2, \ldots \tag{5.17}$$

*for some constants $M \geq 1$ and $\omega \geq 0$. If for some $\lambda$ with $\operatorname{Re} \lambda > \omega$ the range of $\lambda I - (A_1 + A_2 + \cdots + A_k)$ is dense in $X$ then $\overline{A_1 + A_2 + \cdots + A_k} \in G(M, \omega)$. If $S(t)$ is the semigroup generated by $\overline{A_1 + A_2 + \cdots + A_k}$ we have*

$$S(t)x = \lim_{n \to \infty} \left( S_1\!\left(\frac{t}{n}\right) S_2\!\left(\frac{t}{n}\right) \cdots S_k\!\left(\frac{t}{n}\right) \right)^n x \qquad \text{for } x \in X \quad (5.18)$$

*and the limit is uniform on bounded $t$ intervals.*

PROOF. Set $F(t) = S_1(t)S_2(t) \cdots S_k(t)$ and $\prod_{i=1}^{j} S_i(t) = S_1(t)S_2(t) \cdots S_j(t)$. For $x \in \cap_{j=1}^{k} D(A_j)$ and $t \to 0$ we have

$$\frac{F(t)x - x}{t} = \sum_{j=1}^{k} \left( \prod_{i=1}^{j-1} S_i(t) \right) \frac{S_j(t)x - x}{t} \to (A_1 + A_2 + \cdots + A_k)x.$$

$$(5.19)$$

The result follows now directly from Theorem 5.3.                    □

Corollary 5.5 is an abstract version of the method of fractional steps which is used in solving partial differential equations. The idea behind this method is the following; in order to solve the initial value problem

$$\frac{du}{dt} = (A_1 + A_2 + \cdots + A_k)u, \qquad u(0) = x, \qquad (5.20)$$

we have only to solve the $k$ simpler problems

$$\frac{du_j}{dt} = A_j u_j, \qquad u_j(0) = u_0, \qquad j = 1, 2, \ldots, k \qquad (5.21)$$

and obtain the solution of (5.20) by combining the solutions of (5.21) according to (5.18).

The method of "alternating directions" is also a special case of this general abstract result.

We conclude this section with the analogue of Corollary 5.5 for the backwards difference approximations of (5.20).

**Corollary 5.6.** *Let $A_j \in G(M_j, \omega_j)$, $j = 1, 2, \ldots, k$. If $A_1 + A_2 + \cdots + A_k \in G(M, \omega)$ and*

$$\left\| \left( (I - tA_1)^{-1} \cdots (I - tA_k)^{-1} \right)^n \right\| \leq M e^{\omega n t} \qquad n = 1, 2, \ldots \quad (5.22)$$

*then the semigroup $S(t)$ generated by $A_1 + A_2 + \cdots + A_k$ is given by*

$$S(t)x = \lim_{n \to \infty} \left[ \left( I - \frac{t}{n}A_1 \right)^{-1} \left( I - \frac{t}{n}A_2 \right)^{-1} \cdots \left( I - \frac{t}{n}A_k \right)^{-1} \right]^n x$$

$$\text{for every } x \in X. \quad (5.23)$$

PROOF. Set $F(t) = (I - tA_1)^{-1}(I - tA_2)^{-1} \cdots (I - tA_k)^{-1} = \prod_{i=1}^k (I - tA_i)^{-1}$. For $x \in \cap_{j=1}^k D(A_j) = D(A_1 + \cdots + A_k)$ and $t \to 0$ we have

$$\frac{F(t)x - x}{t} = \sum_{j=1}^k \left( \prod_{i=1}^{j-1} (I - tA_i)^{-1} \right) \frac{(I - tA_j)^{-1}x - x}{t}$$

$$\to (A_1 + A_2 + \cdots + A_k)x. \tag{5.24}$$

Here we used that if $A \in G(M, \omega)$ then $(I - tA)^{-1}y \to y$ as $t \to 0$ for every $y \in X$ and $t^{-1}((I - tA)^{-1}x - x) \to Ax$ as $t \to 0$ for $x \in D(A)$. The result follows now directly from Corollary 5.4.                                                   □


## 3.6. Approximation by Discrete Semigroups

In this section we show by means of an example how the results of the previous sections can be applied to obtain solutions of initial value problems for partial differential equations via difference equations.

The results that we present here are rather special in the sense that stronger results of similar nature may be obtained even under somewhat weaker assumptions. Since our goal here is only to demonstrate the method we preferred to make some superfluous assumptions (e.g., part (iv) of Assumption 6.1 below) in order to avoid some of the technicalities.

Let $X$ and $X_n$ be Banach spaces with norms $\| \cdot \|$ and $\| \ \|_n$ respectively. We shall make the following assumption.

**Assumption 6.1.** *For every $n \geq 1$ there exist bounded linear operators $P_n: X \to X_n$ and $E_n: X_n \to X$ such that*

(i) $\|P_n\| \leq N$, $\|E_n\| \leq N'$, $N$ and $N'$ independent of $n$.
(ii) $\|P_n x\|_n \to \|x\|$ as $n \to \infty$ for every $x \in X$.
(iii) $\|E_n P_n x - x\| \to 0$ as $n \to \infty$ for every $x \in X$.
(iv) $P_n E_n = I_n$ where $I_n$ is the identity operator on $X_n$.

EXAMPLE 6.2. Let $X = BU([-\infty, \infty])$ be the space of all bounded uniformly continuous real valued functions defined on $\mathbb{R}^1$. Let $X_n = b$ be the space of all bounded real sequences $\{c_n\}_{n=-\infty}^\infty$. Both spaces $BU([-\infty, \infty])$ and $b$ are normed with the usual supremum norm. We define $P_n f(x) = \{f(k/n)\}_{k=-\infty}^\infty$. Then $P_n$ is obviously linear and $\|P_n\| \leq 1$. From the definitions of the norms and the uniform continuity of the elements of $X$ it is also clear that (ii) is satisfied. Taking for $E_n$ the linear operator which assigns to a sequence $\{c_k\}_{k=-\infty}^\infty$ the function $f(x)$ which is equal to $c_k$ at the point $x = k/n$ and is linear between any two consecutive points $j/n$ and $j + 1/n$ we obtain $\|E_n\| \leq 1$. Obviously $P_n E_n = I_n$ and (iii) follows from the uniform continuity of the elements of $X$ and the definitions of $E_n$ and $P_n$.

**Definition 6.3.** A sequence $x_n \in X_n$ converges to $x \in X$ if

$$\|P_n x - x_n\|_n \to 0 \qquad \text{as} \quad n \to \infty. \tag{6.1}$$

This type of convergence will be denoted, without danger of confusion, by $x_n \to x$ or $\lim_{n \to \infty} x_n = x$.

Note that the limit of such a convergent sequence is unique. Indeed, if $x_n \to x$ and $x_n \to y$ then

$$\|x - y\| = \lim_{n \to \infty} \|P_n(x - y)\|_n$$

$$\leq \lim_{n \to \infty} \|P_n x - x_n\|_n + \lim_{n \to \infty} \|x_n - P_n y\|_n = 0.$$

**Definition 6.4.** A sequence of linear operators $A_n$, $A_n : X_n \to X_n$ converges to an operator $A$, $A : X \to X$ if

$$D(A) = \{ x : P_n x \in D(A_n), A_n P_n x \text{ converges} \} \tag{6.2}$$

and

$$Ax = \lim_{n \to \infty} A_n P_n x \qquad \text{for} \quad x \in D(A). \tag{6.3}$$

We will denote this type of convergence by $A_n \to \to A$.

Note that $A_n \to \to A$ means that for every $x \in D(A)$

$$\|A_n P_n x - P_n A x\|_n \to 0 \qquad \text{as} \quad n \to \infty. \tag{6.4}$$

**Lemma 6.5.** *Let $\{x_n^k\}_{k=1}^\infty$ be a Cauchy sequence in $X_n$. If for every fixed $k$, $x_n^k \to x^k \in X$ as $n \to \infty$ (in the sense of Definition 6.3) and $x_n^k \to x_n \in X_n$ uniformly with respect to $n$ as $k \to \infty$ then the double limit exists and*

$$\lim_{n \to \infty} x_n = \lim_{k \to \infty} x^k = x \in X. \tag{6.5}$$

PROOF. We first prove that $x^k$ is a convergent sequence in $X$. We have

$$\|P_n x^k - P_n x^l\|_n \leq \|P_n x^k - x_n^k\|_n + \|x_n^k - x_n^l\|_n + \|x_n^l - P_n x^l\|_n.$$

Given $\varepsilon > 0$ we choose $k, l \geq K(\varepsilon)$ such that $\|x_n^k - x_n^l\|_n < \varepsilon/6$ for all $n$. Then we choose $n_0 = n_0(k, l)$ so large that $\|P_n x^k - x_n^k\|_n < \varepsilon/6$, $\|P_n x^l - x_n^l\|_n < \varepsilon/6$ and $\|x^k - x^l\| \leq \|P_n(x^k - x^l)\|_n + \varepsilon/2$ for all $n \geq n_0$. So for $k, l \geq K(\varepsilon)$ we have, by choosing $n \geq n_0(k, l)$, $\|x^k - x^l\| < \varepsilon$ and therefore $x^k \to x$ as $k \to \infty$. Next, we show that $\lim_{n \to \infty} x_n = x$. Indeed,

$$\|P_n x - x_n\|_n \leq \|P_n x - P_n x^k\|_n + \|P_n x^k - x_n^k\|_n + \|x_n^k - x_n\|_n$$

$$\leq N\|x - x^k\| + \|P_n x^k - x_n^k\|_n + \|x_n^k - x_n\|_n.$$

Given $\varepsilon > 0$ we first choose and fix $k$ so large that $N\|x - x^k\| < \varepsilon/3$ and $\|x_n^k - x_n\|_n < \varepsilon/3$ for all $n$. Then we choose $n_0(k)$ so large that $\|P_n x^k - x_n^k\|_n < \varepsilon/3$ for all $n > n_0$. Thus $\|P_n x - x_n\|_n < \varepsilon$ for $n > n_0$ and $x_n \to x$. $\square$

**Lemma 6.6.** *Let $A_n$ be a sequence of bounded linear operators $A_n: X_n \to X_n$. If $\|A_n\| \leq M$, $A_n \to \to A$ and $D(A)$ is dense in $X$, then $D(A) = X$ and $\|A\| \leq M$.*

PROOF. Let $x \in X$. Since $D(A)$ is dense in $X$ there is a sequence $x^k \in D(A)$ such that $x^k \to x$. Now,

$$\|A_n P_n x^k - A_n P_n x\|_n \leq M \|P_n x^k - P_n x\|_n \leq MN \|x^k - x\| \quad (6.6)$$

implies $A_n P_n x^k \to A_n P_n x$ as $k \to \infty$ uniformly in $n$. Moreover, since $x^k \in D(A)$ $A_n P_n x^k \to A x^k$ as $n \to \infty$ for each $k$. Applying Lemma 6.5 to the sequence $x_n^k = A_n P_n x^k$ it follows that $A_n P_n x$ converges as $n \to \infty$ which implies $x \in D(A)$. Thus $D(A) = X$. Finally,

$$\|Ax\| = \lim_{n \to \infty} \|P_n A x\|_n = \lim_{n \to \infty} \|A_n P_n x\|_n \leq M \lim_{n \to \infty} \|P_n x\|_n = M\|x\|$$

and the proof is complete.                                                              □

**Theorem 6.7.** *Let $F(\rho_n)$ be a sequence of bounded linear operators from $X_n$ into $X_n$ satisfying*

$$\|F(\rho_n)^k\| \leq M e^{\omega \rho_n k}, \qquad k = 1, 2, \ldots \quad (6.7)$$

*and*

$$A_n = \rho_n^{-1}(F(\rho_n) - I) \to \to A. \quad (6.8)$$

*If $D(A)$ is dense in $X$ and if there is a $\lambda_0$ with $\operatorname{Re} \lambda_0 > \omega$ such that the range of $\lambda_0 I - A$ is dense in $X$ then $\bar{A}$, the closure of $A$, is the infinitesimal generator of a $C_0$ semigroup $S(t)$ on $X$. Moreover, if $k_n \rho_n \to t$ as $n \to \infty$ then*

$$F(\rho_n)^{k_n} \to \to S(t) \quad (6.9)$$

*where $D(S(t)) = X$.*

PROOF. $A_n$ is a bounded linear operator on $X_n$ and therefore generates a uniformly continuous semigroup $S_n(t)$ on $X_n$. It is easy to check (see, e.g., the proof of Theorem 5.3) that

$$\|S_n(t)\| \leq M e^{\omega_n t} \quad (6.10)$$

where $\omega_n = \rho_n^{-1}(e^{\omega \rho_n} - 1)$. Given $\varepsilon > 0$, $\omega_n \leq \omega + \varepsilon$ for all $\rho_n$ small enough.

Set $\tilde{A}_n = E_n A_n P_n$. $\tilde{A}_n$ is a bounded linear operator on $X$ and therefore generates a semigroup $\tilde{S}_n(t)$ on $X$. Using Assumption 6.1, (iv), we have

$$\tilde{S}_n(t) = \sum_{k=0}^{\infty} \frac{t^k}{k!} \tilde{A}_n^k = \sum_{k=0}^{\infty} \frac{t^k}{k!} E_n A_n^k P_n = E_n \left( \sum_{k=0}^{\infty} \frac{t^k}{k!} A_n^k \right) P_n = E_n S_n(t) P_n. \quad (6.11)$$

Therefore, for $n$ large enough we have

$$\|\tilde{S}_n(t)\| \leq M N N' e^{(\omega + \varepsilon)t} = M_1 e^{(\omega + \varepsilon)t}. \quad (6.12)$$

If $x \in D(A)$ then as $n \to \infty$

$$\|\tilde{A}_n x - Ax\| = \|E_n A_n P_n x - Ax\|$$
$$\leq \|E_n A_n P_n x - E_n P_n Ax\| + \|E_n P_n Ax - Ax\|$$
$$\leq N'\|A_n P_n x - P_n Ax\|_n + \|E_n P_n Ax - Ax\| \to 0 \quad (6.13)$$

where the first term on the right of (6.13) tends to zero as $n \to \infty$ since $x \in D(A)$ and $A_n \to \to A$, and the second term tends to zero as $n \to \infty$ by Assumption 6.1 (iii). From Theorem 4.5 it follows that $\bar{A} \in G(M_1, \omega + \varepsilon)$. Since $\varepsilon > 0$ was arbitrary we actually have $\bar{A} \in G(M_1, \omega)$, that is, $\bar{A}$ is the infinitesimal generator of a $C_0$ semigroup $S(t)$ satisfying $\|S(t)\| \leq M_1 e^{\omega t}$. From Theorem 4.5 we also have

$$\|\tilde{S}_n(t)x - S(t)x\| \to 0 \quad \text{as} \quad n \to \infty. \quad (6.14)$$

Therefore,

$$\|S_n(t)P_n x - P_n S(t)x\|_n = \|P_n \tilde{S}_n(t)x - P_n S(t)x\|_n$$
$$\leq N\|\tilde{S}_n(t)x - S(t)x\| \to 0 \quad \text{as} \quad n \to \infty.$$
$$(6.15)$$

From Lemma 5.1 we have a constant $C$ such that

$$\|S_n(\rho_n k_n)P_n x - F(\rho_n)^{k_n} P_n x\|_n$$

$$\leq C\rho_n^{1/2} \left\| \frac{F(\rho_n) - I}{\rho_n} P_n x \right\|_n = C\rho_n^{1/2}\|A_n P_n x\|_n. \quad (6.16)$$

The estimate (6.16) follows from Lemma 5.1 similarly to the way (5.12) follows from this lemma. Choosing $x \in D(A)$ we have

$$\|A_n P_n x\|_n \leq \|A_n P_n x - P_n Ax\|_n + \|P_n Ax\|_n \leq C_1. \quad (6.17)$$

Finally,

$$\left\|F(\rho_n)^{k_n} P_n x - P_n S(t)x\right\|_n \leq \left\|F(\rho_n)^{k_n} P_n x - S_n(\rho_n k_n)P_n x\right\|_n$$

$$+ \|S_n(\rho_n k_n)P_n x - S_n(t)P_n x\|_n + \|S_n(t)P_n x - P_n S(t)x\|_n$$

$$\leq \left(C\rho_n^{1/2} + |\rho_n k_n - t|\right)\|A_n P_n x\|_n + \|S_n(t)P_n x - P_n S(t)x\|_n.$$
$$(6.18)$$

Combining (6.15), (6.17) and (6.18) and letting $\rho_n \to 0$ such that $\rho_n k_n \to t$ we obtain for every $x \in D(A)$,

$$\|F(\rho_n)^{k_n} P_n x - P_n S(t)x\|_n \to 0 \quad \text{as} \quad n \to \infty. \quad (6.19)$$

Since $\|F(\rho_n)^{k_n}\|$ are uniformly bounded, (6.19) holds for every $x \in X$, i.e., $F(\rho_n)^{k_n} \to \to S(t)$. $\qquad \square$

**Remark.** From the proof of Theorem 6.7 it follows that if $k_n = [t/\rho_n]$ the convergence of $F(\rho_n)^{k_n}$ to $S(t)$ is uniform on bounded $t$ intervals.

We now turn to a concrete example. Let $X = BU([-\infty, \infty])$ be the space of all bounded uniformly continuous real valued functions on $\mathbb{R}^1$, and consider the following initial value problem for the classical heat equation

$$\begin{cases} \dfrac{\partial u}{\partial t} = \dfrac{\partial^2 u}{\partial x^2} & \text{for } -\infty < x < \infty, t > 0 \\ u(0, x) = f(x) & \text{for } -\infty < x < \infty \end{cases} \tag{6.20}$$

with $f \in X$. We intend to prove the existence of a solution $u(t, x)$ of (6.20). Furthermore, we will also obtain a numerical approximation of the solution. This will be done by replacing the differential equation in (6.20) by a difference equation.

In order to reduce the differential equation to a difference equation we consider for each given $n$ and $\tau_n$ functions defined on the lattice $(k/n, l\tau_n)$, $k = 0, \pm 1, \pm 2, \ldots, l = 0, 1, 2, \ldots$ in the $(x, t)$ plane. We set $u(k/n, l\tau_n) = u_{k, l}$. A reasonable difference equation that will correspond to the differential equation in (6.20) is

$$\tau_n^{-1}(u_{k, l+1} - u_{k, l}) = n^2(u_{k+1, l} - 2u_{k, l} + u_{k-1, l}). \tag{6.21}$$

Rearranging (6.21) we have

$$u_{k, l+1} = \left(1 - 2n^2\tau_n\right)u_{k, l} + n^2\tau_n(u_{k+1, l} + u_{k-1, l}). \tag{6.22}$$

Thus if $u_{k, 0} = f_k$ are given we can compute all $u_{k, l}$ by the recursion formula (6.22).

In order to use our previous results we consider the Banach space $X_n = b$ (i.e., the space of all bounded real sequences $\{c_n\}_{n=-\infty}^{\infty}$ with the supremum norm) and define operators $P_n$ and $E_n$ as in Example 6.2. We then define an operator $F(\tau_n)$ mapping $X_n$ into $X_n$ as follows

$$F(\tau_n)\{u_{k, l}\} = \{u_{k, l+1}\} \tag{6.23}$$

where $\{u_{k, l+1}\}$ is obtained from $\{u_{k, l}\}$ by (6.22). Set $\alpha_n = 2n^2\tau_n$ and choose $\tau_n$ such that $\alpha_n < 1$. Then

$$\|F(\tau_n)\{u_{k, l}\}\| = \sup_k |u_{k, l+1}|$$

$$\leq (1 - \alpha_n)\sup_k |u_{k, l}| + \alpha_n \sup_k |u_{k, l}| = \sup_k |u_{k, l}|.$$

$$\tag{6.24}$$

Therefore $\|F(\tau_n)\| \leq 1$ and the stability condition (6.7) of Theorem 6.7 holds with $\omega = 0$ and $M = 1$. Let

$$D = \left\{g : g, \frac{dg}{dx}, \frac{d^2g}{dx^2} \in X\right\}.$$

It is clear that $D$ is dense in $X$. For $f \in D$ we have

$$\|\tau_n^{-1}(F(\tau_n) - I)P_n f - P_n f''\|_n$$

$$= \sup_k \left| n^2 \left( f\left(\frac{k+1}{n}\right) - 2f\left(\frac{k}{n}\right) + f\left(\frac{k-1}{n}\right) \right) - f''\left(\frac{k}{n}\right) \right|.$$

(6.25)

Since $f \in D$, $f''(x)$ is uniformly continuous on $\mathbb{R}^1$ and therefore the right-hand side of (6.25) tends to zero as $n \to \infty$. The assumption (6.8) of Theorem 6.7 is thus satisfied with the operator $A$ defined on $D$ by $Af = f''$.

Finally, to apply Theorem 6.7 to our problem we have to show that for some $\lambda > 0$ the range of $\lambda I - A$ is dense in $X$. Set $\lambda = 1$. We then have to show that for a dense set of elements $h \in X$ the differential equation

$$f - f'' = h \tag{6.26}$$

has a solution $f \in D$. We will show that this is true for any $h \in X$. Let $h \in X$ and consider the function

$$f(x) = \frac{1}{2}\left( e^x \int_x^{\infty} h(\xi)e^{-\xi}\,d\xi + e^{-x}\int_{-\infty}^x h(\xi)e^{\xi}\,d\xi \right)$$

$$= \frac{1}{2}\int_{-\infty}^{\infty} h(\xi)e^{-|\xi-x|}\,d\xi. \tag{6.27}$$

It is easy to show that $f \in D$, $\|f\| \leq \|h\|$ and that $f$ is indeed the solution of (6.26).

Thus all the conditions of Theorem 6.7 are satisfied and we deduce that the closure of $A$ is the infinitesimal generator of a $C_0$ semigroup of contractions $S(t)$ on $X$. In our particular case it is not difficult to show that $A$ is closed and therefore $A$ itself is the infinitesimal generator of $S(t)$. This semigroup as we shall see in more detail in the next chapter, is the solution of the initial value problem (6.20).

Also choosing a sequence $k_n$ such that $\tau_n k_n \to t$ and $2n^2\tau_n = \alpha_n \leq \eta < 1$, we obtain from Theorem 6.7 that

$$\|F(\tau_n)^{k_n} P_n f - P_n S(t)f\|_n \to 0 \qquad \text{as} \quad n \to \infty \tag{6.28}$$

that is, the values that are computed recursively by the difference equation (6.22) at the points $(k/n, l\tau_n)$ converge to the solution of the heat equation (6.20) at $(x, t)$ where $k/n \to x$, $l\tau_n \to t$ as $n \to \infty$.

# CHAPTER 4

# The Abstract Cauchy Problem

## 4.1. The Homogeneous Initial Value Problem

Let $X$ be a Banach space and let $A$ be a linear operator from $D(A) \subset X$ into $X$. Given $x \in X$ the abstract Cauchy problem for $A$ with initial data $x$ consists of finding a solution $u(t)$ to the initial value problem

$$\begin{cases} \dfrac{du(t)}{dt} = Au(t), & t > 0 \\ u(0) = x \end{cases} \tag{1.1}$$

where by a solution we mean an $X$ valued function $u(t)$ such that $u(t)$ is continuous for $t \geq 0$, continuously differentiable and $u(t) \in D(A)$ for $t > 0$ and (1.1) is satisfied. Note that since $u(t) \in D(A)$ for $t > 0$ and $u$ is continuous at $t = 0$, (1.1) cannot have a solution for $x \notin \overline{D(A)}$.

From the results of Chapter 1 it is clear that if $A$ is the infinitesimal generator of a $C_0$ semigroup $T(t)$, the abstract Cauchy problem for $A$ has a solution, namely $u(t) = T(t)x$, for every $x \in D(A)$ (see e.g. Theorem 1.2.4). It is not difficult to show that for $x \in D(A)$, $u(t) = T(t)x$ is the only solution of (1.1). Actually, uniqueness of solutions of the initial value problem (1.1) follows from much weaker assumptions as we will see in Theorem 1.2 below.

**Lemma 1.1.** *Let $u(t)$ be a continuous $X$ valued function on $[0, T]$. If*

$$\left\| \int_0^T e^{ns} u(s) \, ds \right\| \leq M \qquad for \quad n = 1, 2, \ldots \tag{1.2}$$

*then $u(t) \equiv 0$ on $[0, T]$.*

PROOF. Let $x^* \in X^*$ and set $\varphi(t) = \langle x^*, u(t) \rangle$ then $\varphi$ is clearly continuous on $[0, T]$ and

$$\left| \int_0^T e^{ns} \varphi(s) \, ds \right| = \left| \left\langle x^*, \int_0^T e^{ns} u(s) \, ds \right\rangle \right| \leq \|x^*\| \cdot M = M_1$$

$$\text{for} \quad n = 1, 2, \ldots . \quad (1.3)$$

We will show that (1.3) implies that $\varphi(t) \equiv 0$ on $[0, T]$ and since $x^* \in X^*$ was arbitrary it follows that $u(t) \equiv 0$ on $[0, T]$.

Consider the series

$$\sum_{k=1}^{\infty} \frac{(-1)^{k-1}}{k!} e^{kn\tau} = 1 - \exp\{-e^{n\tau}\}.$$

This series converges uniformly in $\tau$ on bounded intervals. Therefore,

$$\left| \int_0^T \sum_{k=1}^{\infty} \frac{(-1)^{k-1}}{k!} e^{kn(t-T+s)} \varphi(s) \, ds \right|$$

$$\leq \sum_{k=1}^{\infty} \frac{1}{k!} e^{kn(t-T)} \left| \int_0^T e^{kns} \varphi(s) \, ds \right| \leq M_1 \left( \exp\{e^{n(t-T)}\} - 1 \right). \quad (1.4)$$

For $t < T$ the right-hand side of (1.4) tends to zero as $n \to \infty$. On the other hand we have

$$\int_0^T \sum_{k=1}^{\infty} \frac{(-1)^{k-1}}{k!} e^{kn(t-T+s)} \varphi(s) \, ds = \int_0^T \left( 1 - \exp\{-e^{n(t-T+s)}\} \right) \varphi(s) \, ds.$$

$$(1.5)$$

Using Lebesgue's dominated convergence theorem we see that the right-hand side of (1.5) converges to $\int_{T-t}^T \varphi(s) \, ds$ as $n \to \infty$. Combining this together with (1.4) we find that for every $0 \leq t < T$, $\int_{T-t}^T \varphi(s) \, ds = 0$ which implies $\varphi(t) \equiv 0$ on $[0, T]$. □

**Theorem 1.2.** *Let $A$ be a densely defined linear operator. If $R(\lambda: A)$ exists for all real $\lambda \geq \lambda_0$ and*

$$\limsup_{\lambda \to \infty} \lambda^{-1} \log \| R(\lambda: A) \| \leq 0 \quad (1.6)$$

*then the initial value problem (1.1) has at most one solution for every $x \in X$.*

PROOF. Note first that $u(t)$ is a solution of (1.1) if and only if $e^{zt} u(t)$ is a solution of the initial value problem

$$\frac{dv}{dt} = (A + zI)v, \qquad v(0) = x.$$

Thus we may translate $A$ by a constant multiple of the identity and assume that $R(\lambda: A)$ exists for all real $\lambda$, $\lambda \geq 0$ and that (1.6) is satisfied.

Let $u(t)$ be a solution of (1.1) satisfying $u(0) = 0$. We prove that $u(t) \equiv 0$. To this end consider the function $t \to R(\lambda : A)u(t)$ for $\lambda > 0$. Since $u(t)$ is a solution of (1.1) we have

$$\frac{d}{dt} R(\lambda : A)u(t) = R(\lambda : A) Au(t) = \lambda R(\lambda : A)u(t) - u(t)$$

which implies

$$R(\lambda : A)u(t) = -\int_0^t e^{\lambda(t-\tau)} u(\tau) \, d\tau. \qquad (1.7)$$

From the assumption (1.6) it follows that for every $\sigma > 0$

$$\lim_{\lambda \to \infty} e^{-\sigma\lambda} \|R(\lambda : A)\| = 0$$

and therefore it follows from (1.7) that

$$\lim_{\lambda \to \infty} \int_0^{t-\sigma} e^{\lambda(t-\sigma-\tau)} u(\tau) \, d\tau = 0. \qquad (1.8)$$

From Lemma 1.1 we deduce that $u(\tau) \equiv 0$ for $0 \le \tau \le t - \sigma$. Since $t$ and $\sigma$ were arbitrary, $u(t) \equiv 0$ for $t \ge 0$. $\qquad \square$

From Theorem 1.2 it follows that in order to obtain the uniqueness of the solutions of the initial value problem (1.1) it is not necessary to assume that $A$ is the infinitesimal generator of a $C_0$ semigroup or equivalently, that for some $\omega \in \mathbb{R}^1$, $\rho(A) \supset ]\omega, \infty[$ and $\|(\lambda - \omega)^n R(\lambda : A)^n\| \le M$ for $\lambda > \omega$, much less than this suffices for the uniqueness. Also to obtain existence of solutions of (1.1) for some dense subsets $D$ of initial values it is not necessary to assume that $A$ is the infinitesimal generator of a $C_0$ semigroup. Depending on the set $D$ of initial values, existence results can be obtained under weaker assumptions. However, in order to obtain existence and uniqueness for all $x \in D(A)$ as well as differentiability of the solution on $[0, \infty[$ one has to assume that $A$ is the infinitesimal generator of a $C_0$ semigroup. This is the contents of our next theorem.

**Theorem 1.3.** *Let $A$ be a densely defined linear operator with a nonempty resolvent set $\rho(A)$. The initial value problem (1.1) has a unique solution $u(t)$, which is continuously differentiable on $[0, \infty[$, for every initial value $x \in D(A)$ if and only if $A$ is the infinitesimal generator of a $C_0$ semigroup $T(t)$.*

PROOF. If $A$ is the infinitesimal generator of a $C_0$ semigroup $T(t)$ then from Theorem 1.2.4 it follows that for every $x \in D(A)$, $T(t)x$ is the unique solution of (1.1) with the initial value $x \in D(A)$. Moreover, $T(t)x$ is continuously differentiable for $0 \le t < \infty$.

On the other hand, if (1.1) has a unique continuously differentiable solution on $[0, \infty[$ for every initial data $x \in D(A)$ then we will see that $A$ is the infinitesimal generator of a $C_0$ semigroup $T(t)$. We now assume that for every $x \in D(A)$ the initial value problem (1.1) has a unique continuously differentiable solution on $[0, \infty[$ which we denote by $u(t; x)$.

For $x \in D(A)$ we define the graph norm by $|x|_G = \|x\| + \|Ax\|$. Since $\rho(A) \neq \varnothing$ $A$ is closed and therefore $D(A)$ endowed with the graph norm is a Banach space which we denote by $[D(A)]$. Let $X_{t_0}$ be the Banach space of continuous functions from $[0, t_0]$ into $[D(A)]$ with the usual supremum norm. We consider the mapping $S:[D(A)] \to X_{t_0}$ defined by $Sx = u(t; x)$ for $0 \leq t \leq t_0$. From the linearity of (1.1) and the uniqueness of the solutions it is clear that $S$ is a linear operator defined on all of $[D(A)]$. The operator $S$ is closed. Indeed, if $x_n \to x$ in $[D(A)]$ and $Sx_n \to v$ in $X_{t_0}$ then from the closedness of $A$ and

$$u(t; x_n) = x_n + \int_0^t Au(\tau; x_n) \, d\tau$$

it follows that as $n \to \infty$

$$v(t) = x + \int_0^t Av(\tau) \, d\tau$$

which implies $v(t) = u(t : x)$ and $S$ is closed. Therefore, by the closed graph theorem, $S$ is bounded, and

$$\sup_{0 \leq t \leq t_0} |u(t; x)|_G \leq C|x|_G. \tag{1.9}$$

We now define a mapping $T(t):[D(A)] \to [D(A)]$ by $T(t)x = u(t; x)$. From the uniqueness of the solutions of (1.1) it follows readily that $T(t)$ has the semigroup property. From (1.9) it follows that for $0 \leq t \leq t_0$, $T(t)$ is uniformly bounded. This implies (see, e.g., the proof of Theorem 1.2.2) that $T(t)$ can be extended by, $T(t)x = T(t - nt_0)T(t_0)^n x$ for $nt_0 \leq t < (n + 1)t_0$ to a semigroup on $[D(A)]$ satisfying $|T(t)x|_G \leq Me^{\omega t}|x|_G$.

Next we show that

$$T(t)Ay = AT(t)y \qquad \text{for} \quad y \in D(A^2). \tag{1.10}$$

Setting

$$v(t) = y + \int_0^t u(s; Ay) \, ds \tag{1.11}$$

we have

$$v'(t) = u(t; Ay) = Ay + \int_0^t \frac{d}{ds} u(s; Ay) \, ds$$

$$= A\left(y + \int_0^t u(s; Ay) \, ds\right) = Av(t). \tag{1.12}$$

Since $v(0) = y$ we have by the uniqueness of the solution of (1.1), $v(t) = u(t; y)$ and therefore $Au(t; y) = v'(t) = u(t; Ay)$ which is the same as (1.10).

Now, since $D(A)$ is dense in $X$ and by our assumption $\rho(A) \neq \varnothing$ also $D(A^2)$ is dense in $X$. Let $\lambda_0 \in \rho(A)$, $\lambda_0 \neq 0$, be fixed and let $y \in D(A^2)$. If

$x = (\lambda_0 I - A)y$ then, by (1.10), $T(t)x = (\lambda_0 I - A)T(t)y$ and therefore

$$\|T(t)x\| = \|(\lambda_0 I - A)T(t)y\| \leq C|T(t)y|_G \leq C_1 e^{\omega' t}|y|_G. \quad (1.13)$$

But

$$|y|_G = \|y\| + \|Ay\| \leq C_2\|x\|$$

which implies

$$\|T(t)x\| \leq C_2 e^{\omega' t}\|x\|. \quad (1.14)$$

Therefore $T(t)$ can be extended to all of $X$ by continuity. After this extension $T(t)$ becomes a $C_0$ semigroup on $X$. To complete the proof we have to show that $A$ is the infinitesimal generator of $T(t)$. Denote by $A_1$ the infinitesimal generator of $T(t)$. If $x \in D(A)$ then by the definition of $T(t)$ we have $T(t)x = u(t, x)$ and therefore by our assumptions

$$\frac{d}{dt}T(t)x = AT(t)x \qquad \text{for} \quad t \geq 0$$

which implies in particular that $(d/dt)T(t)x|_{t=0} = Ax$ and therefore $A_1 \supset A$.

Let $\operatorname{Re} \lambda > \omega$ and let $y \in D(A^2)$. It follows from (1.10) and from $A_1 \supset A$ that

$$e^{-\lambda t}AT(t)y = e^{-\lambda t}T(t)Ay = e^{-\lambda t}T(t)A_1 y. \quad (1.15)$$

Integrating (1.15) from 0 to $\infty$ yields

$$AR(\lambda : A_1)y = R(\lambda : A_1)A_1 y. \quad (1.16)$$

But $A_1 R(\lambda : A_1)y = R(\lambda : A_1)A_1 y$ and therefore $AR(\lambda : A_1)y = A_1 R(\lambda : A_1)y$ for every $y \in D(A^2)$. Since $A_1 R(\lambda : A_1)$ are uniformly bounded, $A$ is closed and $D(A^2)$ is dense in $X$, it follows that $AR(\lambda : A_1)y = A_1 R(\lambda : A_1)y$ for every $y \in X$. This implies $D(A) \supset \operatorname{Range} R(\lambda : A_1) = D(A_1)$ and $A \supset A_1$. Therefore $A = A_1$ and the proof is complete.                                           $\square$

Our next theorem describes a situation in which the initial value problem (1.1) has a unique solution for every $x \in X$.

**Theorem 1.4.** *If $A$ is the infinitesimal generator of a differentiable semigroup then for every $x \in X$ the initial value problem (1.1) has a unique solution.*

PROOF. The uniqueness follows from Theorem 1.2. If $x \in D(A)$ the existence follows from Theorem 1.3. If $x \in X$ then by the differentiability of $T(t)x$ and the results of Section 2.2.4 it follows that for every $x \in X$, $(d/dt)T(t)x = AT(t)x$ for $t > 0$ and $AT(t)x$ is Lipschitz continuous for $t > 0$. Thus $T(t)x$ is the solution of (1.1).                                           $\square$

**Corollary 1.5.** *If $A$ is the infinitesimal generator of an analytic semigroup then for every $x \in X$ the initial value problem (1.1) has a unique solution.*

If $A$ is the infinitesimal generator of a $C_0$ semigroup which is not differentiable then, in general, if $x \notin D(A)$, the initial value problem (1.1) does not have a solution. The function $t \to T(t)x$ is then a "generalized solution" of the initial value problem (1.1) which we will call a *mild solution*. There are many different ways to define generalized solutions of the initial value problem (1.1). All lead eventually to $T(t)x$. One such way of defining a generalized solution of (1.1) is the following: A continuous function $u$ on $[0, \infty[$ is a generalized solution of (1.1) if there are $x_n \in D(A)$ such that $x_n \to u(0)$ as $n \to \infty$ and $T(t)x_n \to u(t)$ uniformly on bounded intervals. It is obvious that the generalized solution thus defined is independent of the sequence $\langle x_n \rangle$, is unique and if $u(0) \in D(A)$ it gives the solution of (1.1). Clearly, with this definition of generalized solution, (1.1) has a generalized solution for every $x \in X$ and this generalized solution is $T(t)x$.

## 4.2. The Inhomogeneous Initial Value Problem

In this section we consider the inhomogeneous initial value problem

$$\begin{cases} \dfrac{du(t)}{dt} = Au(t) + f(t) & t > 0 \\ u(0) = x \end{cases} \tag{2.1}$$

where $f:[0, T[ \to X$. We will assume throughout this section that $A$ is the infinitesimal generator of a $C_0$ semigroup $T(t)$ so that the corresponding homogeneous equation, i.e., the equation with $f \equiv 0$, has a unique solution for every initial value $x \in D(A)$.

**Definition 2.1.** A function $u:[0, T[ \to X$ is a (classical) solution of (2.1) on $[0, T[$ if $u$ is continuous on $[0, T[$, continuously differentiable on $]0, T[$, $u(t) \in D(A)$ for $0 < t < T$ and (2.1) is satisfied on $[0, T[$.

Let $T(t)$ be the $C_0$ semigroup generated by $A$ and let $u$ be a solution of (2.1). Then the $X$ valued function $g(s) = T(t - s)u(s)$ is differentiable for $0 < s < t$ and

$$\frac{dg}{ds} = -AT(t - s)u(s) + T(t - s)u'(s)$$

$$= -AT(t - s)u(s) + T(t - s)Au(s) + T(t - s)f(s)$$

$$= T(t - s)f(s). \tag{2.2}$$

If $f \in L^1(0, T: X)$ then, $T(t - s)f(s)$ is integrable and integrating (2.2) from 0 to $t$ yields

$$u(t) = T(t)x + \int_0^t T(t - s)f(s)\, ds. \tag{2.3}$$

Consequently we have

**Corollary 2.2.** *If* $f \in L^1(0, T: X)$ *then for every* $x \in X$ *the initial value problem* (2.1) *has at most one solution. If it has a solution, this solution is given by* (2.3).

For every $f \in L^1(0, T: X)$ the right-hand side of (2.3) is a continuous function on $[0, T]$. It is natural to consider it as a generalized solution of (2.1) even if it is not differentiable and does not strictly satisfy the equation in the sense of Definition 2.1. We therefore define,

**Definition 2.3.** Let $A$ be the infinitesimal generator of a $C_0$ semigroup $T(t)$. Let $x \in X$ and $f \in L^1(0, T: X)$. The function $u \in C([0, T]: X)$ given by

$$u(t) = T(t)x + \int_0^t T(t - s)f(s) \, ds, \qquad 0 \le t \le T,$$

is the *mild solution* of the initial value problem (2.1) on $[0, T]$.

The definition of the mild solution of the initial value problem (2.1) coincides when $f \equiv 0$ with the definition of $T(t)x$ as the mild solution of the corresponding homogeneous equation. It is therefore clear that not every mild solution of (2.1) is indeed a (classical) solution even in the case $f \equiv 0$.

For $f \in L^1(0, T: X)$ the initial value problem (2.1) has by Definition 2.3 a unique mild solution. We will now be interested in imposing further conditions on $f$ so that for $x \in D(A)$, the mild solution becomes a (classical) solution and thus proving, under these conditions, the existence of solutions of (2.1) for $x \in D(A)$.

We start by showing that the continuity of $f$, in general, is not sufficient to ensure the existence of solutions of (2.1) for $x \in D(A)$. Indeed, let $A$ be the infinitesimal generator of a $C_0$ semigroup $T(t)$ and let $x \in X$ be such that $T(t)x \notin D(A)$ for any $t \ge 0$. Let $f(s) = T(s)x$. Then $f(s)$ is continuous for $s \ge 0$. Consider the initial value problem

$$\begin{cases} \dfrac{du(t)}{dt} = Au(t) + T(t)x \\ u(0) = 0 \ . \end{cases} \qquad (2.4)$$

We claim that (2.4) has no solution even though $u(0) = 0 \in D(A)$. Indeed, the mild solution of (2.4) is

$$u(t) = \int_0^t T(t - s)T(s)x \, ds = tT(t)x,$$

but $tT(t)x$ is not differentiable for $t > 0$ and therefore cannot be the solution of (2.4).

Thus in order to prove the existence of solutions of (2.1) we have to require more than just the continuity of $f$. We start with a general criterion for the existence of solutions of the initial value problem (2.1).

**Theorem 2.4.** *Let $A$ be the infinitesimal generator of a $C_0$ semigroup $T(t)$, let $f \in L^1(0, T: X)$ be continuous on $]0, T]$ and let*

$$v(t) = \int_0^t T(t-s)f(s)\, ds, \qquad 0 \le t \le T. \tag{2.5}$$

*The initial value problem (2.1) has a solution $u$ on $[0, T[$ for every $x \in D(A)$ if one of the following conditions is satisfied;*

(i) *$v(t)$ is continuously differentiable on $]0, T[$.*
(ii) *$v(t) \in D(A)$ for $0 < t < T$ and $Av(t)$ is continuous on $]0, T[$.*

*If (2.1) has a solution $u$ on $[0, T[$ for some $x \in D(A)$ then $v$ satisfies both (i) and (ii).*

PROOF. If the initial value problem (2.1) has a solution $u$ for some $x \in D(A)$ then this solution is given by (2.3). Consequently $v(t) = u(t) - T(t)x$ is differentiable for $t > 0$ as the difference of two such differentiable functions and $v'(t) = u'(t) - T(t)Ax$ is obviously continuous on $]0, T[$. Therefore (i) is satisfied. Also if $x \in D(A)$ $T(t)x \in D(A)$ for $t \ge 0$ and therefore $v(t) = u(t) - T(t)x \in D(A)$ for $t > 0$ and $Av(t) = Au(t) - AT(t)x = u'(t) - f(t) - T(t)Ax$ is continuous on $]0, T[$. Thus also (ii) is satisfied.

On the other hand, it is easy to verify for $h > 0$ the identity

$$\frac{T(h) - I}{h}v(t) = \frac{v(t+h) - v(t)}{h} - \frac{1}{h}\int_t^{t+h} T(t+h-s)f(s)\, ds. \tag{2.6}$$

From the continuity of $f$ it is clear that the second term on the right-hand side of (2.6) has the limit $f(t)$ as $h \to 0$. If $v(t)$ is continuously differentiable on $]0, T[$ then it follows from (2.6) that $v(t) \in D(A)$ for $0 < t < T$ and $Av(t) = v'(t) - f(t)$. Since $v(0) = 0$ it follows that $u(t) = T(t)x + v(t)$ is the solution of the initial value problem (2.1) for $x \in D(A)$. If $v(t) \in D(A)$ it follows from (2.6) that $v(t)$ is differentiable from the right at $t$ and the right derivative $D^+v(t)$ of $v$ satisfies $D^+v(t) = Av(t) + f(t)$. Since $D^+v(t)$ is continuous, $v(t)$ is continuously differentiable and $v'(t) = Av(t) + f(t)$. Since $v(0) = 0$, $u(t) = T(t)x + v(t)$ is the solution of (2.1) for $x \in D(A)$ and the proof is complete.                                                    □

From Theorem 2.4 we draw the following two useful corollaries.

**Corollary 2.5.** *Let $A$ be the infinitesimal generator of a $C_0$ semigroup $T(t)$. If $f(s)$ is continuously differentiable on $[0, T]$ then the initial value problem (2.1) has a solution $u$ on $[0, T[$ for every $x \in D(A)$.*

PROOF. We have

$$v(t) = \int_0^t T(t-s)f(s)\, ds = \int_0^t T(s)f(t-s)\, ds. \tag{2.7}$$

It is clear from (2.7) that $v(t)$ is differentiable for $t > 0$ and that its derivative

$$v'(t) = T(t)f(0) + \int_0^t T(s)f'(t-s)\,ds = T(t)f(0) + \int_0^t T(t-s)f'(s)\,ds$$

is continuous on $]0, T[$. The result therefore follows from Theorem 2.4 (i).  $\square$

**Corollary 2.6.** *Let $A$ be the infinitesimal generator of a $C_0$ semigroup $T(t)$. Let $f \in L^1(0, T: X)$ be continuous on $]0, T[$. If $f(s) \in D(A)$ for $0 < s < T$ and $Af(s) \in L^1(0, T: X)$ then for every $x \in D(A)$ the initial value problem (2.1) has a solution on $[0, T[$.*

PROOF. From the conditions it follows that for $s > 0$, $T(t-s)f(s) \in D(A)$ and that $AT(t-s)f(s) = T(t-s)Af(s)$ is integrable. Therefore $v(t)$ defined by (2.5) satisfies $v(t) \in D(A)$ for $t > 0$ and

$$Av(t) = A\int_0^t T(t-s)f(s)\,ds = \int_0^t T(t-s)Af(s)\,ds$$

is continuous. The result follows now from Theorem 2.4 (ii).                       $\square$

As a consequence of the previous results we can prove,

**Theorem 2.7.** *Let $f \in L^1(0, T: X)$. If $u$ is the mild solution of (2.1) on $[0, T]$ then for every $T' < T$, $u$ is the uniform limit on $[0, T']$ of solutions of (2.1).*

PROOF. Assume that $\|T(t)\| \le Me^{\omega t}$. Let $x_n \in D(A)$ satisfy $x_n \to x$ and let $f_n \in C^1([0, T]: X)$ satisfy $f_n \to f$ in $L^1(0, T: X)$. From Corollary 2.5 it follows that for each $n \ge 1$ the initial value problem

$$\begin{cases} \dfrac{du_n(t)}{dt} = Au_n(t) + f_n(t) \\ u_n(0) = x_n \end{cases} \tag{2.8}$$

has a solution $u_n(t)$ on $[0, T[$ satisfying

$$u_n(t) = T(t)x_n + \int_0^t T(t-s)f_n(s)\,ds.$$

If $u$ is the mild solution of (2.1) on $[0, T]$ then

$$\|u_n(t) - u(t)\| \le Me^{\omega t}\|x_n - x\| + \int_0^t Me^{\omega(t-s)}\|f_n(s) - f(s)\|\,ds$$

$$\le Me^{\omega T}\left(\|x_n - x\| + \int_0^T \|f_n(s) - f(s)\|\,ds\right) \tag{2.9}$$

and the result follows readily from (2.9).                                         $\square$

We conclude this section with a few remarks concerning still another notion of solution of the initial value problem (2.1) namely the strong solution.

**Definition 2.8.** A function $u$ which is differentiable almost everywhere on $[0, T]$ such that $u' \in L^1(0, T: X)$ is called a *strong solution* of the initial value problem (2.1) if $u(0) = x$ and $u'(t) = Au(t) + f(t)$a.e. on $[0, T]$.

We note that if $A = 0$ and $f \in L^1(0, T: X)$ the initial value problem (2.1) has usually no solution unless $f$ is continuous. It has however always a strong solution given by $u(t) = u(0) + \int_0^t f(s)$. It is easy to show that if $u$ is a strong solution of (2.1) and $f \in L^1(0, T: X)$ then $u$ is given by (2.3) and therefore is a mild solution of (2.1) and is the unique strong solution of (2.1). A natural problem is to determine when is a mild solution a strong solution. It is not difficult to show, essentially with the same proof as the proof of Theorem 2.4 that we have:

**Theorem 2.9.** *Let $A$ be the infinitesimal generator of a $C_0$ semigroup $T(t)$, let $f \in L^1(0, T: X)$ and let*

$$v(t) = \int_0^t T(t - s) f(s) \, ds, \qquad 0 \le t \le T.$$

*The initial value problem* (2.1) *has a strong solution $u$ on $[0, T]$ for every $x \in D(A)$ if one of the following conditions is satisfied;*

 (i) *$v(t)$ is differentiable a.e. on $[0, T]$ and $v'(t) \in L^1(0, T: X)$*
(ii) *$v(t) \in D(A)$ a.e. on $[0, T]$ and $Av(t) \in L^1(0, T: X)$.*

*If* (2.1) *has a strong solution $u$ on $[0, T]$ for some $x \in D(A)$ then $v$ satisfies both* (i) *and* (ii).

As a consequence of Theorem 2.9 we have:

**Corollary 2.10.** *Let $A$ be the infinitesimal generator of a $C_0$ semigroup $T(t)$. If $f$ is differentiable a.e. on $[0, T]$ and $f' \in L^1(0, T: X)$ then for every $x \in D(A)$ the initial value problem* (2.1) *has a unique strong solution on $[0, T]$.*

In general, the Lipschitz continuity of $f$ on $[0, T]$ is not sufficient to assure the existence of a strong solution of (2.1) for $x \in D(A)$. However, if $X$ is reflexive and $f$ is Lipschitz continuous on $[0, T]$ that is

$$\|f(t_1) - f(t_2)\| \le C|t_1 - t_2| \qquad \text{for} \quad t_1, t_2 \in [0, T]$$

then by a classical result $f$ is differentiable a.e. and $f' \in L^1(0, T: X)$. Corollary 2.10 therefore implies:

**Corollary 2.11.** *Let $X$ be a reflexive Banach space and let $A$ be the infinitesimal generator of a $C_0$ semigroup $T(t)$ on $X$. If $f$ is Lipschitz continuous on $[0, T]$ then for every $x \in D(A)$ the initial value problem* (2.1) *has a unique solution $u$ on $[0, T]$ given by*

$$u(t) = T(t)x + \int_0^t T(t - s) f(s) \, ds.$$

PROOF. From the previous remarks it is obvious that (2.1) has a strong solution. Therefore by Theorem 2.9, $v(t)$ given by (2.5), is differentiable a.e. on $[0, T]$ and

$$v'(t) = T(t)f(0) + \int_0^t T(t - s)f'(s)\, ds = g(t).$$

It is easy to verify that $g(t)$ is continuous on $[0, T]$ and the result follows from Theorem 2.4. $\qquad\qquad\square$

## 4.3. Regularity of Mild Solutions for Analytic Semigroups

Let $A$ be the infinitesimal generator of a $C_0$ semigroup $T(t)$ and let $f \in L^1(0, T: X)$. In the previous section we defined the mild solution of the initial value problem

$$\begin{cases} \dfrac{du(t)}{dt} = Au(t) + f(t) \\ u(0) = x \end{cases} \tag{3.1}$$

to be the continuous function

$$u(t) = T(t)x + \int_0^t T(t - s)f(s)\, ds. \tag{3.2}$$

We saw that if one imposes further conditions on $f$, e.g., $f \in C^1([0, T]: X)$ then the mild solution (3.2) becomes a (classical) solution, i.e., a continuously differentiable solution of (3.1). If $A$ is the infinitesimal generator of an analytic semigroup we have stronger results. For example we will see (Corollary 3.3) that in this case Hölder continuity of $f$ already implies that the mild solution (3.2) is a solution of (3.1).

We start by showing that if $T(t)$ is an analytic semigroup and $f \in L^p(0, T: X)$ with $p > 1$ then the mild solution (3.2) is Hölder continuous. More precisely we have:

**Theorem 3.1.** *Let $A$ be the infinitesimal generator of an analytic semigroup $T(t)$ and let $f \in L^p(0, T: X)$ with $1 < p < \infty$. If $u$ is the mild solution of (3.1) then $u$ is Hölder continuous with exponent $(p - 1)/p$ on $[\varepsilon, T]$ for every $\varepsilon > 0$. If moreover $x \in D(A)$ then $u$ is Hölder continuous with the same exponent on $[0, T]$.*

PROOF. Let $\|T(t)\| \le M$ on $[0, T]$. Since $T(t)$ is analytic there is a constant $C$ such that $\|AT(t)\| \le Ct^{-1}$ on $]0, T]$. This implies that $T(t)x$ is Lipschitz continuous on $[\varepsilon, T]$ for every $\varepsilon > 0$. If $x \in D(A)$, $T(t)x$ is Lipschitz continuous on $[0, T]$. It suffices therefore to show that if $f \in L^p(0, T: X)$, $1 < p < \infty$ then $v(t) = \int_0^t T(t - s)f(s)\, ds$ is Hölder continuous with exponent $(p - 1)/p$ on $[0, T]$. For $h > 0$ we have

$$v(t + h) - v(t) = \int_t^{t+h} T(t + h - s)f(s)\, ds$$

$$+ \int_0^t (T(t + h - s) - T(t - s))f(s)\, ds = I_1 + I_2.$$

We estimate $I_1$ and $I_2$ separately. For $I_1$ we use Hölder's inequality to obtain,

$$\|I_1\| \le M\int_t^{t+h}\|f(s)\|ds \le Mh^{(p-1)/p}\left(\int_t^{t+h}\|f(s)\|^p\,ds\right)^{1/p} \le M|f|_p h^{(p-1)/p}$$
(3.3)

where $|f|_p = (\int_0^T\|f(s)\|^p\,ds)^{1/p}$ is the norm of $f$ in $L^p(0, T: X)$. In order to estimate $I_2$ we note that for $h > 0$

$$\|T(t+h) - T(t)\| \le 2M \qquad \text{for} \quad t, t+h \in [0, T]$$

and

$$\|T(t+h) - T(t)\| \le C\frac{h}{t} \qquad \text{for} \quad t, t+h \in\ ]0, T].$$

Therefore,

$$\|T(t+h) - T(t)\| \le C_1\mu(h, t) = C_1\min\left(1, \frac{h}{t}\right)$$
$$\text{for} \quad t, t+h \in [0, T] \quad (3.4)$$

where $C_1$ is a constant satisfying $C_1 \ge \max(2M, C)$. Using (3.4) and Hölder's inequality we find

$$\|I_2\| \le C_1\int_0^t\mu(h, t-s)\|f(s)\|ds \le C_1|f|_p\left(\int_0^t\mu(h, t-s)^{p/(p-1)}\,ds\right)^{(p-1)/p}.$$
(3.5)

But since $\mu \ge 0$ we have

$$\int_0^t\mu(h, t-s)^{p/(p-1)}\,ds = \int_0^t\mu(h, \tau)^{p/(p-1)}\,d\tau \le \int_0^\infty\mu(h, \tau)^{p/(p-1)}\,d\tau = ph$$

and combining (3.5) with the last inequality we find $\|I_2\| \le \text{const}\cdot h^{(p-1)/p}$.
$\square$

We turn now to conditions on $f$ that will ensure that the mild solution of (3.1) is a strong solution.

**Theorem 3.2.** *Let $A$ be the infinitesimal generator of an analytic semigroup $T(t)$. Let $f \in L^1(0, T: X)$ and assume that for every $0 < t < T$ there is a $\delta_t > 0$ and a continuous real value function $W_t(\tau)$: $[0, \infty[ \to [0, \infty[$ such that*

$$\|f(t) - f(s)\| \le W_t(|t - s|) \tag{3.6}$$

*and*

$$\int_0^{\delta_t}\frac{W_t(\tau)}{\tau}\,d\tau < \infty. \tag{3.7}$$

*Then for every $x \in X$ the mild solution of (3.1) is a classical solution.*

PROOF. Since $T(t)$ is an analytic semigroup, $T(t)x$ is the solution of the homogeneous equation with initial data $x$ for every $x \in X$. To prove the theorem it is therefore sufficient, by Theorem 2.4, to show that $v(t) = \int_0^t T(t-s)f(s)\,ds \in D(A)$ for $0 < t < T$ and that $Av(t)$ is continuous on this interval. To this end we write

$$v(t) = v_1(t) + v_2(t)$$
$$= \int_0^t T(t-s)(f(s) - f(t)) \, ds + \int_0^t T(t-s)f(t) \, ds. \quad (3.8)$$

From Theorem 1.2.4 (b) it follows that $v_2(t) \in D(A)$ and that $Av_2(t) = (T(t) - I)f(t)$. Since the assumptions of our theorem imply that $f$ is continuous on $]0, T[$ it follows that $Av_2(t)$ is continuous on $]0, T[$. To prove the same conclusion for $v_1$ we define

$$v_{1,\varepsilon}(t) = \int_0^{t-\varepsilon} T(t-s)(f(s) - f(t)) \, ds \qquad \text{for} \quad t \ge \varepsilon \quad (3.9)$$

and

$$v_{1,\varepsilon}(t) = 0 \qquad \text{for} \quad t < \varepsilon. \quad (3.10)$$

From this definition it is clear that $v_{1,\varepsilon}(t) \to v_1(t)$ as $\varepsilon \to 0$. It is also clear that $v_{1,\varepsilon}(t) \in D(A)$ and for $t \ge \varepsilon$

$$Av_{1,\varepsilon}(t) = \int_0^{t-\varepsilon} AT(t-s)(f(s) - f(t)) \, ds. \quad (3.11)$$

From (3.6) and (3.7) it follows that for $t > 0$ $Av_{1,\varepsilon}(t)$ converges as $\varepsilon \to 0$ and that

$$\lim_{\varepsilon \to 0} Av_{1,\varepsilon}(t) = \int_0^t AT(t-s)(f(s) - f(t)) \, ds.$$

The closedness of $A$ then implies that $v_1(t) \in D(A)$ for $t > 0$ and

$$Av_1(t) = \int_0^t AT(t-s)(f(s) - f(t)) \, ds. \quad (3.12)$$

To conclude the proof we have to show that $Av_1(t)$ is continuous on $]0, T[$. For $0 < \delta < t$ we have

$$Av_1(t) = \int_0^\delta AT(t-s)(f(s) - f(t)) \, ds + \int_\delta^t AT(t-s)(f(s) - f(t)) \, ds.$$
$$(3.13)$$

For fixed $\delta > 0$ the second integral on the right of (3.13) is a continuous function of $t$ while the first integral is $O(\delta)$ uniformly in $t$. Thus, $Av_1(t)$ is continuous and the proof is complete.                                     □

Let $I$ be an interval. A function $f: I \to X$ is Hölder continuous with exponent $\vartheta$, $0 < \vartheta < 1$ on $I$ if there is a constant $L$ such that

$$\|f(t) - f(s)\| \le L|t - s|^\vartheta \qquad \text{for} \quad s, t \in I. \quad (3.14)$$

It is locally Hölder continuous if every $t \in I$ has a neighborhood in which $f$ is Hölder continuous. It is easy to check that if $I$ is compact then $f$ is Hölder continuous on $I$ if it is locally Hölder continuous. We denote the family of all Hölder continuous functions with exponent $\vartheta$ on $I$ by $C^\vartheta(I : X)$.

An immediate consequence of Theorem 3.2 is,

**Corollary 3.3.** *Let $A$ be the infinitesimal generator of an analytic semigroup $T(t)$. If $f \in L^1(0, T: X)$ is locally Hölder continuous on $]0, T]$ then for every $x \in X$ the initial value problem (3.1) has a unique solution $u$.*

More can be said on the regularity of the solution $u$ under the assumptions of Corollary 3.3. This will be seen in Theorem 3.5. In the proof of Theorem 3.5 we will need the following lemma.

**Lemma 3.4.** *Let $A$ be the infinitesimal generator of an analytic semigroup $T(t)$ and let $f \in C^\vartheta([0, T]: X)$. If*

$$v_1(t) = \int_0^t T(t - s)(f(s) - f(t))\, ds \tag{3.15}$$

*then $v_1(t) \in D(A)$ for $0 \le t \le T$ and $Av_1(t) \in C^\vartheta([0, T]: X)$.*

PROOF. The fact that $v_1(t) \in D(A)$ for $0 \le t \le T$ is an immediate consequence of the proof of Theorem 3.2, so we have only to prove the Hölder continuity of $Av_1(t)$. Assume that $\|T(t)\| \le M$ on $[0, T]$ and that

$$\|AT(t)\| \le Ct^{-1} \qquad \text{for} \quad 0 < t \le T. \tag{3.16}$$

Then, for every $0 < s < t \le T$ we have

$$\|AT(t) - AT(s)\| = \left\| \int_s^t A^2 T(\tau)\, d\tau \right\| \le \int_s^t \|A^2 T(\tau)\|\, d\tau$$

$$\le 4C \int_s^t \tau^{-2}\, d\tau = 4Ct^{-1}s^{-1}(t - s). \tag{3.17}$$

Let $t \ge 0$ and $h > 0$ then

$$Av_1(t + h) - Av_1(t) = A \int_0^t (T(t + h - s) - T(t - s))(f(s) - f(t))\, ds$$

$$+ A \int_0^t T(t + h - s)(f(t) - f(t + h))\, ds$$

$$+ A \int_t^{t+h} T(t + h - s)(f(s) - f(t + h))\, ds$$

$$= I_1 + I_2 + I_3. \tag{3.18}$$

We estimate each of the three terms separately. First from (3.14) and (3.17) we have

$$\|I_1\| \le \int_0^t \|AT(t + h - s) - AT(t - s)\|\, \|f(s) - f(t)\|\, ds$$

$$\le 4CLh \int_0^t \frac{ds}{(t - s + h)(t - s)^{1 - \vartheta}} \le C_1 h^\vartheta. \tag{3.19}$$

To estimate $I_2$ we use Theorem 1.2.4(b) and (3.14),

$$\|I_2\| = \|(T(t + h) - T(h))(f(t) - f(t + h))\|$$

$$\le \|T(t + h) - T(h)\|\, \|f(t) - f(t + h)\| \le 2MLh^\vartheta. \tag{3.20}$$

Finally, to estimate $I_3$ we use (3.16) and (3.14) to find

$$\|I_3\| \leq \int_t^{t+h} \|AT(t + h - s)\| \, \|f(s) - f(t + h)\| \, ds$$

$$\leq CL \int_t^{t+h} (t + h - s)^{\vartheta - 1} \, ds \leq C_2 h^{\vartheta}. \tag{3.21}$$

Combining (3.18) with the estimates (3.19), (3.20) and (3.21) we see that $Av_1(t)$ is Hölder continuous with exponent $\vartheta$ on $[0, T]$. $\qquad\square$

The main regularity result for the case where $A$ generates an analytic semigroup and $f$ is Hölder continuous comes next.

**Theorem 3.5.** *Let $A$ be the infinitesimal generator of an analytic semigroup $T(t)$ and let $f \in C^{\vartheta}([0, T]: X)$. If $u$ is the solution of the initial value problem (3.1) on $[0, T]$ then,*

(i) *For every $\delta > 0$, $Au \in C^{\vartheta}([\delta, T]: X)$ and $du/dt \in C^{\vartheta}([\delta, T]: X)$.*
(ii) *If $x \in D(A)$ then $Au$ and $du/dt$ are continuous on $[0, T]$.*
(iii) *If $x = 0$ and $f(0) = 0$ then $Au$, $du/dt \in C^{\vartheta}([0, T]: X)$.*

PROOF. We have,

$$u(t) = T(t)x + \int_0^t T(t - s)f(s) \, ds = T(t)x + v(t).$$

Since by (3.17) $AT(t)x$ is Lipschitz continuous on $\delta \leq t \leq T$ for every $\delta > 0$ it suffices to show that $Av(t) \in C^{\vartheta}([\delta, T]: X)$. To this end we decompose $v$ as before to

$$v(t) = v_1(t) + v_2(t) = \int_0^t T(t - s)(f(s) - f(t)) \, ds + \int_0^t T(t - s)f(t) \, ds.$$

From Lemma 3.4 it follows that $Av_1(t) \in C^{\vartheta}([0, T]: X)$ so it remains only to show that $Av_2(t) \in C^{\vartheta}([\delta, T]: X)$ for every $\delta > 0$. But $Av_2(t) = (T(t) - I)f(t)$ and since $f \in C^{\vartheta}([0, T]: X)$ we have only to show that $T(t)f(t) \in C^{\vartheta}([\delta, T]: X)$ for every $\delta > 0$. Let $t \geq \delta$ and $h > 0$ then

$$\|T(t + h)f(t + h) - T(t)f(t)\|$$

$$\leq \|T(t + h)\| \, \|f(t + h) - f(t)\| + \|T(t + h) - T(t)\| \, \|f(t)\|$$

$$\leq MLh^{\vartheta} + \frac{C}{\delta} h \|f\|_{\infty} \leq C_1 h^{\vartheta} \tag{3.22}$$

where we used (3.4), (3.14) and denoted $\|f\|_{\infty} = \max_{0 \leq t \leq T} \|f(t)\|$. This completes the proof of (i). To prove (ii) we note first that if $x \in D(A)$ then $AT(t)x \in C([0, T]: X)$. By Lemma 3.4 $Av_1(t) \in C^{\vartheta}([0, T]: X)$ and since $f$ is continuous on $[0, T]$ it remains only to show that $T(t)f(t)$ is continuous on $[0, T]$. From (i) it is clear that $T(t)f(t)$ is continuous on $]0, T]$. The continuity at $t = 0$ follows readily from,

$$\|T(t)f(t) - f(0)\| \leq \|T(t)f(0) - f(0)\| + M\|f(t) - f(0)\|$$

and this completes the proof of (ii). Finally, to prove (iii) we have again only to show that in this case $T(t)f(t) \in C^{\vartheta}([0, T]: X)$ and this follows from

$$\|T(t + h)f(t + h) - T(t)f(t)\|$$

$$\leq \|T(t + h)\| \, \|f(t + h) - f(t)\| + \|(T(t + h) - T(t))f(t)\|$$

$$\leq MLh^{\vartheta} + \left\| \int_t^{t+h} AT(\tau)f(t) \, d\tau \right\|$$

$$\leq MLh^{\vartheta} + \int_t^{t+h} \|AT(\tau)(f(t) - f(0))\| \, d\tau$$

$$\leq MLh^{\vartheta} + CL\int_t^{t+h} \tau^{-1}t^{\vartheta} \, d\tau \leq MLh^{\vartheta} + CL\int_t^{t+h} \tau^{\vartheta-1} \, d\tau \leq Ch^{\vartheta}$$

and the proof is complete.                                                                           □

We conclude this section with a result which is analogous to Theorem 3.2 in which the condition on the modulus of continuity of $f$ is replaced by another regularity condition.

**Theorem 3.6.** *Let $A$ be the infinitesimal generator of an analytic semigroup $T(t)$ and let $0 \in \rho(A)$. If $f(s)$ is continuous, $f(s) \in D((-A)^{\alpha})$, $0 < \alpha \leq 1$ and $\|(-A)^{\alpha}f(s)\|$ is bounded, then for every $x \in X$ the mild solution of (3.1) is a classical solution.*

PROOF. As in the proof of Theorem 3.2 it suffices to show that $v(t) \in D(A)$ for $t > 0$ and that $Av(t)$ is continuous for $t > 0$. Since $T(t)$ is analytic $T(t - s)f(s) \in D(A)$ for $t > s$ and by Theorem 2.6.13(c),

$$\|AT(t - s)f(s)\| = \|(-A)^{1-\alpha}T(t - s)(-A)^{\alpha}f(s)\|$$

$$\leq C|t - s|^{\alpha-1}\|(-A)^{\alpha}f(s)\|.$$

Therefore $AT(t - s)f(s)$ is integrable and $v(t) \in D(A)$ as well as

$$Av(t) = \int_0^t AT(t - s)f(s) \, ds.$$

The continuity of $Av(t)$ for $t > 0$ is proved exactly as the continuity of $A_1v(t)$ is proved in Theorem 3.2.                                                          □

## 4.4. Asymptotic Behavior of Solutions

In this section we intend to study the asymptotic behavior of solutions of the initial value problem

$$\frac{du(t)}{dt} = Au(t) + f(t), \qquad u(0) = x. \tag{4.1}$$

We start with the solutions of the homogeneous problem i.e., $f \equiv 0$ and look for conditions that guarantee their exponential decay.

**Theorem 4.1.** *Let $A$ be the infinitesimal generator of a $C_0$ semigroup $T(t)$. If for some $p$, $1 \leq p < \infty$*

$$\int_0^\infty \|T(t)x\|^p \, dt < \infty \qquad \text{for every} \quad x \in X \tag{4.2}$$

*then there are constants $M \geq 1$ and $\mu > 0$ such that $\|T(t)\| \leq Me^{-\mu t}$.*

PROOF. We start by showing that (4.2) implies the boundedness of $t \to \|T(t)\|$. Let $\|T(t)\| \leq M_1 e^{\omega t}$ where $M_1 \geq 1$ and $\omega \geq 0$. If $\omega = 0$ there is nothing to prove so we assume $\omega > 0$. From (4.2) it then follows that $T(t)x \to 0$ as $t \to \infty$ for every $x \in X$. Indeed, if this were false we could find $x \in X$, $\delta > 0$ and $t_j \to \infty$ such that $\|T(t_j)x\| \geq \delta$. Without loss of generality we can assume that $t_{j+1} - t_j > \omega^{-1}$. Set $\Delta_j = [t_j - \omega^{-1}, t_j]$, then $m(\Delta_j) = \omega^{-1} > 0$ and the intervals $\Delta_j$ do not overlap. For $t \in \Delta_j$ we have $\|T(t)x\| \geq \delta(M_1 e)^{-1}$ and therefore

$$\int_0^\infty \|T(t)x\|^p \, dt \geq \sum_{j=1}^\infty \int_{\Delta_j} \|T(t)x\|^p \, dt \geq \left(\frac{\delta}{M_1 e}\right)^p \sum_{j=1}^\infty m(\Delta_j) = \infty$$

contradicting (4.2). Thus $T(t)x \to 0$ as $t \to \infty$ for every $x \in X$ and the uniform boundedness theorem implies $\|T(t)\| \leq M$ for $t \geq 0$.

Next, consider the mapping $S : X \to L^p(\mathbb{R}^+ : X)$ defined by $Sx = T(t)x$. From (4.2) it follows that $S$ is defined on all of $X$. It is not difficult to see that $S$ is closed and therefore, by the closed graph theorem, $S$ is bounded, i.e.,

$$\int_0^\infty \|T(t)x\|^p \, dt \leq M_2^p \|x\|^p. \tag{4.3}$$

Let $0 < \rho < M^{-1}$ where $\|T(t)\| \leq M$. Define $t_x(\rho)$ by

$$t_x(\rho) = \sup\{t : \|T(s)x\| \geq \rho\|x\| \qquad \text{for} \quad 0 \leq s \leq t\}.$$

Since $\|T(t)x\| \to 0$ as $t \to \infty$, $t_x(\rho)$ is finite and positive for every $x \in X$. Moreover,

$$t_x(\rho)\rho^p\|x\|^p \leq \int_0^{t_x(\rho)} \|T(t)x\|^p \, dt \leq \int_0^\infty \|T(t)x\|^p \, dt \leq M_2^p \|x\|^p$$

and therefore $t_x(\rho) \leq (M_2/\rho)^p = t_0$. For $t > t_0$ we have

$$\|T(t)x\| \leq \|T(t - t_x(\rho))\| \, \|T(t_x(\rho))x\| \leq M\rho\|x\| \leq \beta\|x\|$$

where $0 \leq \beta = M\rho < 1$. Finally, let $t_1 > t_0$ be fixed and let $t = nt_1 + s$, $0 \leq s < t_1$. Then

$$\|T(t)\| \leq \|T(s)\| \, \|T(nt_1)\| \leq M\|T(t_1)\|^n \leq M\beta^n \leq M'e^{-\mu t}$$

where $M' = M\beta^{-1}$ and $\mu = -(1/t_1)\log \beta > 0$.                    $\square$

Theorem 4.1 shows that if $T(t)x \in L^p(\mathbb{R}^+ : X)$ for every $x \in X$ then $\|T(t)\| \leq Me^{-\mu t}$ for some $M \geq 1$ and $\mu > 0$. We are now interested in conditions on the infinitesimal generator $A$ of $T(t)$ which will insure a similar behavior.

For a Banach space $X$ of finite dimension it is well known that if $\sup\{\text{Re } \lambda : \lambda \in \sigma(A)\} = \sigma < 0$ then $\|T(t)\|$ decays exponentially. This behavior is a consequence of the fact that linear operators in finite dimensional Banach spaces have only point spectrum. Since this is not the case in general Banach spaces one does not expect this result to be true in general Banach spaces.

EXAMPLE 4.2.  For a measurable function $f$ on $[0, \infty[$ set

$$|f|_1 = \int_0^\infty e^s |f(s)| ds$$

and let $E$ be the space of all measurable functions $f$ on $[0, \infty[$ for which $|f|_1 < \infty$. Let $X = E \cap L^p(0, \infty)$, $1 < p < \infty$. $X$ endowed with the norm $\|f\| = |f|_1 + \|f\|_{L^p}$ is easily seen to be a Banach space. In $X$ we define a semigroup $\{T(t)\}$ by;

$$T(t)f(x) = f(x + t) \qquad \text{for} \quad t \geq 0. \tag{4.4}$$

It follows readily from its definition that $\{T(t)\}$ is a $C_0$ semigroup on $X$ and that $\|T(t)\| \leq 1$. Choosing $f \in X$ to be the characteristic function of the interval $[t, t + \varepsilon^p]$, $\varepsilon > 0$, and letting $\varepsilon \downarrow 0$ shows that $\|T(t)\| \geq 1$ and thus $\|T(t)\| = 1$ for $t \geq 0$.

The infinitesimal generator $A$ of $\{T(t)\}$ is given by

$$D(A) = \{u : u \text{ is absolutely continuous, } u' \in X\} \tag{4.5}$$

and

$$Au = u' \qquad \text{for} \quad u \in D(A). \tag{4.6}$$

Let $f \in X$ and consider the equation

$$\lambda u - Au = \lambda u - u' = f. \tag{4.7}$$

A simple computation shows that

$$u(t) = \int_0^\infty e^{-\lambda s} f(t + s) \, ds = e^{\lambda t} \int_t^\infty e^{-\lambda s} f(s) \, ds \tag{4.8}$$

is a solution of (4.7). We will show that if $\lambda$ satisfies $\text{Re } \lambda > -1$ then $u$, given by (4.8), is in $D(A)$ and thus $\{\lambda : \text{Re } \lambda > -1\} \subset \rho(A)$. To show that $u \in D(A)$ it suffices by (4.7) to show that $u \in X$ and this follows from

$$|u(t)| \leq e^{\text{Re } \lambda t} \int_t^\infty e^{-(\text{Re } \lambda + 1)s} e^s |f(s)| ds \leq e^{-t} \int_t^\infty e^s |f(s)| ds \leq e^{-t} |f|_1$$

which implies that $u \in L^p(0, \infty)$, and

$$|u|_1 \le \int_0^\infty \int_t^\infty e^{(\text{Re}\,\lambda+1)(t-s)} e^s |f(s)| \, ds \, dt$$

$$= \int_0^\infty \left( \int_0^s e^{(\text{Re}\,\lambda+1)(t-s)} \, dt \right) e^s |f(s)| \, ds$$

$$= (\text{Re}\,\lambda + 1)^{-1} \int_0^\infty (1 - e^{-(\text{Re}\,\lambda+1)s}) e^s |f(s)| \, ds$$

$$\le (\text{Re}\,\lambda + 1)^{-1} |f|_1.$$

The set $\{\lambda : \text{Re}\,\lambda > -1\}$ is therefore a subset of $\rho(A)$, $\sigma = \sup\{\text{Re}\,\lambda : \lambda \in \sigma(A)\} \le -1$ while $\|T(t)\|$ does not decay exponentially.                                              $\square$

From Example 4.2 we conclude that in order to obtain exponential decay of $\|T(t)\|$ from the spectral condition $\sup\{\text{Re}\,\lambda : \lambda \in \sigma(A)\} = \sigma < 0$ one has to supplement it with some further conditions on $T(t)$ or $A$. There are many possible assumptions that imply the result. We choose here a simple but rather useful such assumption, namely that $A$ is the infinitesimal generator of an analytic semigroup.

**Theorem 4.3.** *Let $A$ be the infinitesimal generator of an analytic semigroup $T(t)$. If*

$$\sigma = \sup\{\text{Re}\,\lambda : \lambda \in \sigma(A)\} < 0$$

*then there are constants $M \ge 1$ and $\mu > 0$ such that $\|T(t)\| \le Me^{-\mu t}$.*

PROOF. From the results of Section 2.5 it follows easily that there are constants $\omega \ge 0$, $M \ge 1$, $\delta > 0$ and a neighborhood $U$ of $\lambda = \omega$ such that

$$\rho(A) \supset \Sigma = \left\{ \lambda : |\arg(\lambda - \omega)| < \frac{\pi}{2} + \delta \right\} \cup U \qquad (4.9)$$

and

$$\|R(\lambda : A)\| \le \frac{M}{|\lambda - \omega|} \qquad \text{for} \quad \lambda \in \Sigma. \qquad (4.10)$$

Moreover,

$$T(t) = \frac{1}{2\pi i} \int_\Gamma e^{\lambda t} R(\lambda : A) \, d\lambda \qquad (4.11)$$

where $\Gamma$ consists of the two rays $\Gamma_1 = \{\lambda = \rho e^{i\vartheta} + \omega : \rho \ge 0, \pi/2 < \vartheta < \pi/2 + \delta\}$ and $\Gamma_2 = \{\lambda = \rho e^{-i\vartheta} + \omega : \rho \ge 0, \pi/2 < \vartheta < \pi/2 + \delta\}$ and is oriented such that $\text{Im}\,\lambda$ increases along $\Gamma$. The convergence in (4.11) for $t > 0$ is in the uniform operator topology. By our assumption $R(\lambda : A)$ is analytic in a neighborhood of the triangle

$$\Delta = \{\lambda : \text{Re}\,\lambda > \sigma_1, |\arg(\lambda - \omega)| \ge \vartheta\}$$

where $0 > \sigma_1 > \sigma$. From Cauchy's theorem it follows that $\Gamma$ in (4.11) can be shifted without changing the value of the integral to the path $\Gamma'$ where $\Gamma'$ is composed of

$$\Gamma_1' = \left\{ \lambda = \rho e^{i\vartheta} + \omega : \rho \geq \frac{\omega - \sigma_1}{|\cos \vartheta|} \right\},$$

$$\Gamma_2' = \{ \operatorname{Re} \lambda = \sigma_1 : |\operatorname{Im} \lambda| \leq (\omega - \sigma_1)|\tan \vartheta| \},$$

$$\Gamma_3' = \left\{ \lambda = \rho e^{-i\vartheta} + \omega : \rho \geq \frac{\omega - \sigma_1}{|\cos \vartheta|} \right\}$$

and is oriented so that $\operatorname{Im} \lambda$ increases along $\Gamma'$. Thus

$$T(t) = \frac{1}{2\pi i} \int_{\Gamma'} e^{\lambda t} R(\lambda : A) \, d\lambda.$$

Estimating $\| T(t) \|$, on $\Gamma_i'$ $i = 1, 2, 3$ one finds easily that for $t \geq 1$ and some constant $M_1$, $\| T(t) \| \leq M_1 e^{\sigma_1 t}$. Since $\| T(t) \| \leq M_2$ for $0 \leq t \leq 1$ we have $\| T(t) \| \leq M e^{\sigma_1 t}$ for $t \geq 0$ and the proof is complete. $\qquad \square$

We turn now to some simple results on the asymptotic behavior of mild solutions of the inhomogeneous initial value problem (4.1).

**Theorem 4.4.** *Let* $\mu > 0$ *and let* $A$ *be the infinitesimal generator of a* $C_0$ *semigroup* $T(t)$ *satisfying* $\| T(t) \| \leq M e^{-\mu t}$. *Let* $f$ *be bounded and measurable on* $[0, \infty[$. *If*

$$\lim_{t \to \infty} f(t) = f_0 \tag{4.12}$$

*then,* $u(t)$, *the mild solution of* (4.1) *satisfies*

$$\lim_{t \to \infty} u(t) = -A^{-1} f_0. \tag{4.13}$$

PROOF. Since $\| T(t) \| \leq M e^{-\mu t}$ it follows that $0 \in \rho(A)$ (see Theorem 1.5.3) and $\| T(t) x \| \to 0$ as $t \to \infty$. Now

$$v(t) = \int_0^t T(t - s) f(s) \, ds = \int_0^t T(t - s) [f(s) - f_0] \, ds$$

$$+ \int_0^t T(t - s) f_0 \, ds = v_1(t) + v_2(t).$$

Clearly, (see proof of Theorem 1.3.1),

$$\lim_{t \to \infty} v_2(t) = \int_0^\infty T(t) f_0 \, dt = R(0 : A) f_0 = -A^{-1} f_0.$$

To complete the proof we have to show that $v_1(t) \to 0$ as $t \to \infty$.
Given $\varepsilon > 0$ we choose $t_0$ such that for $t > t_0$

$$\| f(t) - f_0 \| < \frac{\varepsilon \mu}{2M}. \tag{4.14}$$

Then, setting $\|f\|_\infty = \sup_{t \geq 0} \|f(t)\|$ we have,

$$\|v_1(t)\| \leq \int_0^{t_0} \|T(t-s)\|\ \|f(s) - f_0\|ds$$

$$+ \int_{t_0}^t \|T(t-s)\|\ \|f(s) - f_0\|ds \leq 2\|f\|_\infty M\mu^{-1} e^{-\mu(t-t_0)} + \frac{\varepsilon}{2}.$$

Choosing $t > t_0$ large enough, the first term on the right becomes less than $\varepsilon/2$ and thus for $t$ large enough $\|v_1(t)\| < \varepsilon$ and the proof is complete. $\quad\square$

A result of similar nature is the following:

**Theorem 4.5.** *Let $\mu > 0$ and let $A$ be the infinitesimal generator of a $C_0$ semigroup $T(t)$ satisfying $\|T(t)\| \leq Me^{-\mu t}$. Let $f$ be continuous and bounded on $[0, \infty[$. If $u_\varepsilon(t)$ is the mild solution of*

$$\varepsilon \frac{du_\varepsilon(t)}{dt} = Au_\varepsilon(t) + f(t), \qquad u_\varepsilon(0) = x, \qquad \varepsilon > 0 \qquad (4.15)$$

*then*

$$\lim_{\varepsilon \to 0} u_\varepsilon(t) = -A^{-1}f(t) \qquad (4.16)$$

*and the limit is uniform on every interval $[\delta, T]$ where $0 < \delta < T$.*

PROOF. The operator $\varepsilon^{-1}A$ is clearly the infinitesimal generator of the $C_0$ semigroup $T_\varepsilon(t) = T(t/\varepsilon)$. We have

$$u_\varepsilon(t) = T_\varepsilon(t)x + \varepsilon^{-1}\int_0^t T_\varepsilon(t-s)f(s)\,ds. \qquad (4.17)$$

Since $\|T_\varepsilon(t)\| \leq Me^{-(\mu/\varepsilon)t}$ it follows that $\|T_\varepsilon(t)x\| \to 0$ as $\varepsilon \to 0$ uniformly on every interval $[\delta, T]$. Now,

$$v_\varepsilon(t) = \varepsilon^{-1}\int_0^t T_\varepsilon(t-s)f(s)\,ds$$

$$= \varepsilon^{-1}\int_0^t T_\varepsilon(t-s)[f(s) - f(t)]\,ds + \varepsilon^{-1}\int_0^t T_\varepsilon(t-s)f(t)\,ds$$

$$= v_{\varepsilon 1}(t) + v_{\varepsilon 2}(t).$$

For $v_{\varepsilon 1}(t)$ we have

$$\|v_{\varepsilon 1}(t)\| \leq \varepsilon^{-1}\int_0^t \|T_\varepsilon(t-s)\|\ \|f(s) - f(t)\|ds$$

$$\leq M\varepsilon^{-1}\int_0^t e^{-\mu\tau/\varepsilon}\|f(t-\tau) - f(t)\|d\tau$$

$$\leq M\varepsilon^{-1}\int_0^{r\varepsilon} e^{-\mu\tau/\varepsilon}\|f(t-\tau) - f(t)\|d\tau + 2\|f\|_\infty M\mu^{-1}e^{-\mu r}$$

$$= M\int_0^r e^{-\mu\sigma}\|f(t-\varepsilon\sigma) - f(t)\|d\sigma + 2e^{-\mu r}M\|f\|_\infty\mu^{-1}$$

where $\|f\|_\infty = \sup_{t\geq 0}\|f(t)\|$ and $r > 0$. Given $\rho > 0$ we first choose $r$ so large that the second term on the right-hand side becomes less than $\rho/2$ and then choose $\varepsilon$ so small that, by the continuity of $f$, the first term on the right-hand side is less than $\rho/2$. Thus $v_{\varepsilon 1}(t) \to 0$ as $\varepsilon \to 0$. Finally,

$$v_{\varepsilon 2}(t) = \varepsilon^{-1}\int_0^t T_\varepsilon(t-s)f(t)\,ds = \varepsilon^{-1}\int_0^t T_\varepsilon(\tau)f(t)\,d\tau$$

$$= \varepsilon^{-1}\int_0^\infty T_\varepsilon(\tau)f(t)\,d\tau - \varepsilon^{-1}\int_t^\infty T_\varepsilon(\tau)f(t)\,d\tau$$

$$= -A^{-1}f(t) + T_\varepsilon(t)A^{-1}f(t).$$

Letting $\varepsilon \to 0$ we therefore have $v_{\varepsilon 2}(t) \to -A^{-1}f(t)$ uniformly on $[\delta, T]$. $\square$

**Remark.** In Theorem 4.5 if $x \in D(A)$ and $f$ is continuously differentiable on $[0, \infty[$ then it is not difficult to show that

$$\frac{du_\varepsilon(t)}{dt} \to -A^{-1}f'(t) \qquad \text{as} \quad \varepsilon \to 0 \tag{4.18}$$

and the limit is uniform on compact subsets of $]0, T[$.

## 4.5. Invariant and Admissible Subspaces

Let $X$ be a Banach space, $Y$ a subspace (not necessarily closed) of $X$ and let $S: D(S) \subset X \to X$ be a linear operator in $X$. The subspace $Y$ of $X$ is an *invariant subspace* of $S$ if $S: D(S) \cap Y \to Y$.

Given a $C_0$ semigroup $T(t)$ on $X$ we will be interested in conditions for a subspace $Y$ of $X$ to be an invariant subspace of $T(t)$ for all $t \geq 0$. Such a subspace will be called an *invariant subspace* of the semigroup $T(t)$. If $Y$ is a closed subspace of $X$ we have:

**Theorem 5.1.** *Let $T(t)$ be a $C_0$ semigroup on $X$ with infinitesimal generator $A$. If $Y$ is a closed subspace of $X$ then $Y$ is an invariant subspace of $T(t)$ if and only if there is a real number $\omega$ such that for every $\lambda > \omega$, $Y$ is an invariant subspace of $R(\lambda : A)$, the resolvent of $A$.*

PROOF. From the results of Chapter 1 it follows that there is an $\omega$ such that

$$R(\lambda : A)x = \int_0^\infty e^{-\lambda t}T(t)x\,dt \tag{5.1}$$

for $\lambda > \omega$. Thus, $\lambda R(\lambda : A)x$ is in the closed convex hull of the trajectory $\{T(t)x : t \geq 0\}$. If $T(t)Y \subset Y$ for every $t \geq 0$ it follows from (5.1) that $\lambda R(\lambda : A)Y \subset Y$ for every $\lambda > \omega$.

Conversely, by the exponential formula (Theorem 1.8.3) we have

$$T(t)x = \lim_{n \to \infty} \left( \frac{n}{t} R\left( \frac{n}{t} : A \right) \right)^n x \qquad \text{for } x \in X. \tag{5.2}$$

If $R(\lambda : A)Y \subset Y$ for $\lambda > \omega$ then for all $n$ large enough $((n/t)R(n/t : A))^n Y \subset Y$ and (5.2) implies that $T(t)Y \subset Y$ for every $t > 0$.                    $\square$

**Remark.** From the proof of Theorem 5.1 it is clear that the result holds also if $Y$ is a closed convex cone with vertex at zero, rather than a closed subspace of $X$.

In the sequel we will be interested in invariant subspaces $Y$ which are not closed in $X$. In order to state such results we need some preliminaries. We start by recalling (see Definition 1.10.3) that if $S : D(S) \subset X \to X$ and $Y$ is a subspace of $X$ then the *part of $S$ in $Y$* is the linear operator $\tilde{S}$ with the domain $D(\tilde{S}) = \{x \in D(S) \cap Y : Sx \in Y\}$ and for $x \in D(\tilde{S})$. $\tilde{S}x = Sx$. The restriction $S_{|Y}$ of $S$ to $Y$ clearly satisfies $S_{|Y} \supset \tilde{S}$. If $Y$ is an invariant subspace of $S$ then $S_{|Y} = \tilde{S}$.

**Lemma 5.2.** *Let $S : D(S) \subset X \to X$ be invertible and let $Y$ be a subspace of $X$. If $S^{-1}Y \subset Y$ then $\tilde{S}$, the part of $S$ in $Y$ is invertible and $\tilde{S}^{-1} = (S^{-1})_{|Y}$.*

PROOF. Let $x \in Y$ and $z = S^{-1}x$. Then $z \in D(S) \cap Y$ and $Sz = x \in Y$. Therefore $z \in D(\tilde{S})$ and $\tilde{S}z = Sz = x$. This shows that the range of $\tilde{S}$ is all of $Y$ and that $\tilde{S}^{-1}$ is well defined and $\tilde{S}^{-1}x = S^{-1}x$ for all $x \in Y$, i.e., $\tilde{S}^{-1} = (S^{-1})_{|Y}$.                    $\square$

In the rest of this section we will assume that $X$ is a Banach space, $Y$ is a subspace of $X$ which is closed with respect to a norm $\| \ \|_Y$ (and hence is itself a Banach space). We will further assume that the norm $\| \ \|_Y$ is stronger than the original norm $\| \ \|$ of $X$. This means that there is a constant $C$ such that

$$\|y\| \leq C\|y\|_Y \qquad \text{for } y \in Y. \tag{5.3}$$

Note that by assumption $Y$ is closed in the norm $\| \ \|_Y$ but in general it is not closed in the norm $\| \ \|$.

**Definition 5.3.** Let $T(t)$ be a $C_0$ semigroup and let $A$ be its infinitesimal generator. A subspace $Y$ of $X$ is called *$A$-admissible* if it is an invariant subspace of $T(t)$, $t \geq 0$, and the restriction of $T(t)$ to $Y$ is a $C_0$ semigroup in $Y$ (i.e., it is strongly continuous in the norm $\| \ \|_Y$).

EXAMPLE 5.4. Let $X$ be the space of bounded uniformly continuous real valued functions on $[0, \infty[$ with the usual supremum norm and let $Y' = X \cap C^1([0, \infty[)$. Set

$$T(t)f(x) = f(x + t) \qquad \text{for } f \in X, t \geq 0. \tag{5.4}$$

$T(t)$ is obviously a $C_0$ semigroup of contractions on $X$. Its infinitesimal generator $A$ is given by $D(A) = \{f \in Y' : f' \in X\}$ and $Af = f'$ for $f \in D(A)$. Denoting the norm in $X$ by $\| \ \|$, we consider the space $Y$ of elements $g \in Y'$ for which $g' \in X$. We equip $Y$ with the norm $\|g\|_Y = \|g\| + \|g'\|$ for $g \in Y$. The norm $\| \ \|_Y$ is stronger than $\| \ \|$, $Y$ is closed in the norm $\| \ \|_Y$ and it is easy to see that the semigroup $T(t)$ defined by (5.4) leaves $Y$ invariant and is a $C_0$ semigroup in $Y$. Thus $Y$ is $A$-admissible.

**Theorem 5.5.** *Let $T(t)$ be a $C_0$ semigroup on $X$ and let $A$ be its infinitesimal generator.*

  *A subspace $Y$ of $X$ is $A$-admissible if and only if*

(i) *$Y$ is an invariant subspace of $R(\lambda : A)$ for all $\lambda > \omega$.*
(ii) *$\tilde{A}$, the part of $A$ in $Y$, is the infinitesimal generator of a $C_0$ semigroup on $Y$.*

*Moreover, if $Y$ is $A$-admissible then $\tilde{A}$ is the infinitesimal generator of the restriction of $T(t)$ to $Y$.*

PROOF. Assume that $Y$ is $A$-admissible. Since $T(t)Y \subset Y$ for $t \geq 0$ and since $\| \ \|_Y$ is stronger than $\| \ \|$ and the restriction of $T(t)$ to $Y$ is a $C_0$ semigroup in $Y$ it follows from (5.1) that there is an $\omega$ such that for $\lambda > \omega$ $R(\lambda : A)Y \subset Y$. Let $A_1$ be the infinitesimal generator of the restriction of $T(t)$ to $Y$. From the definition of the infinitesimal generator it follows readily that $D(A_1) \subset D(A) \cap Y$ and that for $x \in D(A_1)$, $A_1 x = Ax$ and so $\tilde{A} \supset A_1$. On the other hand if $x \in D(\tilde{A})$ then $Ax \in Y$ and the equality

$$T(t)x - x = \int_0^t T(s) Ax \, ds \tag{5.5}$$

holds in $Y$. Dividing (5.5) by $t > 0$ and letting $t \downarrow 0$ it follows that $x \in D(A_1)$ and so $D(A_1) \supset D(\tilde{A})$. Thus $\tilde{A} = A_1$ and $\tilde{A}$ is the infinitesimal generator of a $C_0$ semigroup on $Y$, namely, the restriction of $T(t)$ to $Y$.

  Conversely assume that (i) and (ii) are satisfied and denote by $S(t)$ the $C_0$ semigroup generated by $\tilde{A}$ on $Y$. From the assumption (i) and Lemma 5.2 it follows that $R(\lambda : \tilde{A})x = R(\lambda : A)x$ for every $x \in Y$ and therefore also

$$\left[ \frac{n}{t} R\left( \frac{n}{t} : \tilde{A} \right) \right]^n x = \left[ \frac{n}{t} R\left( \frac{n}{t} : A \right) \right]^n x \tag{5.6}$$

for all $n$ large enough and $x \in Y$. Passing to the limit as $n \to \infty$ it follows from the exponential formula (Theorem 1.8.3) that the left-hand side of (5.6) converges in $Y$, and hence also in $X$, to $S(t)x$ while the right-hand side converges in $X$ to $T(t)x$. Therefore $S(t)x = T(t)x$ for every $x \in Y$ which implies both that $Y$ is an invariant subspace for $T(t)$ and that $T(t)$ is a $C_0$ semigroup on $Y$. $\square$

**Corollary 5.6.** *Y is A-admissible if and only if*

(i) *For sufficiently large $\lambda$, Y is an invariant subspace of $R(\lambda : A)$.*
(ii) *There exist constants M and $\beta$ such that*

$$\|R(\lambda : A)^n\|_Y \le M(\lambda - \beta)^{-n}, \qquad \lambda > \beta, \qquad n = 1, 2, \ldots . \quad (5.7)$$

(iii) *For $\lambda > \beta$, $R(\lambda : A)Y$ is dense in Y.*

PROOF. Condition (i) is the same as in Theorem 5.5. From (i) and Lemma 5.2 it follows that $R(\lambda : \tilde{A})x = R(\lambda : A)x$ for $x \in Y$ and $\lambda > \omega$. Therefore we can replace $A$ by $\tilde{A}$ in (5.7) and in Condition (iii). Condition (iii) is then equivalent to the fact that $D(\tilde{A}) = R(\lambda : \tilde{A})Y = R(\lambda : A)Y$ is dense in $Y$ and from Theorem 1.5.3 it follows that $\tilde{A}$ generates a $C_0$ semigroup on $Y$ if and only if (ii) and (iii) are satisfied. From Theorem 5.5 it then follows that $Y$ is $A$-admissible if and only if (i)–(iii) are satisfied.                    □

**Remark 5.7.** If in Corollary 5.6 $Y$ is reflexive the Condition (iii) follows from (i) and (ii). Indeed, for $\lambda, \mu \in \rho(A)$ we have the resolvent identity

$$R(\lambda : A) - R(\mu : A) = (\mu - \lambda)R(\lambda : A)R(\mu : A)$$

which implies directly that $D = R(\lambda : A)Y$ is independent of $\lambda \in \rho(A)$. From (5.7) with $n = 1$ it follows that for $x \in Y$, $\lambda R(\lambda : A)x$ is bounded in $Y$ as $\lambda \to \infty$. The reflexivity of $Y$ then implies that there is a sequence $\lambda_n \to \infty$ such $\lambda_n R(\lambda_n : A)x$ converges weakly in $Y$ to some $y \in Y$. Since $A$ is the infinitesimal generator of a $C_0$ semigroup on $X$, $\lambda R(\lambda : A)x \to x$ strongly in $X$ as $\lambda \to \infty$ (Lemma 1.3.2) so $y = x$. Since for large values $\lambda_n$, $\lambda_n R(\lambda_n : A)x \in D$ we conclude that the weak closure of $D$ in $Y$ is all of $Y$. But the weak and strong closures of a linear subspace of a Banach space are the same and so $D$ is dense in $Y$.

We conclude this section with a useful criterion for a subspace $Y$ of $X$ to be $A$-admissible.

**Theorem 5.8.** *Let $\overline{Y}$ be the closure of Y in the norm of X. Let S be an isomorphism of Y onto $\overline{Y}$. Y is A-admissible if and only if $A_1 = SAS^{-1}$ is the infinitesimal generator of a $C_0$ semigroup on $\overline{Y}$. In this case we have in $\overline{Y}$*

$$T_1(t) = ST(t)S^{-1}$$

*where $T_1(t)$ is the semigroup generated by $A_1$.*

PROOF. Let $\tilde{A}$ be the part of $A$ in $Y$. From the definition of $A_1$ we have

$$D(A_1) = \{x \in \overline{Y} : S^{-1}x \in D(A), AS^{-1}x \in Y\}$$

$$= \{x \in \overline{Y} : S^{-1}x \in D(\tilde{A})\} = SD(\tilde{A}).$$

It follows that $D(A_1)$ is dense in $\overline{Y}$ if and only if $D(\tilde{A})$ is dense in $Y$.

Moreover for $x \in D(A_1)$ we have

$$(\lambda I - A_1)x = (\lambda I - SAS^{-1})x = S(\lambda I - A)S^{-1}x = S(\lambda I - \tilde{A})S^{-1}x.$$
(5.8)

By assumption for $\lambda > \omega$, $R(\lambda : A)$ is a bounded operator on $X$. We claim that $R(\lambda : A_1)$ exists as a bounded operator on $\overline{Y}$ if and only if $R(\lambda : A)Y \subset Y$ and then

$$R(\lambda : A_1) = SR(\lambda : A)S^{-1} = SR(\lambda : \tilde{A})S^{-1}$$
(5.9)

in $\overline{Y}$. Indeed, if $R(\lambda : A)Y \subset Y$, $SR(\lambda : A)S^{-1}$ is a bounded linear operator on $\overline{Y}$ which is the inverse of $S(\lambda I - A)S^{-1}$ and (5.9) follows from (5.8). On the other hand if $R(\lambda : A_1)$ exists in $\overline{Y}$, $S(\lambda I - A)S^{-1}$ is invertible and its inverse $SR(\lambda : A)S^{-1}$ is a bounded linear operator satisfying (5.9) and therefore also $S^{-1}R(\lambda : A_1) = R(\lambda : A)S^{-1}$ which implies $R(\lambda : A)Y \subset Y$.

Now if $A_1$ is the infinitesimal generator of a $C_0$ semigroup on $\overline{Y}$, $D(A_1)$ is dense in $\overline{Y}$ and therefore $D(\tilde{A})$ is dense in $Y$. Moreover for $\lambda > \omega$ $R(\lambda : A_1)$ exists and therefore by the first part of the proof $R(\lambda : A)Y \subset Y$ and (5.9) holds. Theorem 1.5.3 then implies that $\tilde{A}$ generates a $C_0$ semigroup on $Y$ and by Theorem 5.5, $Y$ is $A$-admissible. On the other hand if $Y$ is $A$-admissible, $R(\lambda : A)Y \subset Y$ (Theorem 5.5) and by the first part of the proof (5.9) holds. Since $D(\tilde{A})$ is then dense in $Y$, $D(A_1)$ is dense in $\overline{Y}$ and Theorem 1.5.3 implies that $A_1$ is the infinitesimal generator of a $C_0$ semigroup on $\tilde{Y}$. Finally, (5.9) together with the exponential formula (Theorem 1.8.3) imply that $T_1(t) = ST(t)S^{-1}$ and the proof is complete. $\quad\square$

# CHAPTER 5

# Evolution Equations

## 5.1. Evolution Systems

Let $X$ be a Banach space. For every $t$, $0 \le t \le T$ let $A(t): D(A(t)) \subset X \to X$ be a linear operator in $X$ and let $f(t)$ be an $X$ valued function. In this chapter we will study the initial value problem

$$\begin{cases} \dfrac{du(t)}{dt} = A(t)u(t) + f(t) & \text{for } s < t \le T \\ u(s) = x . \end{cases} \tag{1.1}$$

The initial value problem (1.1) is called an evolution problem. An $X$ valued function $u:[s, T] \to X$ is a classical solution of (1.1) if $u$ is continuous on $[s, T]$, $u(t) \in D(A(t))$ for $s < t \le T$, $u$ is continuously differentiable on $s < t \le T$ and satisfies (1.1).

The previous chapter was dedicated to the special case of (1.1) where $A(t) = A$ is independent of $t$. We saw that in this case, the solution of the inhomogeneous initial value problem, i.e., the problem with $f \ne 0$, can be represented in terms of the solutions of the homogeneous initial value problem via the formula of "variations of constants"

$$u(t) = T(t - s)u(s) + \int_s^t T(t - \tau)f(\tau)\, d\tau \tag{1.2}$$

where $T(t)x$ is the solution of the initial value problem

$$\frac{du(t)}{dt} = Au(t), \qquad u(0) = x. \tag{1.3}$$

We will see later that a similar result is also true when $A(t)$ depends on $t$.

Therefore we concentrate at the beginning on the homogeneous initial value problem:

$$\begin{cases} \dfrac{du(t)}{dt} = A(t)u(t) & 0 \le s < t \le T \\ u(s) = x. \end{cases} \tag{1.4}$$

In order to obtain some feeling for the behavior of the solutions of (1.4) we consider first the simple case where for $0 \le t \le T$, $A(t)$ is a bounded linear operator on $X$ and $t \to A(t)$ is continuous in the uniform operator topology. For this case we have:

**Theorem 5.1.** *Let $X$ be a Banach space and for every $t, 0 \le t \le T$ let $A(t)$ be a bounded linear operator on $X$. If the function $t \to A(t)$ is continuous in the uniform operator topology then for every $x \in X$ the initial value problem (1.4) has a unique classical solution $u$.*

PROOF. The proof of this theorem is standard using Picard's iterations method. Let $\alpha = \max_{0 \le t \le T} \|A(t)\|$ and define a mapping $S$ from $C([s, T]: X)$ into itself by

$$(Su)(t) = x + \int_s^t A(\tau)u(\tau)\, d\tau. \tag{1.5}$$

Denoting $\|u\|_\infty = \max_{s \le t \le T} \|u(t)\|$ it is easy to check that

$$\|Su(t) - Sv(t)\| \le \alpha(t - s)\|u - v\|_\infty, \qquad s \le t \le T. \tag{1.6}$$

Using (1.5) and (1.6) it follows by induction that

$$\|S^n u(t) - S^n v(t)\| \le \frac{\alpha^n (t - s)^n}{n!} \|u - v\|_\infty, \qquad s \le t \le T,$$

and therefore,

$$\|S^n u - S^n v\|_\infty \le \frac{\alpha^n (T - s)^n}{n!} \|u - v\|_\infty.$$

For $n$ large enough $\alpha^n (T - s)^n / n! < 1$ and by a well known generalization of the Banach contraction principle, $S$ has a unique fixed point $u$ in $C([s, T]: X)$ for which

$$u(t) = x + \int_s^t A(\tau)u(\tau)\, d\tau. \tag{1.7}$$

Since $u$ is continuous, the right hand side of (1.7) is differentiable. Thus $u$ is differentiable and its derivative, obtained by differentiating (1.7), satisfies $u'(t) = A(t)u(t)$. So, $u$ is a solution of the initial value problem (1.4). Since every solution of (1.4) is also a solution of (1.7), the solution of (1.4) is unique. $\qquad\square$

We define the "solution operator" of the initial value problem (1.4) by

$$U(t, s)x = u(t) \qquad \text{for} \quad 0 \le s \le t \le T \tag{1.8}$$

where $u$ is the solution of (1.4). $U(t, s)$ is a two parameter family of operators. From the uniqueness of the solution of the initial value problem (1.4) it follows readily that if $A(t) = A$ is independent of $t$ then $U(t, s) = U(t - s)$ and the two parameter family of operators reduces to the one parameter family $U(t)$, $t \geq 0$, which is of course the semigroup generated by $A$. The main properties of $U(t, s)$, in our special case where $A(t)$ is a bounded linear operator on $X$ for $0 \leq t \leq T$ and $t \to A(t)$ is continuous in the uniform operator topology, are given in the next theorem.

**Theorem 5.2.** *For every $0 \leq s \leq t \leq T$, $U(t, s)$ is a bounded linear operator and*

(i)   $\|U(t, s)\| \leq \exp(\int_s^t \|A(\tau)\| d\tau)$.
(ii)  $U(t, t) = I$, $U(t, s) = U(t, r)U(r, s)$ *for* $0 \leq s \leq r \leq t \leq T$.
(iii) $(t, s) \to U(t, s)$ *is continuous in the uniform operator topology for* $0 \leq s \leq t \leq T$.
(iv)  $\partial U(t, s)/\partial t = A(t)U(t, s)$ *for* $0 \leq s \leq t \leq T$.
(v)   $\partial U(t, s)/\partial s = -U(t, s)A(s)$ *for* $0 \leq s \leq t \leq T$.

PROOF. Since the problem (1.4) is linear it is obvious that $U(t, s)$ is a linear operator defined on all of $X$. From (1.7) it follows that

$$\|u(t)\| \leq \|x\| + \int_s^t \|A(\tau)\| \, \|u(\tau)\| \, d\tau$$

which by Gronwall's inequality implies

$$\|U(t, s)x\| = \|u(t)\| \leq \|x\| \exp\left(\int_s^t \|A(\tau)\| \, d\tau\right) \qquad (1.9)$$

and so $U(t, s)$ is bounded and satisfies (i).

From (1.8) it follows readily that $U(t, t) = I$ and from the uniqueness of the solution of (1.4) the relation $U(t, s) = U(t, r)U(r, s)$ for $0 \leq s \leq r \leq t \leq T$ follows. Combining (i) and (ii), (iii) follows. Finally, from (1.7) and (iii) it follows that $U(t, s)$ is the unique solution of the integral equation

$$U(t, s) = I + \int_s^t A(\tau)U(\tau, s) \, d\tau \qquad (1.10)$$

in $B(X)$ (the space of all bounded linear operators on $X$). Differentiating (1.10) with respect to $t$ yields (iv). Differentiating (1.10) with respect to $s$ we find

$$\frac{\partial}{\partial s} U(t, s) = -A(s) + \int_s^t A(\tau)\frac{\partial}{\partial s} U(\tau, s) \, d\tau. \qquad (1.11)$$

From the uniqueness of the solution of (1.10) it follows that

$$\frac{\partial}{\partial s} U(t, s) = -U(t, s)A(s) \qquad (1.12)$$

and the proof is complete.                                                            □

The two parameter family of operators $U(t, s)$ replaces in the non-autonomous case, i.e., in the case where $A(t)$ depends on $t$, the one parameter semigroup $U(t)$ of the autonomous case. This motivates the following definition.

**Definition 5.3.** A two parameter family of bounded linear operators $U(t, s)$, $0 \le s \le t \le T$, on $X$ is called an *evolution system* if the following two conditions are satisfied:

(i)  $U(s, s) = I$, $U(t, r)U(r, s) = U(t, s)$ for $0 \le s \le r \le t \le T$.
(ii) $(t, s) \to U(t, s)$ is strongly continuous for $0 \le s \le t \le T$.

Note that by analogy to the autonomous case, since we are not really interested in uniform continuity of solutions, we have replaced the continuity of $U(t, s)$ in the uniform operator topology by strong continuity.

In the next sections we will give conditions on a given family of linear, usually unbounded, operators $\{A(t)\}$, $0 \le t \le T$ that guarantee the existence of a unique classical solution of the initial value problem

$$\frac{du(t)}{dt} = A(t)u(t), \qquad u(s) = x \qquad (1.13)$$

for a dense set of initial values $x \in X$. The existence of such a unique solution will provide us with an evolution system associated with the family $\{A(t)\}$, $0 \le t \le T$. The uniqueness of the solution of (1.13) will imply the property (i) of evolution systems while the continuity of the solution at the initial data will imply the property (ii). The relations between $A(t)$ and $U(t, s)$ will be determined by some generalized versions of the equations

$$\frac{\partial U(t, s)}{\partial t} = A(t)U(t, s) \qquad (1.14)$$

$$\frac{\partial U(t, s)}{\partial s} = -U(t, s)A(s). \qquad (1.15)$$

We conclude this section with a remark concerning the inhomogeneous initial value problem (1.1) where $f \in L^1(0, T: X)$. If there is an evolution system $U(t, s)$ associated with this initial value problem such that for every $v \in D(A(s))$, $U(t, s)v \in D(A(t))$ and $U(t, s)v$ is differentiable both in $t$ and $s$ satisfying

$$\frac{\partial}{\partial t} U(t, s)v = A(t)U(t, s)v \qquad (1.16)$$

$$\frac{\partial}{\partial s} U(t, s)v = -U(t, s)A(s)v \qquad (1.17)$$

then every classical solution $u$ of (1.1) with $x \in D(A(s))$ is given by

$$u(t) = U(t, s)x + \int_s^t U(t, r)f(r)\, dr. \qquad (1.18)$$

Indeed, in this case the function $r \to U(t, r)u(r)$ is differentiable on $[s, T]$ and

$$\frac{\partial}{\partial r} U(t, r)u(r) = -U(t, r)A(r)u(r) + U(t, r)A(r)u(r) + U(t, r)f(r)$$

$$= U(t, r)f(r). \tag{1.19}$$

Integrating (1.19) from $s$ to $t$ yields (1.18). Thus, in this case, the inhomogeneous initial value problem (1.1) has at most one classical solution $u$ which, if it exists, is given by (1.18). However, for any evolution system $U(t, s)$ and $f \in L^1(0, T: X)$ the right-hand side of (1.18) is a well defined continuous function satisfying $u(s) = x$. As in the autonomous case (Section 4.5.2) we will often consider this function as a generalized solution of the initial value problem (1.1).

## 5.2. Stable Families of Generators

This section is devoted to some preliminaries that will be needed in the construction of an evolution system for the initial value problem

$$\begin{cases} \dfrac{du(t)}{dt} = A(t)u(t) & 0 \le s \le t \\ u(s) = x \end{cases} \tag{2.1}$$

in the "hyperbolic" case.

**Definition 2.1.** Let $X$ be a Banach space. A family $\{A(t)\}_{t \in [0, T]}$ of infinitesimal generators of $C_0$ semigroups on $X$ is called *stable* if there are constants $M \ge 1$ and $\omega$ (called the stability constants) such that

$$\rho(A(t)) \supset \,]\omega, \infty[ \qquad \text{for} \quad t \in [0, T] \tag{2.2}$$

and

$$\left\| \prod_{j=1}^{k} R(\lambda : A(t_j)) \right\| \le M(\lambda - \omega)^{-k} \qquad \text{for} \quad \lambda > \omega \tag{2.3}$$

and every finite sequence $0 \le t_1 \le t_2, \ldots, t_k \le T$, $k = 1, 2, \ldots$ .

Note that in general the operators $R(\lambda : A(t_j))$ do not commute and therefore the order of terms in (2.3) is important. In (2.3) and in the sequel products containing $\{t_j\}$ will always be "time-ordered", i.e., a factor with a larger $t_j$ stands to the left of ones with smaller $t_j$.

From the definition of stability it is clear that the stability of a family of infinitesimal generators $\{A(t)\}$ is preserved when the norm in $X$ is replaced by an equivalent norm. The constants of stability however, depend on the particular norm in $X$.

If for $t \in [0, T]$, $A(t) \in G(1, \omega)$, i.e., $A(t)$ is the infinitesimal generator of a $C_0$ semigroup $S_t(s)$, $s \geq 0$, satisfying $\| S_t(s) \| \leq e^{\omega s}$ then the family $\{ A(t) \}_{t \in [0, T]}$ is clearly stable with constants $M = 1$ and $\omega$. In particular any family $\{ A(t) \}_{t \in [0, T]}$ of infinitesimal generators of $C_0$ semigroups of contractions is stable.

**Theorem 2.2.** *For $t \in [0, T]$ let $A(t)$ be the infinitesimal generator of a $C_0$ semigroup $S_t(s)$ on the Banach space $X$. The family of generators $\{ A(t) \}_{t \in [0, T]}$ is stable if and only if there are constants $M \geq 1$ and $\omega$ such that $\rho(A(t)) \supset ]\omega, \infty[$ for $t \in [0, T]$ and either one of the following conditions is satisfied*

$$\left\| \prod_{j=1}^{k} S_{t_j}(s_j) \right\| \leq M \exp \left\{ \omega \sum_{j=1}^{k} s_j \right\} \qquad for \quad s_j \geq 0 \qquad (2.4)$$

*and any finite sequence $0 \leq t_1 \leq t_2 \leq \cdots \leq t_k \leq T$, $k = 1, 2, \ldots$ or*

$$\left\| \prod_{j=1}^{k} R(\lambda_j : A(t_j)) \right\| \leq M \prod_{j=1}^{k} (\lambda_j - \omega)^{-1} \qquad for \quad \lambda_j > \omega \qquad (2.5)$$

*and any finite sequence $0 \leq t_1 \leq t_2 \leq \cdots \leq t_k \leq T$, $k = 1, 2, \ldots$.*

PROOF. From the statement of the theorem it is clear that it suffices to prove that for a family $\{ A(t) \}_{t \in [0, T]}$ of infinitesimal generators for which $\rho(A(t)) \supset ]\omega, \infty[$ the estimates (2.3), (2.4) and (2.5) are equivalent.

Assume that (2.3) holds and let $s_j$, $1 \leq j \leq k$ be positive rational numbers. Let $\lambda = N$ be a positive integer such that $Ns_j = m_j$ is a positive integer for $1 \leq j \leq k$. In (2.3) we take $m = \sum_{j=1}^{k} m_j$ terms and subdivide them into $k$ subsets containing $m_j$, $1 \leq j \leq k$, terms. All values of $t$ in the $j$-th subset are taken to be equal to $t_j$. After dividing both sides of the inequality by $N^m$ we find

$$\left\| \prod_{j=1}^{k} \left[ \frac{m_j}{s_j} R\left( \frac{m_j}{s_j} : A(t_j) \right) \right]^{m_j} \right\| \leq M \left( 1 - \frac{\omega}{N} \right)^{-m}. \qquad (2.6)$$

Letting $N \to \infty$, such that $Ns_j$, $1 \leq j \leq k$, stay integers, each one of the $m_j$ tends to infinity and by the exponential formula (Theorem 1.8.3) we obtain

$$\left\| \prod_{j=1}^{k} S_{t_j}(s_j) \right\| \leq M \exp \left\{ \omega \sum_{j=1}^{k} s_j \right\}$$

and therefore (2.4) holds for all positive rationals $s_j$. The general case of non-negative real $s_j$ follows from the strong continuity of $S_t(s)$ in $s$ and thus (2.3) implies (2.4).

In Chapter 1 we saw that

$$R(\lambda_j : A(t_j)) x = \int_0^{\infty} e^{-\lambda_j s} S_{t_j}(s) x \, ds \qquad for \quad \lambda_j > \omega. \qquad (2.7)$$

Iterating (2.7) a finite number of times yields

$$\prod_{j=1}^{k} R(\lambda_j : A(t_j)) x = \int_0^{\infty} \cdots \int_0^{\infty} \exp\left\{ - \sum_{j=1}^{k} \lambda_j s_j \right\} \prod_{j=1}^{k} S_{t_j}(s_j) x \, ds_1 \cdots ds_k.$$

(2.8)

Using (2.4) to estimate the norm of the right-hand side of (2.8) we find

$$\left\| \prod_{j=1}^{k} R(\lambda_j : A(t_j)) x \right\| \leq M\|x\| \prod_{j=1}^{k} \int_0^{\infty} e^{(\omega - \lambda_j)s_j} \, ds_j = M\|x\| \prod_{j=1}^{k} (\lambda_j - \omega)^{-1}$$

and therefore (2.4) implies (2.5). Finally, choosing all $\lambda_j$ equal to $\lambda$ in (2.5) shows that (2.5) implies (2.3) and the proof is complete.                     $\square$

We have noted above that if $\{A(t)\}_{t \in [0, T]}$ is a family of infinitesimal generators satisfying $A(t) \in G(1, \omega)$ for $t \in [0, T]$ then it is a stable family. In general however, it is not always easy to decide whether or not a given family of infinitesimal generators is stable. The following perturbation theorem is a useful criterion for this.

**Theorem 2.3.** Let $\{A(t)\}_{t \in [0, T]}$ be a stable family of infinitesimal generators with stability constants $M$ and $\omega$. Let $B(t)$, $0 \leq t \leq T$ be bounded linear operators on $X$. If $\|B(t)\| \leq K$ for all $0 \leq t \leq T$ then $\{A(t) + B(t)\}_{t \in [0, T]}$ is a stable family of infinitesimal generators with stability constants $M$ and $\omega + KM$.

**Proof.** From Theorem 3.1.1 it follows that for every $t \in [0, T]$, $A(t) + B(t)$ is the infinitesimal generator of a $C_0$ semigroup. It is easy to check that if $\lambda > \omega + KM$ then $\lambda$ is in the resolvent set of $A(t) + B(t)$ and

$$R(\lambda : A(t) + B(t)) = \sum_{n=0}^{\infty} R(\lambda : A(t))[B(t) R(\lambda : A(t))]^n.$$

Therefore,

$$\prod_{j=1}^{k} R(\lambda : A(t_j) + B(t_j)) = \prod_{j=1}^{k} \left( \sum_{n=0}^{\infty} R(\lambda : A(t_j))[B(t_j) R(\lambda : A(t_j))]^n \right).$$

(2.9)

Expanding the right-hand side of (2.9) we find a series whose general term is of the form

$$R(\lambda : A(t_k))[B(t_k) R(\lambda : A(t_k))]^{n_k}$$
$$\cdots R(\lambda : A(t_1))[B(t_1) R(\lambda : A(t_1))]^{n_1}$$

where $n_j \geq 0$. If $\sum_{j=1}^{k} n_j = n$ then estimating this term, using the stability of the family $\{A(t)\}_{t \in [0, T]}$, yields the estimate $M^{n+1} K^n (\lambda - \omega)^{-n-k}$. The number of terms in which $\sum_{j=1}^{k} n_j = n$ in this series is $\binom{n+k-1}{k-1}$ and therefore

$$\left\| \prod_{j=1}^{k} R(\lambda : A(t_j) + B(t_j)) \right\| \leq M(\lambda - \omega)^{-k} \sum_{n=0}^{\infty} \binom{n+k-1}{k-1} (MK(\lambda - \omega)^{-1})^n$$

$$= M(\lambda - \omega - MK)^{-k}$$

and the proof is complete.                                                                          □

Let $X$ and $Y$ be Banach spaces and assume that $Y$ is densely and continuously embedded in $X$. Let $\{A(t)\}_{t \in [0, T]}$ be a stable family of infinitesimal generators in $X$ and let $\{\tilde{A}(t)\}_{t \in [0, T]}$ be the family of parts $\tilde{A}(t)$ of $A(t)$ in $Y$. Our last result gives a useful sufficient condition for $\{\tilde{A}(t)\}_{t \in [0, T]}$ to be stable in $Y$.

**Theorem 2.4.** *Let $Q(t)$, $0 \leq t \leq T$, be a family of isomorphisms of $Y$ onto $X$ with the following properties.*

(i) *$\|Q(t)\|_{Y \to X}$ and $\|Q(t)^{-1}\|_{X \to Y}$ are uniformly bounded by a constant $C$.*
(ii) *The map $t \to Q(t)$ is of bounded variation in the $B(Y, X)$ norm $\| \cdot \|_{Y \to X}$.*

*Let $\{A(t)\}_{t \in [0, T]}$ be a stable family of infinitesimal generators in $X$ and let $A_1(t) = Q(t)A(t)Q(t)^{-1}$. If $\{A_1(t)\}_{t \in [0, T]}$ is a stable family of infinitesimal generators in $X$ then $Y$ is $A(t)$-admissible for $t \in [0, T]$ and $\{\tilde{A}(t)\}_{t \in [0, T]}$ is a stable family of infinitesimal generators in $Y$.*

PROOF. From Theorem 4.5.8 it follows readily that $Y$ is $A(t)$-admissible for every $t \in [0, T]$ and therefore by Theorem 4.5.5, $\tilde{A}(t)$ the part of $A(t)$ in $Y$ is the infinitesimal generator of a $C_0$ semigroup in $Y$. From the definition of $A_1(t)$ it follows that

$$D(A_1(t)) = \{x \in X : Q(t)^{-1}x \in D(A(t)), A(t)Q(t)^{-1}x \in Y\}$$

$$= \{x \in X : Q(t)^{-1}x \in D(\tilde{A}(t))\} = Q(t)D(\tilde{A}(t))$$

and therefore $A_1(t) = Q(t)\tilde{A}(t)Q(t)^{-1}$. This implies that for large enough real $\lambda$

$$R(\lambda : \tilde{A}(t)) = Q(t)^{-1}R(\lambda : A_1(t))Q(t)$$

and thus,

$$\prod_{j=1}^{k} R(\lambda : \tilde{A}(t_j)) = \prod_{j=1}^{k} Q(t_j)^{-1}R(\lambda : A_1(t_j))Q(t_j). \qquad (2.10)$$

Setting $P_j = (Q(t_j) - Q(t_{j-1}))Q(t_{j-1})^{-1}$ the right-hand side of (2.10) becomes

$$Q(t_k)^{-1}\{R(\lambda : A_1(t_k))(I + P_k) \cdots (I + P_2)R(\lambda : A_1(t_1))\}Q(t_1). \qquad (2.11)$$

Let $M_1$ and $\omega_1$ be the stability constants of $\{A_1(t)\}_{t \in [0, T]}$. Expanding the expression in the curly bracket into a polynomial in the $P_j$ and noting that similarly to the proof of Theorem 2.3, only $m + 1$ factors of $M_1$ are needed to estimate a term involving $m$ of the $P_j$, we can estimate the $X$ norm of this expression by

$$M_1(\lambda - \omega_1)^{-k} \prod_{j=2}^{k} (1 + M_1 \|P_j\|). \qquad (2.12)$$

From the definition of $P_j$ we have $\|P_j\| \le C \|Q(t_j) - Q(t_{j-1})\|_{Y \to X}$ and therefore

$$(1 + \alpha \|P_j\|) \le \exp\{\alpha C \|Q(t_j) - Q(t_{j-1})\|_{Y \to X}\}. \qquad (2.13)$$

Denoting by $V$ the total variation of $t \to Q(t)$ in the $B(Y, X)$ norm and estimating the $Y$ norm of (2.11) using (2.12) and (2.13) yields

$$\left\| \prod_{j=1}^{k} R(\lambda : \tilde{A}(t_j)) \right\|_Y \le C^2 M_1 (\lambda - \omega_1)^{-k}$$

$$\cdot \exp \left\{ M_1 C \sum_{j=2}^{k} \|Q(t_j) - Q(t_{j-1})\|_{Y \to X} \right\}$$

$$\le C^2 M_1 e^{C M_1 V} (\lambda - \omega_1)^{-k}$$

and thus $\{\tilde{A}(t)\}_{t \in [0, T]}$ is stable in $Y$.                                     $\square$

## 5.3. An Evolution System in the Hyperbolic Case

This section is devoted to the construction of an evolution system for the initial value problem

$$\begin{cases} \dfrac{du(t)}{dt} = A(t)u(t) & 0 \le s \le t \le T \\ u(s) = v \end{cases} \qquad (3.1)$$

where the family $\{A(t)\}_{t \in [0, T]}$ satisfies the conditions $(H_1)$–$(H_3)$ below. The set of conditions $(H_1)$–$(H_3)$ is usually referred to as the "hyperbolic" case in contrast to the "parabolic" case in which each of the operators $A(t)$, $t \ge 0$ is assumed to be the infinitesimal generator of an analytic semigroup. The reason for these names lies in the different applications of the abstract results to partial differential equations.

Let $X$ and $Y$ be Banach spaces with norms $\| \ \|$ and $\| \ \|_Y$ respectively. Throughout this section we will assume that $Y$ is densely and continuously imbedded in $X$, i.e., $Y$ is a dense subspace of $X$ and there is a constant $C$

such that

$$\|w\| \leq C\|w\|_Y \quad \text{for} \quad w \in Y. \tag{3.2}$$

Let $A$ be the infinitesimal generator of a $C_0$ semigroup $S(s)$, $s \geq 0$, on $X$. Recall (Definition 4.5.3) that $Y$ is $A$-admissible if $Y$ is an invariant subspace of $S(s)$ and the restriction $\tilde{S}(s)$ of $S(s)$ to $Y$ is a $C_0$ semigroup on $Y$. Moreover, $\tilde{A}$ the part of $A$ in $Y$ is, in this case, the infinitesimal generator of the semigroup $\tilde{S}(s)$ on $Y$.

For $t \in [0, T]$ let $A(t)$ be the infinitesimal generator of a $C_0$ semigroup $S_t(s)$, $s \geq 0$, on $X$. We will make the following assumptions.

($H_1$) $\{A(t)\}_{t \in [0, T]}$ is a stable family with stability constants $M, \omega$.

($H_2$) $Y$ is $A(t)$-admissible for $t \in [0, T]$ and the family $\{\tilde{A}(t)\}_{t \in [0, T]}$ of parts $\tilde{A}(t)$ of $A(t)$ in $Y$, is a stable family in $Y$ with stability constants $\tilde{M}, \tilde{\omega}$.

($H_3$) For $t \in [0, T]$, $D(A(t)) \supset Y$, $A(t)$ is a bounded operator from $Y$ into $X$ and $t \rightarrow A(t)$ is continuous in the $B(Y, X)$ norm $\| \ \|_{Y \rightarrow X}$.

The principal result of this section, Theorem 3.1, shows that if $\{A(t)\}_{t \in [0, T]}$ satisfies the conditions ($H_1$)–($H_3$) then one can associate a unique evolution system $U(t, s)$, $0 \leq s \leq t \leq T$, with the initial value problem (3.1).

**Theorem 3.1.** *Let $A(t)$, $0 \leq t \leq T$, be the infinitesimal generator of a $C_0$ semigroup $S_t(s)$, $s \geq 0$, on $X$. If the family $\{A(t)\}_{t \in [0, T]}$ satisfies the conditions ($H_1$)–($H_3$) then there exists a unique evolution system $U(t, s)$, $0 \leq s \leq t \leq T$, in $X$ satisfying*

($E_1$) $\quad \|U(t, s)\| \leq M \exp\{\omega(t - s)\} \quad$ *for* $\quad 0 \leq s \leq t \leq T.$

($E_2$) $\quad \dfrac{\partial^+}{\partial t} U(t, s)v \Big|_{t=s} = A(s)v \quad$ *for* $\quad v \in Y, 0 \leq s \leq T.$

($E_3$) $\quad \dfrac{\partial}{\partial s} U(t, s)v = -U(t, s)A(s)v \quad$ *for* $\quad v \in Y, 0 \leq s \leq t \leq T.$

*Where the derivative from the right in ($E_2$) and the derivative in ($E_3$) are in the strong sense in $X$.*

PROOF. We start by approximating the family $\{A(t)\}_{t \in [0, T]}$ by piecewise constant families $\{A_n(t)\}_{t \in [0, T]}$, $n = 1, 2, \ldots$, defined as follows: Let $t_k^n = (k/n)T$, $k = 0, 1, \ldots, n$ and let

$$\begin{cases} A_n(t) = A(t_k^n) & \text{for} \quad t_k^n \leq t < t_{k+1}^n, \quad k = 0, 1, \ldots, n-1 \\ A_n(T) = A(T). \end{cases}$$

$$\tag{3.3}$$

Since $t \rightarrow A(t)$ is continuous in the $B(Y, X)$ norm it follows that

$$\|A(t) - A_n(t)\|_{Y \rightarrow X} \rightarrow 0 \quad \text{as} \quad n \rightarrow \infty \tag{3.4}$$

uniformly in $t \in [0, T]$. From the definition of $A_n(t)$ and the conditions of the theorem it follows readily that for $n \geq 1$, $\{A_n(t)\}_{t \in [0, T]}$ is a stable family in $X$ with constants $M, \omega$ and $\{\tilde{A}_n(t)\}_{t \in [0, T]}$ is a stable family in $Y$ with constants $\tilde{M}, \tilde{\omega}$.

Next we define a two parameter family of operators $U_n(t, s)$, $0 \leq s \leq t \leq T$ by,

$$U_n(t, s) = \begin{cases} S_{t_j^n}(t - s) & \text{for } t_j^n \leq s \leq t \leq t_{j+1}^n \\ S_{t_k^n}(t - t_k^n)\left[\displaystyle\prod_{j=l+1}^{k-1} S_{t_j^n}\left(\frac{T}{n}\right)\right] S_{t_l^n}(t_{l+1}^n - s) \end{cases}$$

$$\text{for } k > l, t_k^n \leq t \leq t_{k+1}^n, t_l^n \leq s \leq t_{l+1}^n. \quad (3.5)$$

It is easy to verify that $U_n(t, s)$ is an evolution system, that is

$$U_n(s, s) = I, U_n(t, s) = U_n(t, r)U_n(r, s) \qquad \text{for } 0 \leq s \leq r \leq t \leq T \quad (3.6)$$

and

$$(t, s) \rightarrow U_n(t, s) \text{ is strongly continuous on } 0 \leq s \leq t \leq T. \quad (3.7)$$

From Theorem 2.2 it follows that

$$\|U_n(t, s)\| \leq Me^{\omega(t-s)} \qquad \text{for } 0 \leq s \leq t \leq T \quad (3.8)$$

and from $(H_2)$ we have

$$U_n(t, s)Y \subset Y \qquad \text{for } 0 \leq s \leq t \leq T. \quad (3.9)$$

Since $D(A(t)) \supset Y$ for $t \in [0, T]$, the definition of $U_n(t, s)$ implies that for $v \in Y$

$$\frac{\partial}{\partial t} U_n(t, s)v = A_n(t)U_n(t, s)v \qquad \text{for } t \neq t_j^n, j = 0, 1, \ldots, n \quad (3.10)$$

$$\frac{\partial}{\partial s} U_n(t, s)v = -U_n(t, s)A_n(s)v \qquad \text{for } s \neq t_j^n, j = 0, 1, \ldots, n. \quad (3.11)$$

Moreover, $(H_2)$ together with Theorem 2.2 imply

$$\|U_n(t, s)\|_Y \leq \tilde{M}e^{\tilde{\omega}(t-s)} \qquad \text{for } 0 \leq s \leq t \leq T. \quad (3.12)$$

Let $v \in Y$ and consider the map $r \rightarrow U_n(t, r)U_m(r, s)v$. From (3.10) and (3.11) it follows that except for a finite number of values of $r$, this map is differentiable in $r$, $s \leq r \leq t$, and

$$U_n(t, s)v - U_m(t, s)v = -\int_s^t \frac{\partial}{\partial r} U_n(t, r)U_m(r, s)v \, dr$$

$$= \int_s^t U_n(t, r)(A_n(r) - A_m(r))U_m(r, s)v \, dr. \quad (3.13)$$

Denoting $\gamma = \max(\omega, \tilde{\omega})$, (3.13) implies

$$\| U_n(t,s)v - U_m(t,s)v \| \leq M\tilde{M}e^{\gamma(t-s)}\|v\|_Y \int_s^t \|A_n(r) - A_m(r)\|_{Y \to X} \, dr.$$

(3.14)

From (3.14) and (3.4) it follows that $U_n(t,s)v$ converges in $X$, uniformly on $0 \leq s \leq t \leq T$, as $n \to \infty$. As $Y$ is dense in $X$, this convergence of $U_n(t,s)v$ together with (3.8) imply that $U_n(t,s)$ converges strongly in $X$, uniformly on $0 \leq s \leq t \leq T$, as $n \to \infty$. Let

$$U(t,s)x = \lim_{n \to \infty} U_n(t,s)x \qquad \text{for} \quad x \in X, 0 \leq s \leq t \leq T. \quad (3.15)$$

From (3.6) and (3.7) it is clear that $U(t,s)$ is an evolution system in $X$ and from (3.8) it follows that $(E_1)$ is satisfied.

To prove $(E_2)$ and $(E_3)$ consider the function $r \to U_n(t,r)S_\tau(r-s)v$ for $v \in Y$. This function is differentiable except for a finite number of values of $r$ and we have

$$U_n(t,s)v - S_\tau(t-s)v = -\int_s^t \frac{\partial}{\partial r} U_n(t,r)S_\tau(r-s)v \, dr$$

$$= \int_s^t U_n(t,r)(A_n(r) - A(\tau))S_\tau(r-s)v \, dr$$

(3.16)

and therefore,

$$\| U_n(t,s)v - S_\tau(t-s)v \| \leq M\tilde{M}e^{\gamma(t-s)}\|v\|_Y \int_s^t \|A_n(r) - A(\tau)\|_{Y \to X} \, dr.$$

Passing to the limit as $n \to \infty$ this yields

$$\| U(t,s)v - S_\tau(t-s)v \| \leq M\tilde{M}e^{\gamma(t-s)}\|v\|_Y \int_s^t \|A(r) - A(\tau)\|_{Y \to X} \, dr.$$

(3.17)

Choosing $\tau = s$ in (3.17), dividing it by $t - s > 0$ and letting $t \downarrow s$ we find

$$\limsup_{t \downarrow s} \frac{1}{t-s} \| U(t,s)v - S_s(t-s)v \| = 0 \qquad (3.18)$$

where we used the continuity of $t \to A(t)$ in the $B(Y, X)$ norm. Since $S_s(t-s)v$ is differentiable from the right at $t = s$, it follows from (3.18) that so is $U(t,s)v$ and that their derivatives from the right at $t = s$ are the same. This implies $(E_2)$.

Choosing $\tau = t$ in (3.17), dividing it by $t - s > 0$ and letting $s \uparrow t$ we find

$$\limsup_{s \uparrow t} \frac{1}{t-s} \| U(t,s)v - S_t(t-s)v \| = 0 \qquad (3.19)$$

which implies, as above, that

$$\frac{\partial^-}{\partial s} U(t,s)v \bigg|_{s=t} = -A(t)v. \qquad (3.20)$$

For $s < t$, $(E_2)$ together with the strong continuity of $U(t, s)$ in $X$ imply

$$\frac{\partial^+}{\partial s} U(t, s)v = \lim_{h \downarrow 0} \frac{1}{h}\{U(t, s + h)v - U(t, s)v\}$$

$$= \lim_{h \downarrow 0} U(t, s + h)\left\{\frac{v - U(s + h, s)v}{h}\right\} = -U(t, s)A(s)v$$

$$(3.21)$$

and for $s \leq t$ we have by (3.20)

$$\frac{\partial^-}{\partial s} U(t, s)v = \lim_{h \downarrow 0} \frac{1}{h}\{U(t, s)v - U(t, s - h)v\}$$

$$= \lim_{h \downarrow 0} U(t, s)\left\{\frac{v - U(s, s - h)v}{h}\right\} = -U(t, s)A(s)v.$$

$$(3.22)$$

Combining (3.21) and (3.22) shows that $U(t, s)$ satisfies $(E_3)$.

To complete the proof it remains to show that $U(t, s)$, $0 \leq s \leq t \leq T$ is the only evolution system satisfying $(E_1), (E_2), (E_3)$. Suppose $V(t, s)$ is an evolution system satisfying $(E_1)$–$(E_3)$. For $v \in Y$ consider the function $r \to V(t, r)U_n(r, s)v$. Since $V(t, s)$ satisfies $(E_3)$ it follows from the construction of $U_n(t, s)$ that this function is differentiable except for a finite number of values of $r$. Integrating its derivative yields

$$V(t, s)v - U_n(t, s)v = \int_s^t V(t, r)(A(r) - A_n(r))U_n(r, s)v \, dr$$

and therefore,

$$\|V(t, s)v - U_n(t, s)v\| \leq M\tilde{M}e^{\gamma(t-s)}\|v\|_Y \int_s^t \|A(r) - A_n(r)\|_{Y \to X} \, dr.$$

$$(3.23)$$

Letting $n \to \infty$ in (3.23) and using (3.4) implies $V(t, s)v = U(t, s)v$ for $v \in Y$. Since $Y$ is dense in $X$ and both $U(t, s)$ and $V(t, s)$ satisfy $(E_1)$, $U(t, s) = V(t, s)$ and the proof is complete.                                    □

The assumption that the family $\{A(t)\}_{t \in [0, T]}$ satisfies $(H_2)$ is not always easy to check. A sufficient condition for $(H_2)$ which can be effectively checked in many applications is given in Theorem 2.4 above. It states that $(H_2)$ holds if there is a family $\{Q(t)\}$ of isomorphisms of $Y$ onto $X$ for which $\|Q(t)\|_{Y \to X}$ and $\|Q(t)^{-1}\|_{X \to Y}$ are uniformly bounded and $t \to Q(t)$ is of bounded variation in the $B(Y, X)$ norm.

**Remark 3.2.** If condition $(H_3)$ in Theorem 3.2 is replaced by the weaker condition:

$(H_3)'$ For $t \in [0, T]$, $D(A(t)) \supset Y$ and $A(t) \in L^1(0, T: B(Y, X))$ we can still construct a unique evolution system $U(t, s)$ for the initial value problem (3.1). Indeed, if $(H_3)'$ is satisfied, there exists a sequence of partitions $\{t_k^n\}_{n=1}^{N(n)}$ of $[0, T]$ for which $\delta_n = \max\{t_{k+1}^n - t_k^n\} \to 0$ as $n \to \infty$ and the corresponding operators $A_n(t)$, constructed as in the proof of Theorem 3.1, satisfy

$$\lim_{n \to \infty} \int_0^T \|A_n(r) - A(r)\|_{Y \to X} \, dr = 0. \tag{3.24}$$

Constructing $U_n(t, s)$ as in the proof of Theorem 3.1, replacing of course the partition $\{(k/n)T\}_{k=1}^n$ by the partition $\{t_k^n\}_{k=1}^{N(n)}$ it follows from (3.14) together with (3.24) that $U_n(t, s)v$ converge uniformly on $0 \leq s \leq t \leq T$ to $U(t, s)v$ and thus $U(t, s)$ exists and satisfies $(E_1)$. Moreover, in this case, (3.19) holds a.e. on $[0, T]$ and hence we have

$(E_2)'$

$$\frac{\partial^+}{\partial t} U(t, s)v \bigg|_{t=s} = A(s)v \qquad \text{for} \quad v \in Y \text{ and a.e. on } 0 \leq s \leq t \leq T$$

and similarly,

$(E_3)'$

$$\frac{\partial}{\partial s} U(t, s)v = -U(t, s)A(s)v \qquad \text{for} \quad v \in Y \text{ and a.e. on } 0 \leq s \leq t \leq T.$$

The properties $(E_2)'$ and $(E_3)'$ together with $(E_1)$ and the strong continuity of $U(t, s)$ suffice to ensure the uniqueness of $U(t, s)$.

## 5.4. Regular Solutions in the Hyperbolic Case

Let $X$ and $Y$ be Banach spaces such that $Y$ is densely and continuously imbedded in $X$ and let $\{A(t)\}_{t \in [0, T]}$ be a family of infinitesimal generators of $C_0$ semigroups on $X$ satisfying the assumptions $(H_1), (H_2), (H_3)$ of the previous section. Let $f \in C([s, T]: X)$ and consider the initial value problem

$$\begin{cases} \dfrac{du(t)}{dt} = A(t)u(t) + f(t) & \text{for} \quad 0 \leq s \leq t \leq T \\ u(s) = v. \end{cases} \tag{4.1}$$

A function $u \in C([s, T]: X)$ is a *classical solution* of (4.1) if $u$ is continuously differentiable in $X$ on $]s, T]$, $u(t) \in D(A(t))$ for $s < t \leq T$ and (4.1) is satisfied in $X$. Unfortunately we do not know any simple conditions that guarantee the existence of classical solutions of the initial value problem (4.1) in the hyperbolic case even if $f \equiv 0$. In order to obtain classical solutions of (4.1) under reasonable conditions, we will restrict ourselves in this section to a rather strong and therefore quite restricted notion of solutions of (4.1) namely the $Y$-valued solutions.

**Definition 4.1.** A function $u \in C([s, T]: Y)$ is a *Y-valued solution* of the initial value problem (4.1) if $u \in C^1(]s, T]: X)$ and (4.1) is satisfied in $X$.

A $Y$-valued solution $u$ of (4.1) differs from a classical solution by satisfying for $s \le t \le T$, $u(t) \in Y \subset D(A(t))$ rather than only $u(t) \in D(A(t))$ and by being continuous in the stronger $Y$-norm rather than merely in the $X$-norm. For $Y$-valued solutions we have:

**Theorem 4.2.** *Let* $\{A(t)\}_{t \in [0, T]}$ *be a family of infinitesimal generators of $C_0$ semigroups on $X$ satisfying the condition* $(H_1), (H_2), (H_3)$ *of Theorem 3.1 and let $f \in C([s, T]: X)$. If the initial value problem (4.1) has a Y-valued solution $u$ then this solution is unique and moreover*

$$u(t) = U(t, s)v + \int_s^t U(t, r)f(r) \, dr \tag{4.2}$$

*where $U(t, s)$ is the evolution system provided by Theorem 3.1.*

PROOF. Let $U_n(t, s)$, $0 \le s \le t \le T$ be the evolution system constructed in the proof of Theorem 3.1 (see (3.5)) and let $u$ be a $Y$-valued solution of (4.1). From the properties of $U_n(t, s)$ and $u$ it follows that the function $r \to U_n(t, r)u(r)$ is continuously differentiable in $X$ except for a finite number of values of $r$ and

$$\frac{\partial}{\partial r} U_n(t, r)u(r) = -U_n(t, r)A_n(r)u(r)$$
$$+ U_n(t, r)A(r)u(r) + U_n(t, r)f(r). \tag{4.3}$$

Integrating (4.3) from $s$ to $t$ we find

$$u(t) = U_n(t, s)v + \int_s^t U_n(t, r)f(r) \, dr$$
$$+ \int_s^t U_n(t, r)(A(r) - A_n(r))u(r) \, dr. \tag{4.4}$$

Denoting $C = \max_{s \le r \le T} \|u(r)\|_Y$ and using (3.8) to estimate (4.4) we find

$$\left\| u(t) - U_n(t, s)v - \int_s^t U_n(t, r)f(r) \, dr \right\|$$
$$\le Me^{\omega(t-s)}C \int_s^t \|A(r) - A_n(r)\|_{Y \to X} \, dr. \tag{4.5}$$

Letting $n \to \infty$ in (4.5) and using (3.4) and (3.15) we find (4.2). The uniqueness of $u$ is a consequence of the representation (4.2). □

We turn now to the problem of the existence of $Y$-valued solutions of the homogeneous initial value problem

$$\begin{cases} \dfrac{du(t)}{dt} = A(t)u(t) & \text{for } 0 \le s < t \le T \\ u(s) = v. \end{cases} \tag{4.6}$$

From Theorem 4.2 it follows that if the family $\{A(t)\}_{t\in[0,T]}$ satisfies the conditions of Theorem 3.1 and the initial value problem (4.6) has a $Y$-valued solution, this solution is given by $u(t) = U(t, s)v$ where $U(t, s)$, $0 \le s \le t \le T$, is the evolution system associated with the family $\{A(t)\}_{t\in[0,T]}$ by Theorem 3.1. In general however, $u(t) = U(t, s)v$ is not a $Y$-valued solution of (4.6) even if $v \in Y$. The reason for this is twofold, $Y$ need not be an invariant subspace for $U(t, s)$ and even if it is such an invariant subspace, $U(t, s)v$ for $v \in Y$ need not be continuous in the $Y$-norm. Both these properties of $U(t, s)$ are needed for $u(t) = U(t, s)v$ to be a $Y$-valued solution of (4.6). Our next result shows that they are also sufficient for this purpose.

**Theorem 4.3.** *Let $\{A(t)\}_{t\in[0,T]}$ satisfy the conditions of Theorem 3.1 and let $U(t, s)$, $0 \le s \le t \le T$ be the evolution system given in Theorem 3.1. If*

$$(E_4) \qquad\qquad U(t, s)Y \subset Y \qquad for \quad 0 \le s \le t \le T$$

*and*

$$(E_5) \qquad For \; v \in Y, \; U(t, s)v \; is \; continuous \; in \; Y \; for \; 0 \le s \le t \le T$$

*then for every $v \in Y$, $U(t, s)v$ is the unique $Y$-valued solution of the initial value problem (4.6).*

PROOF. The uniqueness of $Y$-valued solutions of the initial value problem (4.6) is an immediate consequence of Theorem 4.2. It suffices therefore to prove that if $v \in Y$ then $u(t) = U(t, s)v$ is a $Y$-valued solution of (4.6). From $(E_4)$ and $(E_5)$ it follows that $u(t) \in Y$ for $s \le t \le T$ and that it is continuous in the $Y$-norm for $s \le t \le T$. To complete the proof it remains to show that $u$ satisfies the differential equation in (4.6). Since $u(t) = U(t, s)v \in Y$ for $s \le t \le T$ we have by $(E_2)$ that

$$\frac{\partial^+}{\partial t} U(t, s)v = \lim_{h\downarrow 0} \frac{U(t + h, s)v - U(t, s)v}{h}$$

$$= \lim_{h\downarrow 0} \frac{U(t + h, t) - I}{h} U(t, s)v = A(t)U(t, s)v. \qquad (4.7)$$

The right-hand side of (4.7) is continuous in $X$ since $t \to U(t, s)v$ is continuous in the $Y$-norm and $t \to A(t)$ is continuous in $B(Y, X)$. Therefore, the right-derivative of $U(t, s)v$ is continuous in $X$ and as a consequence $U(t, s)v$ is continuously differentiable in $X$ and by (4.7)

$$\frac{\partial}{\partial t} U(t, s)v = A(t)U(t, s)v \qquad for \quad s \le t \le T. \qquad \square$$

From Theorem 4.3 it follows that if $U(t, s)$, the evolution system given by Theorem 3.1 also satisfies $(E_4)$ and $(E_5)$ then for every $v \in Y$ the initial value problem (4.6) has a unique $Y$-valued solution given by $U(t, s)v$. In order to get an evolution system $U(t, s)$ that satisfies $(E_1)-(E_5)$ we will replace the condition $(H_2)$ of Theorem 3.1 by the following condition:

$(H_2)^+$ There is a family $\langle Q(t) \rangle_{t \in [0, T]}$ of isomorphisms of $Y$ onto $X$ such that for every $v \in Y$, $Q(t)v$ is continuously differentiable in $X$ on $[0, T]$ and

$$Q(t)A(t)Q(t)^{-1} = A(t) + B(t) \qquad (4.8)$$

where $B(t)$, $0 \leq t \leq T$, is a strongly continuous family of bounded operators on $X$.

In the proof of our main result, Theorem 4.6, we will need the following two technical results.

**Lemma 4.4.** *The conditions $(H_1)$ and $(H_2)^+$ imply the condition $(H_2)$.*

PROOF. From $(H_2)^+$ it follows that for every $v \in Y$, $t \to dQ(t)v/dt$ is continuous in $X$ on $[0, T]$ and therefore $\|dQ(t)/dt\|_{Y \to X}$ is bounded on $[0, T]$. This implies that $t \to Q(t)$ is Lipschitz continuous and hence of bounded variation on $[0, T]$ in the $B(Y, X)$ norm and $\|Q(t)\|_{Y \to X}$ is bounded on $[0, T]$. The Lipschitz continuity of $t \to Q(t)$ in $B(Y, X)$ also implies the continuity of $t \to Q(t)^{-1}$ in $B(X, Y)$ and therefore $\|Q(t)^{-1}\|_{X \to Y}$ is bounded on $[0, T]$. Since by $(H_1)$ $\langle A(t) \rangle_{t \in [0, T]}$ is stable in $X$ it follows from Theorem 2.3 that $\langle A(t) + B(t) \rangle_{t \in [0, T]}$ is a stable family in $X$. From Theorem 2.4 it then follows that $Y$ is $A(t)$-admissible for every $t \in [0, T]$ and $\langle \tilde{A}(t) \rangle_{t \in [0, T]}$ is a stable family in $Y$. $\qquad \square$

**Lemma 4.5.** *Let $U(t, s)$, $0 \leq s \leq t \leq T$ be an evolution system in a Banach space $X$ satisfying $\|U(t, s)\| \leq M$ for $0 \leq s \leq t \leq T$. If $H(t)$ is a strongly continuous family of bounded linear operators in $X$ then there exists a unique family of bounded linear operators $V(t, s)$, $0 \leq s \leq t \leq T$ on $X$ such that*

$$V(t, s)x = U(t, s)x + \int_s^t V(t, r)H(r)U(r, s)x \, dr \qquad \text{for} \quad x \in X$$

$$(4.9)$$

*and $V(t, s)x$ is continuous in $s$, $t$ for $0 \leq s \leq t \leq T$.*

PROOF. Let $V^{(0)}(t, s) = U(t, s)$ and define

$$V^{(m)}(t, s)x = \int_s^t V^{(m-1)}(t, r)H(r)U(r, s)x \, dr \qquad \text{for} \quad x \in X.$$

$$(4.10)$$

The integrand in (4.10) is continuous on $0 \leq s \leq r \leq t \leq T$ as is easily seen by induction on $m$. From the uniform boundedness principle it follows that there is an $H > 0$ such that $\|H(t)\| \leq H$ for $t \in [0, T]$ and by induction on $m$ one verifies easily the estimate

$$\|V^{(m)}(t, s)\| \leq M^{m+1}H^m \frac{(t - s)^m}{m!}.$$

The series

$$V(t, s) = \sum_{m=0}^{\infty} V^{(m)}(t, s) \qquad (4.11)$$

therefore converges in the uniform operator topology on $X$ and $V(t, s)$ thus defined, is strongly continuous on $0 \le s \le t \le T$. Moreover it follows from (4.10) and (4.11) that $V(t, s)$ satisfies (4.9). To complete the proof it remains to prove the uniqueness of $V(t, s)$. Let $V_1(t, s)$ satisfy (4.9) and set $W(t, s) = V(t, s) - V_1(t, s)$ then

$$W(t, s)x = \int_s^t W(t, r)H(r)U(r, s)x \, dr \qquad \text{for} \quad x \in X. \quad (4.12)$$

Estimating (4.12) yields

$$\|W(t, s)x\| \le MH \int_s^t \|W(t, r)x\| dr \qquad \text{for} \quad x \in X$$

which by Gronwall's inequality implies $W(t, s)x = 0$ for $0 \le s \le t \le T$ and $x \in X$ whence $V(t, s) = V_1(t, s)$ and the proof is complete. $\qquad \square$

The main result of this section is:

**Theorem 4.6.** *Let $A(t)$, $0 \le t \le T$ be the infinitesimal generator of a $C_0$ semigroup on $X$. If the family $\{A(t)\}_{t \in [0, T]}$ satisfies the conditions $(H_1)$, $(H_2)^+$ and $(H_3)$ then there exists a unique evolution system $U(t, s)$, $0 \le s \le t \le T$, in $X$ satisfying $(E_1)-(E_5)$.*

PROOF. From Lemma 4.4 it follows that $\{A(t)\}_{t \in [0, T]}$ satisfies the conditions $(H_1)$, $(H_2)$, $(H_3)$ and therefore, by Theorem 3.1, there exists a unique evolution system $U(t, s)$ satisfying $(E_1)-(E_3)$.

Let $v \in Y$ and denote the derivative of $Q(t)v$ by $\dot{Q}(t)v$. Set

$$C(t) = \dot{Q}(t)Q(t)^{-1}. \quad (4.13)$$

$C(t)$, $0 \le t \le T$, is clearly a strongly continuous family of bounded operators on $X$. Let $W(t, s)$ be the unique solution of the integral equation

$$W(t, s)x = U(t, s)x + \int_s^t W(t, r)[B(r) + C(r)]U(r, s)x \, dr$$

$$\text{for} \quad x \in X. \quad (4.14)$$

The existence, uniqueness and properties of $W(t, s)$ follow from Lemma 4.5. Below we will prove

$$U(t, s) = Q(t)^{-1}W(t, s)Q(s). \quad (4.15)$$

From (4.15) it follows that $U(t, s)Y \subset Y$ since $W(t, s) \in B(X)$. Thus $U(t, s)$ satisfies $(E_4)$. Moreover, from the continuity of $W(t, s)x$ on $0 \le s \le t \le T$ and the properties of $Q(s)$ and $Q(t)^{-1}$ it follows that $U(t, s)$ is strongly continuous in $Y$ for $0 \le s \le t \le T$ and therefore satisfies $(E_5)$.

We turn now to the proof of (4.15). First we note that from our assumptions on $Q(t)$ it follows easily that for every $x \in X$ $Q(t)^{-1}x$ is differentiable in $Y$ and

$$\frac{d}{dt}(Q(t)^{-1}x) = -Q(t)^{-1}\dot{Q}(t)Q(t)^{-1}x. \quad (4.16)$$

Set

$$Q(t, r) = U(t, r)Q(r)^{-1}. \tag{4.17}$$

From $(E_3)$ and (4.16) it follows that for every $x \in X$, $r \to Q(t, r)x$ is differentiable in $X$ and

$$\frac{\partial}{\partial r} Q(t, r)x = -U(t, r)A(r)Q(r)^{-1}x - U(t, r)Q(r)^{-1}\dot{Q}(r)Q(r)^{-1}x$$

$$= -U(t, r)A(r)Q(r)^{-1}x - Q(t, r)C(r)x.$$

But for every $v \in Y$ we have by $(H_2)^+$

$$A(r)Q(r)^{-1}v = Q(r)^{-1}(A(r) + B(r))v$$

and therefore for $v \in Y$

$$\frac{\partial}{\partial r} Q(t, r)v = -Q(t, r)[A(r) + B(r) + C(r)]v. \tag{4.18}$$

Let $U_n(t, s)$ be the operators constructed in the proof of Theorem 3.1 (see (3.5)) then by (3.10)

$$\frac{\partial}{\partial r} U_n(r, s)v = A_n(r)U_n(r, s)v \qquad \text{for} \quad v \in Y \tag{4.19}$$

where (4.19) holds for all $s \le r$ except for a finite number of values of $r$. Combining (4.18) and (4.19) we find

$$\frac{\partial}{\partial r} Q(t, r)U_n(r, s)v$$

$$= -Q(t, r)(A(r) + B(r) + C(r) - A_n(r))U_n(r, s)v. \tag{4.20}$$

Integrating (4.20) from $r = s$ to $r = t$ yields

$$Q(t)^{-1}U_n(t, s)v - Q(t, s)v$$

$$= -\int_s^t Q(t, r)(A(r) + B(r) + C(r) - A_n(r))U_n(r, s)v \, dr. \tag{4.21}$$

From $(E_1)$ and (3.12) we deduce

$$\left\| \int_s^t Q(t, r)(A(r) - A_n(r))U_n(r, s)v \, dr \right\|$$

$$\le M\tilde{M}e^{\gamma(t-s)} \sup_r \|Q^{-1}(r)\|_{X \to Y} \|v\|_Y \int_s^t \|A(r) - A_n(r)\|_{Y \to X} \, dr \tag{4.22}$$

where $\gamma = \max(\omega, \tilde{\omega})$. Passing to the limit as $n \to \infty$ in (4.21) and using (4.22) and (3.15) we obtain for $v \in Y$

$$Q(t)^{-1}U(t, s)v - Q(t, s)v = -\int_s^t Q(t, r)(B(r) + C(r))U(r, s)v \, dr. \tag{4.23}$$

Since all operators in (4.23) are bounded in $X$ and since $Y$ is dense in $X$, (4.23) holds for every $v \in X$ and hence after rearrangement we have

$$Q(t, s)x = Q(t)^{-1}U(t, s)x + \int_s^t Q(t, r)(B(r) + C(r))U(r, s)x\, dr.$$

$$(4.24)$$

On the other hand, multiplying (4.14) from the left by $Q(t)^{-1}$ yields

$$Q(t)^{-1}W(t, s)x = Q(t)^{-1}U(t, s)x + \int_s^t Q(t)^{-1}W(t, r)(B(r)$$

$$+ C(r))U(r, s)x\, dr. \qquad (4.25)$$

From (4.24) and the uniqueness of the solution of (4.25) it follows that

$$U(t, s)Q(s)^{-1} = Q(t, s) = Q(t)^{-1}W(t, s)$$

which implies (4.15) and the proof is complete. $\qquad\qquad\square$

From Theorems 4.6 and 4.3 we obtain,

**Corollary 4.7.** *Let $\{A(t)\}_{t \in [0, T]}$ be a family of infinitesimal generators of $C_0$ semigroups on $X$. If $\{A(t)\}_{t \in [0, T]}$ satisfies the conditions $(H_1)$, $(H_2)^+$ and $(H_3)$ then for every $v \in Y$ the initial value problem*

$$\begin{cases} \dfrac{du(t)}{dt} = A(t)u(t) & \text{for}\quad s < t \le T \\ u(s) = v \end{cases} \qquad (4.26)$$

*has a unique $Y$-valued solution $u$ on $s \le t \le T$.*

One special case in which the conditions of Theorem 4.6 can be easily verified is the case where $D(A(t)) = D$ is independent of $t$. In this case we define on $D$ a norm $\| \ \|_Y$ by

$$\|v\|_Y = \|v\| + \|A(0)v\| \qquad \text{for}\quad v \in Y = D \qquad (4.27)$$

and it is not difficult to see, using the closedness of $A(0)$, that $D$ equipped with this norm is a Banach space which we denote by $Y$. This $Y$ is clearly densely and continuously imbedded in $X$ and we have:

**Theorem 4.8.** *Let $\{A(t)\}_{t \in [0, T]}$ be a stable family of infinitesimal generators of $C_0$ semigroups on $X$. If $D(A(t)) = D$ is independent of $t$ and for $v \in D$, $A(t)v$ is continuously differentiable in $X$ then there exists a unique evolution system $U(t, s)$, $0 \le s \le t \le T$, satisfying $(E_1)$–$(E_5)$ where $Y$ is $D$ equipped with the norm $\| \ \|_Y$ given by (4.27).*

**PROOF.** We will show that $\{A(t)\}_{t \in [0, T]}$ satisfies the conditions $(H_1)$, $(H_2)^+$ and $(H_3)$. Condition $(H_1)$ is explicitly assumed in our theorem. The continuous differentiability of $A(t)v$ in $X$ clearly implies that $t \to A(t)$ is continuous in the $B(Y, X)$ norm so $(H_3)$ is satisfied. To prove $(H_2)^+$ note that for $\lambda_0 > \omega$ the operator $Q(t) = \lambda_0 I - A(t)$ is an isomorphism of $Y$ onto $X$ and by our assumption on $A(t)u$ it follows that $Q(t)v$ is continuously differentiable in $X$ for every $v \in Y$. Finally,

$$Q(t)A(t)Q(t)^{-1} = A(t)$$

and therefore (4.8) is satisfied with $B(t) \equiv 0$, so $(H_2)^+$ holds and the proof is complete.                                                                              □

## 5.5. The Inhomogeneous Equation in the Hyperbolic Case

This section is devoted to a few remarks concerning the solutions of the inhomogeneous initial value problem

$$\begin{cases} \dfrac{du(t)}{dt} = A(t)u(t) + f(t) & \text{for } 0 \le s < t \le T \\ u(s) = v \end{cases} \tag{5.1}$$

in the hyperbolic case. In Section 5.3 we have considered the corresponding homogeneous initial value problem and under the assumptions $(H_1)$, $(H_2)$, $(H_3)$ we have constructed (Theorem 3.1) a unique evolution system $U(t, s)$, $0 \le s \le t \le T$, satisfying the properties $(E_1)$–$(E_3)$. Motivated by the autonomous case (see Section 4.2) we make the following definition.

**Definition 5.1.** Let $\{A(t)\}_{t \in [0,T]}$ satisfy the conditions of Theorem 3.1 and let $U(t, s)$, $0 \le s \le t \le T$ be the evolution system given by Theorem 3.1. For every $f \in L^1(s, T: X)$ and $v \in X$ the continuous function

$$u(t) = U(t, s)v + \int_s^t U(t, r)f(r)\, dr \tag{5.2}$$

is called the *mild solution* of the initial value problem (5.1).

From the concluding remarks of Section 5.1 it follows that if the evolution system $U(t, s)$ is regular enough and $f \in C^1([s, T]: X)$ then the initial value problem (5.1) has a unique classical solution for every $v \in D(A(s))$ and this solution coincides with the mild solution (5.2). A similar result (Theorem 4.2) holds for $Y$-valued solutions of (5.1).

Existence of $Y$-valued solutions for the inhomogeneous initial value problem is provided by:

**Theorem 5.2.** Let $\{A(t)\}_{t \in [0,T]}$ satisfy the condition of Theorem 4.3. If $f \in C([s, T]: Y)$ then for every $v \in Y$ the initial value problem (5.1) possesses a unique $Y$-valued solution $u$ given by (5.2).

PROOF. It has been shown in Theorem 4.3 that $U(t, s)v$ is a $Y$-valued solution of the homogeneous initial value problem

$$\begin{cases} \dfrac{du(t)}{dt} = A(t)u(t) & \text{for} \quad 0 \le s < t \le T \\ u(s) = v. \end{cases} \tag{5.3}$$

To prove that $u$ given by (5.2) is a $Y$-valued solution of (5.1) we will show that

$$w(t) = \int_s^t U(t, r)f(r)\, dr \tag{5.4}$$

is a $Y$-valued solution of (5.1) with the initial value $w(s) = v = 0$. From our assumptions on $f$ and $(E_4)$ it follows readily that $w(t) \in Y$ for $s \le t \le T$. From $(E_5)$ it follows that $r \to U(t, r)f(r)$ is continuous in $Y$ which implies that $t \to w(t)$ is continuous in $Y$ and that $r \to A(t)U(t, r)f(r)$ is continuous in $X$ for $s \le t \le T$. The continuity of $r \to A(t)U(t, r)f(r)$ implies that $w(t)$ is continuously differentiable in $X$ and that

$$\frac{d}{dt} w(t) = A(t)w(t) + f(t) \qquad \text{for} \quad s \le t \le T$$

holds in $X$ as desired. Finally, the uniqueness of $Y$-valued solutions of (5.1) is a direct consequence of Theorem 4.2. $\qquad\square$

Theorem 5.2 shows that if the family $\{A(t)\}_{t \in [0, T]}$ of infinitesimal generators of $C_0$ semigroups on $X$ satisfies the conditions $(H_1)$, $(H_2)^+$ and $(H_3)$ then for every $v \in y$ and $f \in C([s, T]: Y)$ the initial value problem 5.1 possesses a unique $Y$-valued solution $u$ given by (5.2). This result is reminiscent of Corollary 4.2.6.

Our next result, for the special case where all the operators $A(t)$, $0 \le t \le T$, have a common domain $D$ independent of $t$ is reminiscent of Corollary 4.2.5.

**Theorem 5.3.** *Let $\{A(t)\}_{t \in [0, T]}$ be a stable family of infinitesimal generators of $C_0$ semigroups on $X$ such that $D(A(t)) = D$ is independent of $t$ and for every $v \in D$, $A(t)v$ is continuously differentiable in $X$. If $f \in C^1([s, T]: X)$ then for every $v \in D$ the initial value problem (5.1) has a unique classical solution $u$ given by*

$$u(t) = U(t, s)v + \int_s^t U(t, r)f(r)\, dr. \tag{5.5}$$

PROOF. As in Theorem 4.8 we endow $D$ with the graph norm of $A(0)$ and denote this Banach space by $Y$. From our assumptions it then follows that for $\lambda_0$ large enough and every $t \in [0, T]$, $Q(t) = \lambda_0 I - A(t)$ is an isomorphism of $Y$ onto $X$ such that $Q(t)v$ is continuously differentiable in $X$ for every $v \in Y$. We denote the derivative of $Q(t)v$ by $\dot{Q}(t)v$ and note that $\dot{Q}(t) \in B(Y, X)$ and that $\{\|\dot{Q}(t)\|_{Y \to X}\}$ is uniformly bounded. From Theorem 4.8 it follows that $U(t, s)v$ is the $Y$-valued solution of the homogeneous initial value problem (5.3). To show that $u$ given by (5.5) is a classical solution of (5.1) it is, therefore sufficient to show that

$$w(t) = \int_s^t U(t, r) f(r) \, dr$$

is a classical solution of (5.1) satisfying $w(s) = 0$. To this end we note first that $Q(r)^{-1} f(r)$ is differentiable in $Y$ and that

$$\frac{d}{dr} \left( Q(r)^{-1} f(r) \right) = -Q(r)^{-1} \dot{Q}(r) Q(r)^{-1} f(r) + Q(r)^{-1} f'(r)$$

$$= Q(r)^{-1} g(r) \tag{5.6}$$

where $f'(r)$ is the derivative of $f(r)$ and $g(r) = f'(r) - \dot{Q}(r) Q(r)^{-1} f(r)$. Differentiating $U(t, r) Q(r)^{-1} f(r)$ with respect to $r$ using $(E_3)$ and (5.6) we find

$$\frac{\partial}{\partial r} U(t, r) Q(r)^{-1} f(r)$$

$$= -U(t, r) A(r) Q(r)^{-1} f(r) + U(t, r) Q(r)^{-1} g(r)$$

$$= U(t, r) f(r) + U(t, r) Q(r)^{-1} (g(r) - \lambda_0 f(r)).$$

Integrating this equality from $r = s$ to $r = t$ we obtain after rearrangement

$$w(t) = Q(t)^{-1} f(t)$$

$$- \left[ U(t, s) Q(s)^{-1} f(s) + \int_s^t U(t, r) Q(r)^{-1} (g(r) - \lambda_0 f(r)) \, dr \right]$$

$$= Q(t)^{-1} f(t) - v(t) \tag{5.7}$$

where $v(t)$ is defined by the second equality of (5.7). Since $Q(s)^{-1} f(s) \in Y$ and $r \to Q(r)^{-1}(g(r) - \lambda_0 f(r))$ is continuous in $Y$ on $[s, T]$ it follows from Theorem 5.2 that

$$\frac{dv(t)}{dt} = A(t) v(t) + Q(t)^{-1} (g(t) - \lambda_0 f(t)) \qquad \text{for} \quad 0 \le s \le t \le T. \tag{5.8}$$

Therefore, using (5.7) we have

$$\frac{dw(t)}{dt} = \frac{d}{dt} \left( Q(t)^{-1} f(t) \right) - \frac{dv(t)}{dt}$$

$$= Q(t)^{-1} g(t) - Q(t)^{-1} (g(t) - \lambda_0 f(t)) - A(t) v(t)$$

$$= A(t) w(t) + \lambda_0 Q(t)^{-1} f(t) - A(t) Q(t)^{-1} f(t)$$

$$= A(t) w(t) + f(t).$$

Since $dv(t)/dt$ and $Q(t)^{-1} g(t)$ are continuous in $X$ it follows that $dw(t)/dt$

is continuous in $X$ and $w$ is a classical solution of (5.1) with $v = 0$. To prove the uniqueness of the classical solution $u$, let $v_1$ be a classical solution of (5.1). From our assumptions and the properties of $U(t, s)$ (see Theorem 4.8) it follows that $r \rightarrow U(t, r)v_1(r)$ is continuously differentiable in $X$ and that

$$\frac{\partial}{\partial r} U(t, r)v_1(r) = U(t, r)f(r).$$

Integrating this equality from $s$ to $t$ yields $v_1(t) = u(t)$.                                      □


## 5.6. An Evolution System for the Parabolic Initial Value Problem

This section starts the second part of Chapter 5 in which we study the initial value problem

$$\begin{cases} \dfrac{du(t)}{dt} + A(t)u(t) = f(t) & 0 \le s < t \le T \\ u(s) = x \end{cases} \tag{6.1}$$

in the parabolic case.[1] The results of this part are independent of the results of Sections 5.2–5.5 in which the corresponding hyperbolic case was treated. The evolution system for the parabolic initial value problem

$$\begin{cases} \dfrac{du(t)}{dt} + A(t)u(t) = 0 & 0 \le s < t \le T \\ u(s) = x \end{cases} \tag{6.2}$$

will be constructed below by a method which is entirely different from the method used in the hyperbolic case (Section 5.3). The present section is devoted to this construction and to the study of the main properties of the resulting evolution system.

We start with a formal computation that will lead us to the method of construction of the evolution system. Suppose that for each $t \in [0, T]$, $-A(t)$ is the infinitesimal generator of a $C_0$ semigroup $S_t(s)$, $s \ge 0$ on the Banach space $X$ and let $U(t, s)$ be an evolution system for (6.2). Set

$$U(t, s) = S_s(t - s) + W(t, s) = S_s(t - s) + \int_s^t S_\tau(t - \tau)R(\tau, s)\, d\tau. \tag{6.3}$$

Then (formally)

$$\frac{\partial}{\partial t} U(t, s) = -A(s)S_s(t - s) + R(t, s) - \int_s^t A(\tau)S_\tau(t - \tau)R(\tau, s)\, d\tau$$

[1] In the parabolic case it is customary to write the term $A(t)u(t)$ on the left-hand side of the equation. This is done to overcome some notational difficulties related to the use of fractional powers of $A(t)$.

and

$$\frac{\partial}{\partial t} U(t, s) + A(t)U(t, s) = R(t, s) - R_1(t, s) - \int_s^t R_1(t, \tau)R(\tau, s)\, d\tau$$

$$(6.4)$$

where

$$R_1(t, s) = (A(s) - A(t))S_s(t - s).$$ $$(6.5)$$

Since $U(t, s)$ is an evolution system for (6.2) it follows from (6.4) that the integral equation

$$R(t, s) = R_1(t, s) + \int_s^t R_1(t, \tau)R(\tau, s)\, d\tau$$ $$(6.6)$$

must be satisfied. Given $R_1(t, s)$ we can try to reverse the argument, i.e., solve the integral equation (6.6) for $R(t, s)$ and then define $U(t, s)$ by (6.3). This will indeed be the method of constructing $U(t, s)$ below. To carry out this program rigorously we will need the following assumptions:

$(P_1)$ The domain $D(A(t)) = D$ of $A(t)$, $0 \le t \le T$ is dense in $X$ and independent of $t$.
$(P_2)$ For $t \in [0, T]$, the resolvent $R(\lambda : A(t))$ of $A(t)$ exists for all $\lambda$ with $\operatorname{Re} \lambda \le 0$ and there is a constant $M$ such that

$$\|R(\lambda : A(t))\| \le \frac{M}{|\lambda| + 1} \qquad \text{for} \quad \operatorname{Re} \lambda \le 0, t \in [0, T]. \quad (6.7)$$

$(P_3)$ There exist constants $L$ and $0 < \alpha \le 1$ such that

$$\|(A(t) - A(s))A(\tau)^{-1}\| \le L|t - s|^\alpha \qquad \text{for} \quad s, t, \tau \in [0, T].$$

$$(6.8)$$

We note that the interval for which $(P_1)$–$(P_3)$ are satisfied was chosen as $[0, T]$ only for notational convenience. All the results that will be proven below, hold for any interval $[a, b]$, $0 \le a < b < \infty$, on which these assumptions are satisfied.

The main result of this section is:

**Theorem 6.1.** *Under the assumptions $(P_1)$–$(P_3)$ there is a unique evolution system $U(t, s)$ on $0 \le s \le t \le T$, satisfying:*

$(E_1)'$ $\|U(t, s)\| \le C$ *for* $0 \le s \le t \le T$.
$(E_2)^+$ *For* $0 \le s < t \le T$, $U(t, s): X \to D$ *and* $t \to U(t, s)$ *is strongly differentiable in $X$. The derivative $(\partial/\partial t)U(t, s) \in B(X)$ and it is strongly continuous on $0 \le s < t \le T$. Moreover,*

$$\frac{\partial}{\partial t} U(t, s) + A(t)U(t, s) = 0 \qquad \text{for} \quad 0 \le s < t \le T, \quad (6.9)$$

$$\left\| \frac{\partial}{\partial t} U(t, s) \right\| = \|A(t)U(t, s)\| \le \frac{C}{t - s} \quad (6.10)$$

*and*

$$\|A(t)U(t, s)A(s)^{-1}\| \le C \qquad \text{for} \quad 0 \le s \le t \le T. \quad (6.11)$$

$(E_3)^+$ *For every $v \in D$ and $t \in ]0, T]$, $U(t, s)v$ is differentiable with respect to $s$ on $0 \le s \le t \le T$ and*

$$\frac{\partial}{\partial s} U(t, s)v = U(t, s)A(s)v. \tag{6.12}$$

The proof of Theorem 6.1 will occupy most of this section. It will be divided into three main parts. In the first part we will construct $U(t, s)$ by solving the integral equation (6.6) and using (6.3) to define $U(t, s)$. In the second part we will prove that $U(t, s)$ satisfies the properties stated in $(E_2)^+$ and in the third part the uniqueness of $U(t, s)$ together with $U(t, s) = U(t, r)U(r, s)$, for $0 \le s \le r \le t \le T$ and $(E_3)^+$ will be proved.

Before starting with the proof we derive some direct consequences of the assumptions $(P_1)$–$(P_3)$. First we note that $(P_2)$ and the fact that $D$ is dense in $X$ imply that for every $t \in [0, T]$, $-A(t)$ is the infinitesimal generator of an analytic semigroup $S_t(s)$, $s \ge 0$, satisfying (see Theorem 2.5.2)

$$\|S_t(s)\| \le C \qquad \text{for} \quad s \ge 0 \tag{6.13}$$

$$\|A(t)S_t(s)\| \le \frac{C}{s} \qquad \text{for} \quad s > 0 \tag{6.14}$$

where here, and in the sequel, we denote by $C$ a generic constant. From $(P_2)$ it also follows that there exists an angle $\vartheta \in ]0, \pi/2[$ such that

$$\rho(A(t)) \supset \Sigma = \{\lambda : |\arg \lambda| \ge \vartheta\} \cup \{0\} \tag{6.15}$$

and that (6.7) holds for all $\lambda \in \Sigma$, possibly with a different constant $M$.

Some more consequences of the assumptions $(P_1)$–$(P_3)$ are collected in the next lemma.

**Lemma 6.2.** *Let $(P_1)$–$(P_3)$ be satisfied then*

$$\|(A(t_1) - A(t_2))S_\tau(s)\| \le \frac{C}{s}|t_1 - t_2|^\alpha$$

$$\text{for} \quad s \in ]0, T], t_1, t_2 \in [0, T] \tag{6.16}$$

$$\|A(t)(S_\tau(s_2) - S_\tau(s_1))\| \le \frac{C}{s_1 s_2}|s_2 - s_1|$$

$$\text{for} \quad 0 < s_1, s_2 \in ]0, T], t, \tau \in [0, T] \tag{6.17}$$

$$\|A(t)(S_{\tau_1}(s) - S_{\tau_2}(s))\| \le \frac{C}{s}|\tau_1 - \tau_2|^\alpha$$

$$\text{for} \quad s \in ]0, T], t, \tau_1, \tau_2 \in [0, T]. \tag{6.18}$$

*Moreover, $A(t)S_\tau(s) \in B(X)$ for $s \in ]0, T]$, $\tau, t \in [0, T]$ and the $B(X)$ valued function $A(t)S_\tau(s)$ is uniformly continuous in the uniform operator topology for $s \in [\varepsilon, T], t, \tau \in [0, T]$ for every $\varepsilon > 0$.*

PROOF. Since $S_\tau(s): X \to D$ for $s > 0$ (Theorem 2.5.2) it follows from the closed graph theorem that $A(t)S_\tau(s)$ is a bounded linear operator for $t, \tau \in [0, T]$, $s \in ]0, T]$. From (6.8) and (6.14) we have

$$\|(A(t_1) - A(t_2))S_\tau(s)\| \le \|(A(t_1) - A(t_2))A(\tau)^{-1}\| \, \|A(\tau)S_\tau(s)\|$$

$$\le \frac{C}{s}|t_2 - t_1|^\alpha$$

which proves (6.16). To prove (6.17) let $0 < s_1 \le s_2$ and $x \in X$. From the results of Section 2.5 we have

$$A(\tau)S_\tau(s_2)x - A(\tau)S_\tau(s_1)x = -\int_{s_1}^{s_2} A(\tau)^2 S_\tau(\sigma)x \, d\sigma$$

$$= -\int_{s_1}^{s_2}\left(A(\tau)S_\tau\left(\frac{\sigma}{2}\right)\right)^2 x \, d\sigma$$

and therefore by (6.14)

$$\|A(\tau)S_\tau(s_2)x - A(\tau)S_\tau(s_1)x\| \le C\|x\|\int_{s_1}^{s_2}\frac{1}{\sigma^2}d\sigma = \frac{C\|x\|}{s_1 s_2}|s_2 - s_1|.$$

Since by (6.8), $\|A(t)A(\tau)^{-1}\| \le C$ for $t, \tau \in [0, T]$, we obtain

$$\|A(t)(S_\tau(s_1) - S_\tau(s_2))\| \le \|A(t)A(\tau)^{-1}\| \, \|A(\tau)(S_\tau(s_1) - S_\tau(s_2))\|$$

$$\le \frac{C}{s_2 s_1}|s_2 - s_1|$$

which proves (6.17).

To prove (6.18) note that from $(P_2)$ it follows that

$$\|A(t)R(\lambda : A(t))\| \le C \qquad \text{for} \quad 0 \le t \le T$$

and therefore

$$\|A(t)(R(\lambda : A(\tau_1)) - R(\lambda : A(\tau_2)))\|$$

$$= \|A(t)R(\lambda : A(\tau_1))(A(\tau_1) - A(\tau_2))R(\lambda : A(\tau_2))\|$$

$$\le \|A(t)A(\tau_1)^{-1}\| \, \|A(\tau_1)R(\lambda : A(\tau_1))\|$$

$$\cdot \|(A(\tau_1) - A(\tau_2))A(\tau_2)^{-1}\| \, \|A(\tau_2)R(\lambda : A(\tau_2))\|$$

$$\le C|\tau_1 - \tau_2|^\alpha. \tag{6.19}$$

From the results of Section 2.5 we get the following representation

$$A(t)S_{\tau_1}(s)x - A(t)S_{\tau_2}(s)x$$

$$= \frac{1}{2\pi i}\int_\Gamma e^{-\lambda s}A(t)(R(\lambda : A(\tau_1)) - R(\lambda : A(\tau_2)))x \, d\lambda \tag{6.20}$$

where $\Gamma$ is a smooth path in $\Sigma$ connecting $\infty e^{-i\vartheta}$ to $\infty e^{i\vartheta}$. Using (6.19) to estimate (6.20) we find

$$\|A(t)S_{\tau_1}(s)x - A(t)S_{\tau_2}(s)x\| \leq C|\tau_1 - \tau_2|^\alpha \|x\| \int_\Gamma |e^{-\lambda s}| \, |d\lambda|$$

$$\leq \frac{C}{s}|\tau_1 - \tau_2|^\alpha \|x\|$$

whence (6.18).

Finally, to prove the uniform continuity of $A(t)S_\tau(s)$ in the uniform operator topology for $t, \tau \in [0, T]$, $s \in [\varepsilon, T]$ we have only to combine (6.16), (6.17), (6.18) and use the triangle inequality.                    □

**Corollary 6.3.** *The operator $R_1(t, s) = (A(s) - A(t))S_s(t - s)$ is uniformly continuous in the uniform operator topology on $0 \leq s \leq t - \varepsilon \leq T$ for every $\varepsilon > 0$ and*

$$\|R_1(t, s)\| \leq C|t - s|^{\alpha - 1} \qquad for \quad 0 \leq s < t \leq T. \qquad (6.21)$$

PROOF. The first part of the claim is a direct consequence of the uniform continuity of $A(t)S_\tau(s)$ in $B(X)$ while (6.21) follows from

$$\|R_1(t, s)\| \leq \|(A(t) - A(s))A(s)^{-1}\| \, \|A(s)S_s(t - s)\|$$

$$\leq C|t - s|^\alpha |t - s|^{-1} = C|t - s|^{\alpha - 1}. \qquad □$$

We are now ready to start the construction of $U(t, s)$.

## I. Construction of the Evolution System

We begin by solving the integral equation (6.6) for $R(t, s)$. If $R_1(t, s)$ satisfies (6.21) then (6.6) can be solved by successive approximations as follows: For $m \geq 1$ we define inductively

$$R_{m+1}(t, s) = \int_s^t R_1(t, \tau)R_m(\tau, s) \, d\tau. \qquad (6.22)$$

Then we prove by induction that $R_m(t, s)$ is continuous in the uniform operator topology for $0 \leq s < t \leq T$ and that

$$\|R_m(t, s)\| \leq \frac{(C\Gamma(\alpha))^m}{\Gamma(m\alpha)}(t - s)^{m\alpha - 1} \qquad (6.23)$$

where $\Gamma(\cdot)$ is the classical gamma function. In the inductive proof of (6.23) we use the well known identity

$$\int_s^t (t - \tau)^{\alpha - 1}(\tau - s)^{\beta - 1} \, d\tau = (t - s)^{\alpha + \beta - 1}\frac{\Gamma(\alpha)\Gamma(\beta)}{\Gamma(\alpha + \beta)} \qquad (6.24)$$

which holds for every $\alpha$, $\beta > 0$. We note that the integral defining $R_{m+1}(t, s)$ is an improper integral whose existence is an immediate consequence of (6.23). The continuity of $R_{m+1}(s, t)$ also follows easily from the continuity of $R_m(t, s)$, $R_1(t, s)$ and (6.23).

The estimates (6.23) imply that the series

$$R(t, s) = \sum_{m=1}^{\infty} R_m(t, s)$$

converges uniformly in the uniform operator topology for $0 \le s \le t - \varepsilon \le T$ and every $\varepsilon > 0$. As a consequence $R(t, s)$ is uniformly continuous in $B(X)$ for $0 \le s \le t - \varepsilon \le T$ and every $\varepsilon > 0$.

From (6.22) it follows that

$$R(t, s) = \sum_{m=1}^{\infty} R_m(t, s) = R_1(t, s) + \sum_{m=1}^{\infty} \int_s^t R_1(t, \tau) R_m(\tau, s) \, d\tau.$$

$$(6.25)$$

The continuity of $R_m(t, s)$, $m \ge 1$, (6.21) and (6.23) imply that one can interchange the summation and integral in (6.25) and thus see that $R(t, s)$ is a solution of the integral equation (6.6). Moreover, using Stirling's formula we have

$$\|R(t, s)\| \le \sum_{n=1}^{\infty} \Gamma(m\alpha)^{-1} (C\Gamma(\alpha))^m (t - s)^{m\alpha - 1}$$

$$\le \left( \sum_{m=1}^{\infty} \Gamma(m\alpha)^{-1} (C\Gamma(\alpha))^m T^{\alpha(m-1)} \right) (t - s)^{\alpha - 1}$$

$$\le C(t - s)^{\alpha - 1}. \tag{6.26}$$

Defining $U(t, s)$ by (6.3) it follows readily from the strong continuity of $S_s(\tau)$, (6.13) and (6.26) that $U(t, s)$ is strongly continuous for $0 \le s \le t \le T$ and that

$$\|U(t, s)\| \le \|S_s(t - s)\| + \int_s^t \|S_\tau(t - \tau)\| \, \|R(\tau, s)\| \, d\tau$$

$$\le C_1 + C_2 \int_s^t (\tau - s)^{\alpha - 1} \le C. \tag{6.27}$$

Therefore $(E_1)'$ is satisfied. In order to show that $U(t, s)$, $0 \le s \le t \le T$ is an evolution system it remains to show that $U(t, s) = U(t, r)U(r, s)$ for $s \le r \le t$. This will follow from the uniqueness of the solution of the initial value problem (6.2) that will be proved below (Theorem 6.8), and the fact that by $(E_2)^+$ the solution of (6.2) is $U(t, s)x$.

## II. Differentiability of $U(t, s)$.

We turn now to the proof that $U(t, s)$, constructed above, has the properties stated in $(E_2)^+$. For this we need a few preliminaries.

**Lemma 6.4.** *For every $\beta$, $0 < \beta \le \alpha$, there is a constant $C_\beta$ such that*

$$\|R_1(t, s) - R_1(\tau, s)\| \le C_\beta (t - \tau)^\beta (\tau - s)^{\alpha - \beta - 1}$$

$$\text{for} \quad 0 \le s < \tau < t \le T. \quad (6.28)$$

PROOF. We have

$$R_1(t, s) - R_1(\tau, s) = (A(\tau) - A(t))S_s(t - s)$$
$$+ (A(s) - A(\tau))(S_s(t - s) - S_s(\tau - s)).$$

From (6.16) it follows that

$$\|(A(\tau) - A(t))S_s(t - s)\| \le C(t - \tau)^\alpha (t - s)^{-1}$$

$$\le C(t - \tau)^\alpha (\tau - s)^{-1}.$$

Also,

$$\|(A(s) - A(\tau))(S_s(t - s) - S_s(\tau - s))\|$$
$$\le \|(A(s) - A(\tau))A(s)^{-1}\| \cdot \|A(s)(S_s(t - s) - S_s(\tau - s))\|.$$

Estimating the right-hand side of the last inequality using (6.8) and (6.17) we find that it is bounded by $C(\tau - s)^{\alpha - 2}(t - \tau)$ while estimating it using (6.8) and (6.14) we find that it is bounded by $C(\tau - s)^{\alpha - 1}$. Therefore,

$$\|(A(s) - A(\tau))(S_s(t - s) - S_s(\tau - s))\|$$
$$\le C\left[(\tau - s)^{\alpha - 2}(t - \tau)\right]^\alpha \left[(\tau - s)^{\alpha - 1}\right]^{1 - \alpha}$$
$$= C(t - \tau)^\alpha (\tau - s)^{-1}$$

and thus

$$\|R_1(t, s) - R_1(\tau, s)\| \le C(t - \tau)^\alpha (\tau - s)^{-1}.$$

On the other hand we have by (6.21)

$$\|R_1(t, s) - R_1(\tau, s)\| \le \|R_1(t, s)\| + \|R_1(\tau, s)\|$$

$$\le C\left((t - s)^{\alpha - 1} + (\tau - s)^{\alpha - 1}\right) \le C(\tau - s)^{\alpha - 1}.$$

Interpolating the two estimates for $\|R_1(t, s) - R_1(\tau, s)\|$ we find

$$\|R_1(t, s) - R_1(\tau, s)\| \le C\left[(t - \tau)^\alpha (\tau - s)^{-1}\right]^{\beta/\alpha} \left[(\tau - s)^{\alpha - 1}\right]^{1 - \beta/\alpha}$$

$$\le C(t - \tau)^\beta (\tau - s)^{\alpha - \beta - 1}. \qquad \square$$

**Corollary 6.5.** *For every* $\beta$, $0 < \beta < \alpha$, *there is a constant* $C_\beta$ *such that*

$$\|R(t, s) - R(\tau, s)\| \le C_\beta(t - \tau)^\beta(\tau - s)^{\alpha - \beta - 1}$$

$$for \quad 0 \le s < \tau < t \le T. \quad (6.29)$$

PROOF. From the integral equation (6.6) we have

$$R(t, s) - R(\tau, s) = R_1(t, s) - R_1(\tau, s) + \int_\tau^t R_1(t, \sigma)R(\sigma, s)\,d\sigma$$

$$+ \int_s^\tau (R_1(t, \sigma) - R_1(\tau, \sigma))R(\sigma, s)\,d\sigma.$$

The estimates (6.21) and (6.26) imply

$$\left\| \int_\tau^t R_1(t, \sigma)R(\sigma, s)\,d\sigma \right\|$$

$$\le C\int_\tau^t (t - \sigma)^{\alpha - 1}(\sigma - s)^{\alpha - 1}\,d\sigma$$

$$\le C(\tau - s)^{\alpha - 1}\int_\tau^t (t - \sigma)^{\alpha - 1}\,d\sigma$$

$$\le C(\tau - s)^{\alpha - 1}(t - \tau)^\alpha \le C(\tau - s)^{\alpha - \beta - 1}(t - \tau)^\beta$$

while (6.28), (6.26) and (6.24) imply

$$\left\| \int_s^\tau (R_1(t, \sigma) - R_1(\tau, \sigma))R(\sigma, s)\,d\sigma \right\|$$

$$\le C(t - \tau)^\beta \int_s^\tau (\tau - \sigma)^{\alpha - \beta - 1}(\sigma - s)^{\alpha - 1}\,d\sigma$$

$$\le C(t - \tau)^\beta(\tau - s)^{2\alpha - \beta - 1} \le C(t - \tau)^\beta(\tau - s)^{\alpha - \beta - 1}.$$

The estimate (6.29) is now an immediate consequence of (6.28) and the two last inequalities. $\qquad \square$

**Lemma 6.6.** *For every* $x \in X$ *we have*

$$\lim_{\varepsilon \to 0} S_t(\varepsilon)x = x \qquad uniformly\ in \quad 0 \le t \le T. \qquad (6.30)$$

PROOF. For $x \in D$ we have

$$x - S_t(\varepsilon)x = \int_0^\varepsilon A(t)S_t(\sigma)x\,d\sigma = \int_0^\varepsilon S_t(\sigma)A(t)x\,d\sigma.$$

Therefore,

$$\|x - S_t(\varepsilon)x\| \le \int_0^\varepsilon \|S_t(\sigma)\|\,\|A(t)A(0)^{-1}\|\,\|A(0)x\|\,d\sigma \le \varepsilon C\|A(0)x\|$$

and (6.30) holds for every $x \in D$. Since $D$ is dense in $X$ and $\|S_t(s)\| \le C$ the result for every $x \in X$ follows by approximation. $\qquad \square$

We turn now to prove the differentiability of $U(t, s)$. Since $S_s(t - s)$ is differentiable for $t > s$ and $(\partial/\partial t)S_s(t - s) = -A(s)S_s(t - s)$ is a bounded linear operator which is continuous in $B(X)$ for $t > s$ it suffices to prove the differentiability of $W(t, s)$. To this end we set

$$W_\varepsilon(t, s) = \int_s^{t-\varepsilon} S_\tau(t - \tau) R(\tau, s)\, d\tau \qquad \text{for} \quad 0 < \varepsilon < t - s. \quad (6.31)$$

As $\varepsilon \to 0$, $W_\varepsilon(t, s) \to W(t, s)$. Moreover, $W_\varepsilon(t, s)$ is differentiable in $t$ and

$$\frac{\partial}{\partial t} W_\varepsilon(t, s) = S_{t-\varepsilon}(\varepsilon) R(t - \varepsilon, s) - \int_s^{t-\varepsilon} A(\tau) S_\tau(t - \tau) R(\tau, s)\, d\tau.$$

$$(6.32)$$

Using the equality $A(t)S_t(t - \tau) = (\partial/\partial\tau)S_t(t - \tau)$ we can rewrite the last equation as

$$\frac{\partial}{\partial t} W_\varepsilon(t, s) = S_{t-\varepsilon}(\varepsilon) R(t - \varepsilon, s)$$

$$+ \int_s^{t-\varepsilon} (A(t)S_t(t - \tau) - A(\tau)S_\tau(t - \tau)) R(\tau, s)\, d\tau$$

$$+ \int_s^{t-\varepsilon} A(t)S_t(t - \tau)(R(t, s) - R(\tau, s))\, d\tau$$

$$+ (S_t(t - s) - S_t(\varepsilon)) R(t, s). \quad (6.33)$$

From (6.13) and (6.26) it follows that the first and the last terms on the right-hand side of (6.33) are bounded in norm by $C(t - s - \varepsilon)^{\alpha-1}$ while from (6.16) and (6.18) we deduce easily that

$$\|A(t)S_t(t - \tau) - A(\tau)S_\tau(t - \tau)\| \le C(t - \tau)^{\alpha-1}$$

and therefore,

$$\left\| \int_s^{t-\varepsilon} (A(t)S_t(t - \tau) - A(\tau)S_\tau(t - \tau)) R(\tau, s)\, d\tau \right\|$$

$$\le C \int_s^{t-\varepsilon} (t - \tau)^{\alpha-1} (\tau - s)^{\alpha-1}\, d\tau \le C(t - s)^{2\alpha-1}$$

$$\le C(t - s)^\alpha (t - s - \varepsilon)^{\alpha-1} \le C(t - s - \varepsilon)^{\alpha-1}.$$

Finally, from (6.14) and (6.29) we have

$$\left\| \int_s^{t-\varepsilon} A(t)S_t(t - \tau)(R(t, s) - R(\tau, s))\, d\tau \right\|$$

$$\le C \int_s^{t-\varepsilon} (t - \tau)^{\beta-1} (\tau - s)^{\alpha-\beta-1}\, d\tau$$

$$\le C(t - s)^{\alpha-1} \le C(t - s - \varepsilon)^{\alpha-1}.$$

Combining these estimates we find

$$\left\| \frac{\partial}{\partial t} W_\varepsilon(t, s) \right\| \le \frac{C}{(t - s - \varepsilon)^{1-\alpha}} \quad (6.34)$$

where $C$ is a constant which is also independent of $\varepsilon > 0$. Letting $\varepsilon \to 0$ on the right-hand side of (6.33) and using Lemma 6.6 we see that $(\partial / \partial t) W_\varepsilon(t, s)$ converges strongly as $\varepsilon \to 0$. Denoting its limit by $W'(t, s)$ we have,

$$W'(t, s) = S_t(t - s) R(t, s)$$
$$+ \int_s^t (A(t) S_t(t - \tau) - A(\tau) S_\tau(t - \tau)) R(\tau, s) \, d\tau$$
$$+ \int_s^t A(t) S_t(t - \tau)(R(t, s) - R(\tau, s)) \, d\tau \qquad (6.35)$$

which implies that $W'(t, s)$ is strongly continuous for $0 \le s < t \le T$. Passing to the limit as $\varepsilon \to 0$ in (6.34) yields moreover

$$\| W'(t, s) \| \le C(t - s)^{\alpha - 1}. \qquad (6.36)$$

Now, letting $\varepsilon \to 0$ in

$$W_\varepsilon(t_2, s) - W_\varepsilon(t_1, s) = \int_{t_1}^{t_2} \frac{\partial}{\partial \tau} W_\varepsilon(\tau, s) \, d\tau$$

yields

$$W(t_2, s) - W(t_1, s) = \int_{t_1}^{t_2} W'(\tau, s) \, d\tau$$

where $t_2 > t_1 > s + \varepsilon$. Since $W'(t, s)$ is strongly continuous for $0 \le s < t \le T$, it follows that $W(t, s)$ is strongly continuously differentiable with respect to $t$ and that

$$\frac{\partial}{\partial t} W(t, s) = W'(t, s).$$

Therefore, $U(t, s)$ is strongly continuously differentiable,

$$\frac{\partial}{\partial t} U(t, s) = -A(s) S_s(t - s) + \frac{\partial}{\partial t} W(t, s),$$

and by (6.14) and (6.36)

$$\left\| \frac{\partial}{\partial t} U(t, s) \right\| \le \frac{C}{t - s}.$$

Setting

$$U_\varepsilon(t, s) = S_s(t - s) + W_\varepsilon(t, s) \qquad \text{for} \quad \varepsilon > 0, \, t - s > 0,$$

it follows readily that $U_\varepsilon(t, s) : X \to D$ and by (6.31), (6.32),

$$\frac{\partial}{\partial t} U_\varepsilon(t, s) + A(t) U_\varepsilon(t, s)$$
$$= S_{t-\varepsilon}(\varepsilon) R(t - \varepsilon, s) - R_1(t, s) - \int_s^{t-\varepsilon} R_1(t, \tau) R(\tau, s) \, d\tau. \qquad (6.37)$$

Passing to the limit as $\varepsilon \to 0$, the right hand side of (6.37) tends strongly to zero. Since $(\partial / \partial t) U_\varepsilon(t, s) \to (\partial / \partial t) U(t, s)$ strongly, it follows from (6.37) that $A(t) U_\varepsilon(t, s)$ converges strongly as $\varepsilon \to 0$. Let $x \in X$, the closedness of

$A(t)$ together with $U_\varepsilon(t, s)x \to U(t, s)x$ imply that $U(t, s)x \in D$ and that $A(t)U_\varepsilon(t, s)x \to A(t)U(t, s)x$. Thus passing to the strong limit as $\varepsilon \to 0$ in (6.37) yields

$$\frac{\partial}{\partial t} U(t, s) + A(t)U(t, s) = 0 \qquad \text{for} \quad t > s.$$

This concludes the proofs of (6.9) and (6.10).

To prove (6.11) we will need:

**Lemma 6.7.** *Let* $\varphi(t, s) \geq 0$ *be continuous on* $0 \leq s < t \leq T$. *If there are positive constants* $A, B, \alpha$ *such that*

$$\varphi(t, s) \leq A + B\int_s^t (t - \sigma)^{\alpha - 1}\varphi(\sigma, s)\, d\sigma \qquad \text{for} \quad 0 \leq s < t \leq T$$

(6.38)

*then there is a constant* $C$ *such that* $\varphi(t, s) \leq C$ *for* $0 \leq s < t \leq T$.

PROOF. Iterating (6.38) $n - 1$ times using the identity (6.24) and estimating $t - s$ by $T$ we find

$$\varphi(t, s) \leq A \sum_{j=0}^{n-1} \left(\frac{BT^\alpha}{\alpha}\right)^j + \frac{(B\Gamma(\alpha))^n}{\Gamma(n\alpha)} \int_s^t (t - \sigma)^{n\alpha - 1}\varphi(\sigma, s)\, d\sigma.$$

Choosing $n$ sufficiently large so that $n\alpha > 1$ and estimating $(t - \sigma)^{n\alpha - 1}$ by $T^{n\alpha - 1}$ we get

$$\varphi(t, s) \leq c_1 + c_2\int_s^t \varphi(\sigma, s)\, d\sigma$$

which by Gronwall's inequality implies $\varphi(t, s) \leq c_1 e^{c_2(t-s)} \leq c_1 e^{c_2 T} \leq C$. Since $c_1$ and $c_2$ do not depend on $s$ this estimate holds for $0 \leq s < t \leq T$. $\square$

We turn now to the proof of (6.11). Let $x \in X$ and consider the function $\psi(s) = S_t(t - s)U(s, \tau)A(\tau)^{-1}x$ for $0 \leq \tau < s < t \leq T$. It is easy to see that $\psi$ is differentiable with respect to $s$ and that

$$\psi'(s) = S_t(t - s)[A(t) - A(s)]U(s, \tau)A(\tau)^{-1}x.$$

Integrating $\psi'$ from $\tau$ to $t$ and applying $A(t)$ to the result we find

$$Z(t, \tau)x = A(t)S_t(t - \tau)A(\tau)^{-1}x + \int_\tau^t Y(t, s)Z(s, \tau)x\, ds$$

(6.39)

where

$$Y(t, s) = A(t)S_t(t - s)[A(t) - A(s)]A(s)^{-1}$$

and

$$Z(t, \tau) = A(t)U(t, \tau)A(\tau)^{-1}.$$

From (6.8) and (6.13) we have

$$\|A(t)S_t(t-\tau)A(\tau)^{-1}\| = \|S_t(t-\tau)A(t)A(\tau)^{-1}\|$$

$$\leq \|S_t(t-\tau)\| \|A(t)A(\tau)^{-1}\| \leq C_1$$

and from (6.8) and (6.14),

$$\|Y(t,s)\| \leq \|A(t)S_t(t-s)\| \|(A(t)-A(s))A(s)^{-1}\| \leq C_2(t-s)^{\alpha-1}.$$

Estimating now (6.39) we find

$$\|Z(t,\tau)x\| \leq C_1\|x\| + C_2\int_\tau^t (t-s)^{\alpha-1}\|Z(s,\tau)x\|\, ds$$

which implies by Lemma 6.7, $\|Z(t,\tau)x\| \leq C\|x\|$ whence

$$\|Z(t,\tau)\| = \|A(t)U(t,\tau)A(\tau)^{-1}\| \leq C$$

as desired. This completes the proof of $(E_2)^+$.

## III. Uniqueness

The uniqueness of the evolution system $U(t,s)$ satisfying $(E_1)'$, $(E_2)^+$ and $(E_3)^+$ will be a simple consequence of $(E_3)^+$ as we will see below.

We start by proving $(E_3)^+$ under the supplementary assumption that for every $v \in D$, $A(t)v$ is continuously differentiable on $[0,T]$. This assumption and the uniform boundedness theorem imply that $(\partial/\partial t)A(t)A(0)^{-1} = A'(t)A(0)^{-1}$ is uniformly bounded on $[0,T]$. It also implies that for every $\lambda \in \Sigma$, $R(\lambda:A(s))$ is differentiable with respect to $s$ and that

$$\frac{\partial}{\partial s}R(\lambda:A(s)) = R(\lambda:A(s))A'(s)R(\lambda:A(s)). \tag{6.40}$$

From (6.7) and (6.40) we deduce that

$$\left\|\frac{\partial}{\partial s}R(\lambda:A(s))\right\| \leq \frac{c}{|\lambda|+1} \qquad \text{for} \quad \lambda \in \Sigma. \tag{6.41}$$

The assumptions $(P_1)$ and $(P_2)$ imply (see Section 2.5) that

$$S_s(t-s) = \frac{1}{2\pi i}\int_\Gamma e^{-\lambda(t-s)}R(\lambda:A(s))\, d\lambda$$

where $\Gamma$ is a smooth path in $\Sigma$ connecting $\infty e^{-i\vartheta}$ to $\infty e^{i\vartheta}$. From our supplementary assumption it now follows that if $t - s > 0$ then $S_s(t-s)$ is strongly differentiable in $s$ and

$$\frac{\partial}{\partial s} S_s(t - s) = \frac{1}{2\pi i} \int_\Gamma \lambda e^{-\lambda(t-s)} R(\lambda : A(s)) \, d\lambda$$

$$+ \frac{1}{2\pi i} \int_\Gamma e^{-\lambda(t-s)} \frac{\partial}{\partial s} R(\lambda : A(s)) \, d\lambda$$

$$= -\frac{\partial}{\partial t} S_s(t - s) + \frac{1}{2\pi i} \int_\Gamma e^{-\lambda(t-s)} \frac{\partial}{\partial s} R(\lambda : A(s)) \, d\lambda.$$

To prove $(E_3)^+$ we construct an operator valued function $V(t, s)$ satisfying

$$\begin{cases} \dfrac{\partial}{\partial s} V(t, s) v = V(t, s) A(s) v & \text{for} \quad 0 \le s \le t \le T, v \in D \\ V(t, t) = I \end{cases} \tag{6.42}$$

and prove later that $V(t, s) = U(t, s)$. The construction of $V(t, s)$ follows the same lines as the construction of $U(t, s)$ above. We set

$$Q_1(t, s) = \left( \frac{\partial}{\partial t} + \frac{\partial}{\partial s} \right) S_s(t - s) = \frac{1}{2\pi i} \int_\Gamma e^{-\lambda(t-s)} \frac{\partial}{\partial s} R(\lambda : A(s)) \, d\lambda.$$

Using (6.41) and estimating $Q_1(t, s)$ as in the proof of Theorem 1.7.7 we find

$$\|Q_1(t, s)\| = \left\| \frac{1}{2\pi i} \int_\Gamma e^{-\lambda(t-s)} \frac{\partial}{\partial s} R(\lambda : A(s)) \, d\lambda \right\| \le C.$$

Next we solve by successive approximations the integral equation

$$Q(t, s) = Q_1(t, s) + \int_s^t Q(t, \tau) Q_1(\tau, s) \, d\tau. \tag{6.43}$$

This is done in exactly the same way as the solution of the integral equation (6.6). Since in this case $Q_1(t, s)$ is uniformly bounded the solution $Q(t, s)$ of (6.43) will satisfy

$$\|Q(t, s)\| \le C.$$

Setting

$$V(t, s) = S_s(t - s) + \int_s^t Q(t, \tau) S_s(\tau - s) \, d\tau$$

we find that $\|V(t, s)\| \le C$ and for $v \in D$, $V(t, s) v$ is differentiable in $s$. Differentiating $V(t, s) v$ with respect to $s$ yields

$$\frac{\partial}{\partial s} V(t, s) v - V(t, s) A(s) v$$

$$= Q_1(t, s) v + \int_s^t Q(t, \tau) Q_1(\tau, s) v \, d\tau - Q(t, s) v = 0.$$

From the definition of $V(t, s)$ it follows that $V(t, t) = I$ and so $V(t, s)$ is a solution of (6.42).

For $x \in X$ and $s < r < t$ the function $r \to V(t, r)U(r, s)x$ is differentiable in $r$ and

$$\frac{\partial}{\partial r} V(t, r)U(r, s)x = V(t, r)A(r)U(r, s)x - V(t, r)A(r)U(r, s)x = 0.$$

This shows that $V(t, r)U(r, s)x$ is independent of $r$ for $s < r < t$. Letting $r \downarrow s$ and $r \uparrow t$ we find $V(t, s)x = U(t, s)x$ for every $x \in X$. Therefore $U(t, s) = V(t, s)$ and $U(t, s)$ satisfies

$$\frac{\partial}{\partial s} U(t, s)v = U(t, s)A(s)v \qquad \text{for} \quad v \in D \qquad (6.44)$$

as desired.

We continue by showing the validity of (6.44) in general, that is, without assuming the continuous differentiability of $A(t)A(0)^{-1}$ which was assumed above. To do so we approximate $A(t)$ by a sequence of operators $A_n(t)$ for which $A_n(t)A_n(0)^{-1}$ is continuously differentiable. This is done as follows: Let $\rho(t) \geq 0$ be a continuously differentiable real valued function on $\mathbb{R}$ satisfying $\rho(t) = 0$ for $|t| \geq 1$ and $\int_{-\infty}^{\infty} \rho(t)\, dt = 1$. Let $\rho_n(t) = n\rho(nt)$ and extend $A(t)$ to all of $\mathbb{R}$ by defining $A(t) = A(0)$ for $t < 0$ and $A(t) = A(T)$ for $t \geq T$. Let $v \in D$ and set

$$A_n(t)v = \int_{-\infty}^{\infty} \rho_n(t - \sigma)A(\sigma)v\, d\sigma = \int_{-\infty}^{\infty} \rho_n(\sigma)A(t - \sigma)v\, d\sigma.$$

$A_n(t)v$ thus defined is continuously differentiable on $[0, T]$. We will now show that $A_n(t)$ satisfy the conditions $(P_1)$–$(P_3)$.

By defintion we have $D(A_n(t)) = D$ and therefore $(P_1)$ is satisfied. For $\lambda \in \Sigma$ we have

$$x - (\lambda - A_n(t))R(\lambda : A(t))x = -(A(t) - A_n(t))R(\lambda : A(t))x$$

$$= \int_{-\infty}^{\infty} \rho_n(t - \tau)(A(\tau) - A(t))R(\lambda : A(t))x\, d\tau.$$

For $|t - \tau| \leq 1/n$, (6.7) and (6.8) imply

$$\|(A(t) - A(\tau))R(\lambda : A(t))\| \leq Cn^{-\alpha}.$$

Therefore,

$$\|x - (\lambda - A_n(t))R(\lambda : A(t))x\| \leq Cn^{-\alpha}\|x\| \qquad (6.45)$$

and in particular taking $\lambda = 0$ we have

$$\|(A(t) - A_n(t))A(t)^{-1}\| \leq Cn^{-\alpha}. \qquad (6.46)$$

From (6.45) it follows easily that for $v \in D$, we have

$$(1 - Cn^{-\alpha})\|(\lambda - A(t))v\| \leq \|(\lambda - A_n(t))v\|$$

$$\leq (1 + Cn^{-\alpha})\|(\lambda - A(t))v\| \qquad (6.47)$$

and therefore if $n$ is sufficiently large so that $Cn^{-\alpha} < 1$ and $\lambda \in \Sigma$,

$\lambda I - A_n(t)$ is closed, $R(\lambda I - A_n(t)) = X$ and $\lambda I - A_n(t)$ has a bounded inverse $R(\lambda : A_n(t))$ satisfying

$$\|R(\lambda : A_n(t))\| \le \frac{M}{|\lambda| + 1} \qquad \text{for} \quad \lambda \in \Sigma$$

and so $(P_2)$ is satisfied.

Choosing $n$ in (6.46) such that $Cn^{-\alpha} < 1$ we obtain

$$A(t)A_n(t)^{-1} = \sum_{k=0}^{\infty} \left[ I - A_n(t)A(t)^{-1} \right]^k$$

and $\|A(t)A_n(t)^{-1}\| \le C$. From the definition of $A_n(t)$ and (6.8) it follows that

$$\left\| (A_n(t) - A_n(s))A(\sigma)^{-1}v \right\|$$
$$\le \int_{-\infty}^{\infty} \rho_n(\tau) \left\| (A(t - \sigma) - A(s - \sigma))A(\tau)^{-1}v \right\| d\tau \le C|t - s|^{\alpha}\|v\|$$

therefore

$$\|(A_n(t) - A_n(s))A_n(\tau)^{-1}\|$$
$$\le \|(A_n(t) - A_n(s))A(\tau)^{-1}\| \cdot \|A(\tau)A_n(\tau)^{-1}\| \le C|t - s|^{\alpha}$$

and so $(P_3)$ is also satisfied by $A_n(t)$.

From the first part of the proof it follows that there is an operator valued function $U_n(t, s)$ satisfying $\|U_n(t, s)\| \le C$, where $C$ is independent of $n$, and

$$\frac{\partial}{\partial t}U_n(t, s) = -A_n(t)U_n(t, s) \qquad \text{for} \quad 0 \le s < t \le T.$$

Since $A_n(t)v$ is continuously differentiable in $t$ for $v \in D$, it follows that

$$\frac{\partial}{\partial s}U_n(t, s)v = U_n(t, s)A_n(s)v \qquad \text{for} \quad v \in D. \tag{6.48}$$

From (6.48) and the properties of $U(t, s)$ it follows that for every $v \in D$ the function $r \to U_n(t, r)U(r, s)v$ is differentiable and

$$U(t, s)v - U_n(t, s)v = \int_s^t \frac{\partial}{\partial r}\{U_n(t, r)U(r, s)v\} \, dr$$
$$= \int_s^t U_n(t, r)[A_n(r) - A(r)]U(r, s)v \, dr$$
$$= \int_s^t U_n(t, r)[A_n(r) - A(r)]A(r)^{-1}A(r)$$
$$\times U(r, s)A(s)^{-1}A(s)v \, dr. \tag{6.49}$$

Using (6.46) and (6.11) to estimate (6.49) we find

$$\|U(t, s)v - U_n(t, s)v\| \le Cn^{-\alpha}(t - s)\|A(s)v\| \le Cn^{-\alpha}\|A(0)v\|$$

and therefore $U_n(t, s)v \to U(t, s)v$ uniformly in $t$ and $s$. Since $D$ is dense in

$X$ it follows that $U_n(t, s)x \rightarrow U(t, s)x$ uniformly in $t$ and $s$ for every $x \in X$. For $v \in D$ we have

$$\| U_n(t, s)A_n(s)v - U(t, s)A(s)v\|$$
$$\leq \| U_n(t, s)(A_n(s)v - A(s)v)\| + \|(U_n(t, s) - U(t, s))A(s)v\|$$
$$\leq Cn^{-\alpha}\|A(0)v\| + \|(U_n(t, s) - U(t, s))A(s)v\|$$

and therefore $U_n(t, s)A_n(s)v \rightarrow U(t, s)A(s)v$ uniformly in $t$ and $s$. For $r < s < t$ and $v \in D$ we have

$$U_n(t, s)v - U_n(t, r)v = \int_r^s \frac{\partial}{\partial \sigma} U_n(t, \sigma)v \, d\sigma = \int_r^s U_n(t, \sigma)A_n(\sigma)v \, d\sigma$$

which in the limit as $n \rightarrow \infty$ yields

$$U(t, s)v - U(t, r)v = \int_r^s U(t, \sigma)A(\sigma)v \, d\sigma$$

and therefore (6.44) holds in general. This concludes the proof of $(E_3)^+$. To conclude the proof of Theorem 6.1 we still have to show the uniqueness of $U(t, s)$ and that it satisfies $U(t, s) = U(t, r)U(r, s)$ for $0 \leq s \leq r \leq t \leq T$. Both these claims follow from:

**Theorem 6.8.** *Let $A(t)$, $0 \leq t \leq T$ satisfy the conditions $(P_1)$–$(P_3)$. For every $0 \leq s < T$ and $x \in X$ the initial value problem*

$$\begin{cases} \dfrac{du(t)}{dt} + A(t)u(t) = 0 & s < t \leq T \\ u(s) = x \end{cases} \tag{6.50}$$

*has a unique solution $u$ given by $u(t) = U(t, s)x$ where $U(t, s)$ is the evolution system constructed above.*

PROOF. From (6.9) it follows readily that $u(t) = U(t, s)x$ is a solution of (6.50). To prove its uniqueness let $v(t)$ be a solution of (6.50). Since for every $r > s$, $v(r) \in D$, it follows from (6.12) and (6.50) that the function $r \rightarrow U(t, r)v(r)$ is differentiable and

$$\frac{\partial}{\partial r} U(t, r)v(r) = U(t, r)A(r)v(r) - U(t, r)A(r)v(r) = 0.$$

Therefore $U(t, r)v(r)$ is constant on $s < r < t$. Since it is also continuous on $s \leq r \leq t$ we can let $r \rightarrow t$ and $r \rightarrow s$ to obtain $U(t, s)x = v(t)$ and the uniqueness of the solution of (6.50) follows. $\qquad \square$

From Theorem 6.8 it follows readily that for $x \in X$,

$$U(t, s)x = U(t, r)U(r, s)x \qquad \text{for} \quad 0 \leq s \leq t \leq T$$

and $U(t, s)$ is therefore an evolution system satisfying $(E_1)'$, $(E_2)^+$ and $(E_3)^+$. If $V(t, s)$ is an evolution system satisfying $(E_1)'$ and $(E_2)^+$ then $V(t, s)x$ is a solution of (6.50) and from Theorem 6.8 it follows that

$V(t, s)x = U(t, s)x$ and so $V(t, s) = U(t, s)$ and $U(t, s)$ is the unique evolution system satisfying $(E_1)'$, $(E_2)^+$ and $(E_3)^+$. This concludes the proof of Theorem 6.1.                                                          □

In Theorem 6.1 we proved that $-(\partial/\partial t)U(t, s) = A(t)U(t, s)$ is strongly continuous on $0 \le s < t \le T$. Much more is actually true. Indeed we have:

**Theorem 6.9.** *Let the assumptions of Theorem 6.1 be satisfied. Then for every* $\varepsilon > 0$ *the map* $t \to A(t)U(t, s)$ *is Hölder continuous, with exponent* $\beta < \alpha$, *in the uniform operator topology, for* $0 < s + \varepsilon \le t \le T$.

PROOF. We recall that

$$U(t, s) = S_s(t - s) + W(t, s).$$

Since $(\partial/\partial t)S_s(t - s) = A(s)S_s(t - s)$ is Lipschitz continuous in $t$ for $t \in [s + \varepsilon, T]$ it remains to show that $(\partial/\partial t)W(t, s)$ is Hölder continuous as claimed. We fix $\varepsilon > 0$ and assume that $0 \le s < s + \varepsilon \le \tau \le t \le T$. From (6.35) we have

$$\frac{\partial W(t, s)}{\partial t} - \frac{\partial W(\tau, s)}{\partial t}$$

$$= \left[ S_t(t - s)R(t, s) - S_\tau(\tau - s)R(\tau, s) \right]$$

$$+ \int_\tau^t (A(t)S_t(t - \sigma) - A(\sigma)S_\sigma(t - \sigma))R(\sigma, s)\, d\sigma$$

$$+ \int_s^\tau (A(t)S_t(t - \sigma) - A(\sigma)S_\sigma(t - \sigma) - A(\tau)S_\tau(\tau - \sigma)$$

$$+ A(\sigma)S_\sigma(\tau - \sigma))R(\sigma, s)\, d\sigma$$

$$+ \int_\tau^t A(t)S_t(t - \sigma)(R(t, s) - R(\sigma, s))\, d\sigma$$

$$+ \int_s^\tau \left[ A(t)S_t(t - \sigma)(R(t, s) - R(\sigma, s)) \right.$$

$$\left. - A(\tau)S_\tau(\tau - \sigma)(R(\tau, s) - R(\sigma, s)) \right] d\sigma$$

$$= I_1 + I_2 + I_3 + I_4 + I_5.$$

We will now estimate each one of the terms $I_j$, $1 \le j \le 5$ separately. The generic constants appearing in these estimates will usually depend on the $\varepsilon > 0$, chosen above.

$$\|I_1\| \le \|(S_t(t - s) - S_t(\tau - s))R(t, s)\|$$

$$+ \|(S_t(\tau - s) - S_\tau(\tau - s))R(t, s)\|$$

$$+ \|S_\tau(\tau - s)(R(t, s) - R(\tau, s))\|$$

$$\le C_1(t - \tau) + C_2(t - \tau)^\alpha + C_3(t - \tau)^\alpha \le C(t - \tau)^\alpha.$$

Here we used the Lipschitz continuity of $S_t(s)$ for $s > 0$, (6.18) and

Corollary 6.5 with $\alpha = \beta$. The second term is estimated as follows.

$$
\begin{aligned}
\|I_2\| &\leq \int_\tau^t \|(A(t)S_t(t - \sigma) - A(\sigma)S_\sigma(t - \sigma))R(\sigma, s)\| \, d\sigma \\
&\leq C \int_\tau^t (t - \sigma)^{\alpha-1}(\sigma - s)^{\alpha-1} \, d\sigma \\
&\leq C \int_\tau^t (t - \sigma)^{\alpha-1} \, d\sigma \leq C(t - \tau)^\alpha
\end{aligned}
$$

where we used (6.26) and

$$
\|A(t)S_t(t - \sigma) - A(\tau)S_\tau(t - \sigma)\| \leq C(t - \tau)^\alpha (t - \sigma)^{-1} \quad (6.51)
$$

which is a simple consequence of (6.16) and (6.18).

To estimate $I_3$ we note first that from (6.18) and (6.14) we have

$$
\|A(\sigma)S_\sigma(\rho) - A(\tau)S_\tau(\rho)\| \leq \frac{C}{\rho}(\tau - \sigma)^\alpha
$$

and therefore,

$$
\begin{aligned}
\|A(\sigma)^2 &S_\sigma(\rho) - A(\tau)^2 S_\tau(\rho)\| \\
&\leq \left\| A(\sigma)S_\sigma\!\left(\frac{\rho}{2}\right)\!\left( A(\sigma)S_\sigma\!\left(\frac{\rho}{2}\right) - A(\tau)S_\tau\!\left(\frac{\rho}{2}\right)\right)\right\| \\
&\quad + \left\| \left( A(\sigma)S_\sigma\!\left(\frac{\rho}{2}\right) - A(\tau)S_\tau\!\left(\frac{\rho}{2}\right)\right) A(\tau)S_\tau\!\left(\frac{\rho}{2}\right)\right\| \\
&\leq \frac{C}{\rho^2}(\tau - \sigma)^\alpha.
\end{aligned} \quad (6.52)
$$

We rewrite the integrand of $I_3$ as follows,

$$
\begin{aligned}
\big[A(t)&S_t(t - \sigma) - A(\sigma)S_\sigma(t - \sigma) - A(\tau)S_\tau(\tau - \sigma) \\
&+ A(\sigma)S_\sigma(\tau - \sigma)\big]R(\sigma, s) \\
&= \big[A(\sigma)(S_\sigma(\tau - \sigma) - S_\sigma(t - \sigma)) - A(\tau)(S_\tau(\tau - \sigma) - S_\tau(t - \sigma)) \\
&\quad + A(t)S_t(t - \sigma) - A(\tau)S_\tau(t - \sigma)\big]R(\sigma, s) \\
&= \left[\int_{\tau - \sigma}^{t - \sigma}\big(A(\sigma)^2 S_\sigma(\rho) - A(\tau)^2 S_\tau(\rho)\big) \, d\rho + A(t)S_t(t - \sigma) \right. \\
&\quad \left. - A(\tau)S_\tau(t - \sigma)\right]R(\sigma, s).
\end{aligned}
$$

Estimating this integrand using (6.52), (6.51) and (6.26) we find for $0 < \beta < \alpha$

$$
\begin{aligned}
\|I_3\| &\leq C \int_s^\tau \big[(t - \tau)(\tau - \sigma)^{\alpha-1}(t - \sigma)^{-1} \\
&\quad + (t - \tau)^\alpha (t - \sigma)^{-1}\big](\sigma - s)^{\alpha-1} \, d\sigma \\
&\leq C(t - \tau)^\beta \int_s^\tau (\tau - \sigma)^{(\alpha-\beta)-1}(\sigma - s)^{\alpha-1} \, d\sigma \leq C(t - \tau)^\beta
\end{aligned}
$$

where we used that $t - \tau \leq t - \sigma$ and $t - \sigma \geq \tau - \sigma$ and therefore for example,

$$(t - \tau)(\tau - \sigma)^{\alpha-1}(t - \sigma)^{-1} = (t - \tau)^\beta \left(\frac{t - \tau}{t - \sigma}\right)^{1-\beta}(t - \sigma)^{-\beta}(\tau - \sigma)^{\alpha-1}$$

$$\leq (t - \tau)^\beta (\tau - \sigma)^{\alpha-\beta-1}.$$

For $I_4$ we have

$$\|I_4\| \leq \int_\tau^t \|A(t)S_t(t - \sigma)\| \, \|R(t, s) - R(\sigma, s)\| \, d\sigma$$

$$\leq C\int_\tau^t (t - \sigma)^{-1}(t - \sigma)^\alpha (\sigma - s)^{-1} \, d\sigma$$

$$\leq C\int_\tau^t (t - \sigma)^{\alpha-1} \, d\sigma \leq C(t - \tau)^\alpha.$$

Here we used (6.14) and Corollary 6.5. Finally to estimate $I_5$ we rewrite its integrand as follows

$$[A(t)S_t(t - \sigma)(R(t, s) - R(\sigma, s))$$
$$-A(\tau)S_\tau(\tau - \sigma)(R(\tau, s) - R(\sigma, s))]$$
$$= A(t)S_t(t - \sigma)(R(t, s) - R(\tau, s))$$
$$+ [A(t)S_t(t - \sigma) - A(\tau)S_\tau(t - \sigma)$$
$$+ A(\tau)(S_\tau(t - \sigma) - S_\tau(\tau - \sigma))](R(\tau, s) - R(\sigma, s))$$

Taking $\beta < \gamma < \alpha$ and estimating the integrand of $I_5$ using Corollary 6.5, (6.14), (6.51) and (6.17) we find

$$\|I_5\| \leq C\int_s^\tau (t - \sigma)^{-1}\{(t - \tau)^\gamma (\tau - s)^{\alpha-\beta-1}$$

$$+ [(t - \tau)^\gamma + (\tau - \sigma)^{-1}(t - \tau)](\tau - \sigma)^\gamma (\sigma - s)^{\alpha-\beta-1}\} \, d\sigma$$

$$\leq C(t - \tau)^\gamma + C(t - \tau)^\gamma \int_s^\tau (\tau - \sigma)^{\alpha-1}(\sigma - s)^{(\alpha-\beta)-1} \, d\sigma$$

$$+ C(t - \tau)^\beta \int_s^\tau (\tau - \sigma)^{(\gamma-\beta)-1}(\sigma - s)^{(\alpha-\beta)-1} \, d\sigma$$

$$\leq C(t - \tau)^\beta$$

and the proof is complete.                                                          □


## 5.7. The Inhomogeneous Equation in the Parabolic Case

Let $\{A(t)\}_{t\in[0, T]}$ satisfy the conditions $(P_1)$–$(P_3)$ of the previous section and consider the inhomogeneous initial value problem

$$\begin{cases} \dfrac{du(t)}{dt} + A(t)u(t) = f(t) & \text{for } 0 \leq s < t \leq T \\ u(s) = x \end{cases} \qquad (7.1)$$

in the Banach space $X$. From Theorem 6.1 it follows that there is a unique evolution system $U(t, s)$, $0 \leq s \leq t \leq T$, associated with the family $\{A(t)\}_{t \in [0, T]}$. As we have already done in the autonomous and hyperbolic cases, the continuous function

$$u(t) = U(t, s)x + \int_s^t U(t, \sigma)f(\sigma) \, d\sigma \qquad (7.2)$$

with $f \in L^1(s, T: X)$ and $x \in X$ will be called the *mild solution* of the initial value problem (7.1). Thus for every $f \in L^1(s, T: X)$ and $x \in X$ the initial value problem (7.1) possesses a unique mild solution given by (7.2). In this section we will be interested in classical solutions of (7.1) i.e., functions $u : [s, T] \to X$ which are continuous for $s \leq t \leq T$, continuously differentiable for $s < t \leq T$, $u(t) \in D$ for $s < t \leq T$, $u(s) = x$ and $u'(t) + A(t)u(t) = f(t)$ holds for $s < t \leq T$. We will call a function $u$ a solution of the initial value problem (7.1) if it is a classical solution of this problem. In order to obtain (classical) solutions of (7.1) we will have to impose further conditions on the function $f$. The situation here is completely analogous to the situation in the autonomous case with $-A$ being the infinitesimal generator of an analytic semigroup (see Corollary 4.3.3).

**Theorem 7.1.** *Let $\{A(t)\}_{t \in [0, T]}$ satisfy the conditions $(P_1)$–$(P_3)$ of section 5.6 and let $U(t, s)$ be the evolution system provided by Theorem 6.1. If $f$ is Hölder continuous on $[s, T]$ then the initial value problem (7.1) has, for every $x \in X$, a unique solution $u$ given by;*

$$u(t) = U(t, s)x + \int_s^t U(t, \sigma)f(\sigma) \, d\sigma.$$

PROOF. From Theorem 6.8 it follows that $U(t, s)x$ is the unique solution of the homogeneous initial value problem

$$\frac{du(t)}{dt} + A(t)u(t) = 0, \qquad u(s) = x. \qquad (7.3)$$

To prove the existence of a solution of the initial value problem (7.1) it is therefore sufficient to show that

$$v(t) = \int_s^t U(t, \sigma)f(\sigma) \, d\sigma$$

is a solution of the initial value problem

$$\begin{cases} \dfrac{dv(t)}{dt} + A(t)v(t) = f(t) \\ v(s) = 0. \end{cases} \qquad (7.4)$$

To this end we set

$$v_\epsilon(t) = \int_s^{t-\epsilon} U(t, \sigma)f(\sigma) \, d\sigma. \qquad (7.5)$$

From Theorem 6.1 it follows that $v_\varepsilon(t)$ is differentiable in $t$ and

$$\frac{\partial}{\partial t}v_\varepsilon(t) = \frac{\partial}{\partial t}\int_s^{t-\varepsilon}U(t,\sigma)f(\sigma)\,d\sigma$$

$$= U(t,t-\varepsilon)f(t-\varepsilon) - \int_s^{t-\varepsilon}A(t)U(t,\sigma)f(\sigma)\,d\sigma$$

and therefore

$$\frac{\partial}{\partial t}v_\varepsilon(t) + A(t)v_\varepsilon(t) = U(t,t-\varepsilon)f(t-\varepsilon). \tag{7.6}$$

The proof will be concluded by letting $\varepsilon\downarrow 0$ in (7.6). In order to justify this passage to the limit, we will first show that $(\partial/\partial t)v_\varepsilon(t)$ has a limit as $\varepsilon\downarrow 0$.

From the results of Section 6 we have

$$U(t,s) = S_s(t-s) + W(t,s) \tag{7.7}$$

where $S_t(s)$ is the analytic semigroup generated by $-A(t)$ and $W(t,s)$ is strongly continuously differentiable in $t$ for $0 \le s < t \le T$ (see proof of Theorem 6.1) and

$$\frac{\partial}{\partial t}\int_s^{t-\varepsilon}W(t,\sigma)f(\sigma)\,d\sigma = W(t,t-\varepsilon)f(t-\varepsilon)$$

$$+ \int_s^{t-\varepsilon}\frac{\partial}{\partial t}W(t,\sigma)f(\sigma)\,d\sigma.$$

Moreover, $W(t,t) = 0$ and (see (6.36)) $\|(\partial/\partial t)W(t,s)\| \le C(t-s)^{\alpha-1}$. Therefore,

$$\lim_{\varepsilon\downarrow 0}\frac{\partial}{\partial t}\int_s^{t-\varepsilon}W(t,\sigma)f(\sigma)\,d\sigma = \int_s^t\frac{\partial}{\partial t}W(t,\sigma)f(\sigma)\,d\sigma.$$

Next,

$$\frac{\partial}{\partial t}\int_s^{t-\varepsilon}S_\sigma(t-\sigma)f(\sigma)\,d\sigma = S_{t-\varepsilon}(\varepsilon)f(t-\varepsilon) + \int_s^{t-\varepsilon}\frac{\partial}{\partial t}S_\sigma(t-\sigma)f(\sigma)\,d\sigma$$

$$= S_{t-\varepsilon}(\varepsilon)f(t-\varepsilon)$$

$$+ \int_s^{t-\varepsilon}[A(t)S_t(t-\sigma) - A(\sigma)S_\sigma(t-\sigma)]f(\sigma)\,d\sigma$$

$$- \int_s^{t-\varepsilon}A(t)S_t(t-\sigma)f(\sigma)\,d\sigma. \tag{7.8}$$

To show the existence of the limit as $\varepsilon\downarrow 0$, we treat each of the three terms on the right-hand side of (7.8) separately.

From Lemma 6.6 it follows that

$$\lim_{\varepsilon\downarrow 0}S_{t-\varepsilon}(\varepsilon)f(t-\varepsilon) = f(t).$$

The assumptions $(P_1)$–$(P_3)$ imply (see Lemma 6.2) that

$$\|A(t)S_t(t-\sigma) - A(\sigma)S_\sigma(t-\sigma)\| \le C(t-\sigma)^{\alpha-1}$$

and therefore,

$$\lim_{\varepsilon \downarrow 0} \int_s^{t-\varepsilon} [A(t)S_t(t-\sigma) - A(\sigma)S_\sigma(t-\sigma)] f(\sigma) \, d\sigma$$

$$= \int_s^t [A(t)S_t(t-\sigma) - A(\sigma)S_\sigma(t-\sigma)] f(\sigma) \, d\sigma.$$

Finally, for the last term we have

$$\int_s^{t-\varepsilon} A(t)S_t(t-\sigma) f(\sigma) \, d\sigma$$

$$= \int_s^{t-\varepsilon} A(t)S_t(t-\sigma)(f(\sigma) - f(t)) \, d\sigma + \int_s^{t-\varepsilon} A(t)S_t(t-\sigma) f(t) \, d\sigma$$

$$= \int_s^{t-\varepsilon} A(t)S_t(t-\sigma)(f(\sigma) - f(t)) \, d\sigma - [S_t(t-s) - S_t(\varepsilon)] f(t).$$

$$(7.9)$$

The Hölder continuity of $f$ and the estimate $\|A(t)S_t(t-\sigma)\| \leq C(t-\sigma)^{-1}$ imply that the integrand on the first term on the right-hand side of (7.9) is integrable and therefore

$$\lim_{\varepsilon \downarrow 0} \int_s^{t-\varepsilon} A(t)S_t(t-\sigma) f(\sigma) \, d\sigma = \int_s^t A(t)S_t(t-\sigma)(f(\sigma) - f(t)) \, d\sigma$$

$$- S_t(t-s)f(t) + f(t).$$

Combining the last results we conclude that $(\partial/\partial t)\int_s^{t-\varepsilon} S_\sigma(t-\sigma) f(\sigma) \, d\sigma$ converges as $\varepsilon \downarrow 0$ and since we saw that $(\partial/\partial t)\int_s^{t-\varepsilon} W(t,\sigma) f(\sigma) \, d\sigma$ converges as $\varepsilon \downarrow 0$ it follows that $\lim_{\varepsilon \downarrow 0}(\partial/\partial t)v_\varepsilon(t)$ exists. Reviewing the proof it is not difficult to see that this limit is uniform on $[s + \varepsilon, T]$ for every $\varepsilon > 0$.

Returning to (7.6) we can now conclude that $A(t)v_\varepsilon(t)$ converges as $\varepsilon \downarrow 0$ and since by (7.5) $v_\varepsilon(t) \to v(t)$ as $\varepsilon \downarrow 0$ it follows from the closedness of $A(t)$ that $v(t) \in D(A(t)) = D$ and $A(t)v_\varepsilon(t) \to A(t)v(t)$ for $s < t \leq T$. Integrating (7.6) from $t$ to $t + h$ we have

$$v_\varepsilon(t+h) - v_\varepsilon(t) = -\int_t^{t+h} A(\tau)v_\varepsilon(\tau) \, d\tau + \int_t^{t+h} U(\tau, \tau - \varepsilon)f(\tau - \varepsilon) \, d\tau.$$

$$(7.10)$$

Passing to the limit as $\varepsilon \downarrow 0$ yields

$$v(t+h) - v(t) = -\int_t^{t+h} A(\tau)v(\tau) \, d\tau + \int_t^{t+h} f(\tau) \, d\tau. \quad (7.11)$$

Finally, dividing (7.11) by $h$ and letting $h \downarrow 0$ we find that $v(t)$ satisfies (7.4). The uniqueness of the solution of the initial value problem (7.1) is an immediate consequence of the uniqueness of the solution of the homogeneous initial value problem (7.3) which was proved in Theorem 6.8. This concludes the proof of Theorem 7.1.                                                  □

**Remark 7.2.** Let $f$ satisfy

$$\|f(t) - f(\tau)\| \leq L|t - \tau|^{\vartheta}, 0 < \vartheta < 1, \tau, t \in [0, T].$$

It can be shown that if $u$ is the solution of the initial value problem (7.1) then $u'(t)$ is Hölder continuous with exponent $\beta$, $\beta < \min(\alpha, \vartheta)$ on $[s + \varepsilon, T]$ for every $\varepsilon > 0$. Moreover, if $f(s) = 0$ and $x = 0$ then $u'$ is Hölder continuous with exponent $\beta$ on $[s, T]$.

The proof of remark 7.2 is rather long and tedious and we will not give it here. Instead, we will show now how to use Theorem 7.1 in order to obtain higher differentiability of the solution of (7.1) under stronger assumptions. A typical result of this kind is:

**Theorem 7.3.** Let $\{A(t)\}_{t\in[0, T]}$ satisfy the conditions $(P_1)$ and $(P_2)$ and assume that $t \to A(t)A(0)^{-1}$ is differentiable and its derivative $A'(t)A(0)^{-1}$ satisfies a Hölder condition

$$\|A'(t)A(0)^{-1} - A'(\tau)A(0)^{-1}\| \leq C|t - \tau|^a$$

$$\text{for} \quad \tau, t \in [0, T], 0 < \alpha \leq 1.$$

Let $f(t)$ be differentiable on $[0, T]$ and assume that its derivative $f'$ is Hölder continuous on $[0, T]$ i.e.,

$$\|f'(t) - f'(\tau)\| \leq C|t - \tau|^{\beta} \quad \text{for} \quad \tau, t \in [0, T], 0 < \beta \leq 1.$$

If $u$ is the solution of the initial value problem

$$\begin{cases} \dfrac{du(t)}{dt} + A(t)u(t) = f(t) & \text{for} \ \ 0 < t \leq T \\ u(0) = x \end{cases} \tag{7.12}$$

then $u$ is twice continuously differentiable on $]0, T]$.

PROOF. Let $u$ be the solution of (7.12) and set

$$u_h(t) = \frac{u(t + h) - u(t)}{h}, \qquad f_h(t) = \frac{f(t + h) - f(t)}{h}$$

then

$$\frac{du_h(t)}{dt} + A(t)u_h(t) = f_h(t) - \frac{A(t + h) - A(t)}{h}u(t + h) \tag{7.13}$$

The right-hand side of (7.13) is Hölder continuous in $t$ and therefore by Theorem 7.1 we have for $\tau > 0$

$$u_h(t) = U(t, \tau)u_h(\tau)$$

$$+ \int_{\tau}^{t} U(t, \sigma)\left[ f_h(\sigma) - \frac{A(\sigma + h) - A(\sigma)}{h}u(\sigma + h) \right] d\sigma.$$

$$\tag{7.14}$$

Our assumptions on $f(t)$ imply that $f_h(t) \to f'(t)$ as $h \to 0$ uniformly on $[0, T]$. From the differentiability of $A(t)A(0)^{-1}$, the strong continuity of $A(0)A(t)^{-1}$ and the continuity of $A(t)u(t)$ for $t > 0$ it follows that

$$\frac{A(\sigma + h) - A(\sigma)}{h} u(\sigma + h)$$

$$= \frac{A(\sigma + h) - A(\sigma)}{h} A(0)^{-1} A(0)A(\sigma + h)^{-1} A(\sigma + h)u(\sigma + h)$$

converges as $h \to 0$ uniformly on $[\tau, T - \varepsilon]$ for $\varepsilon > 0$, to $A'(\sigma)A(0)^{-1}$ $A(0)u(\sigma)$. Therefore we can let $h \to 0$ in (7.14) and obtain

$$u'(t) = U(t, \tau)u'(\tau) + \int_\tau^t U(t, \sigma)\left[ f'(\sigma) - A'(\sigma)A(0)^{-1}A(0)u(\sigma)\right] d\sigma.$$

From our assumptions and the fact that $u$ is $C^1$ on $[\tau, t]$ it follows that $g(\sigma) = f'(\sigma) - A'(\sigma)A(0)^{-1}A(0)u(\sigma)$ satisfies a Hölder condition on $[\tau, T]$ and therefore by Theorem 7.1 $u'(t)$ is continuously differentiable on $]\tau, t]$ and satisfies the equation

$$\frac{dv(t)}{dt} + A(t)v(t) = f'(t) - A'(t)A(0)^{-1}A(0)u(t) \qquad \text{for} \quad \tau < t \le T.$$

$$(7.15)$$

Since $\tau > 0$ is arbitrary it follows that $u'(t)$ is continuously differentiable on $]0, T]$ as claimed.                                                                                     □

**Remark 7.4.** If in the previous theorem we assume that $A(t)A(0)^{-1}$ is $k$ times continuously differentiable and its $k$-th derivative is Hölder continuous and if we also assume that $f$ is $k$ times continuously differentiable and its $k$-th derivative is Hölder continuous, then we can proceed as in the proof of Theorem 7.3 and show that $u$ is $k + 1$ times continuously differentiable on $]0, T]$. In particular if $A(t)A(0)^{-1}$ and $f(t)$ are $C^\infty$ functions on $[0, T]$ then $u$ is $C^\infty$ on $]0, T]$.

# 5.8. Asymptotic Behavior of Solutions in the Parabolic Case

Let $A(t)$, $t \ge 0$ be a family of operators in a Banach space $X$ satisfying for every $T > 0$ the assumptions $(P_1)$–$(P_3)$, of Section 5. If $f:[0, \infty[ \to X$ is Hölder continuous on $[0, \infty[$ then the initial value problem,

$$\begin{cases} \dfrac{du(t)}{dt} + A(t)u(t) = f(t), & t > 0 \\ u(0) = x \end{cases}$$

$$(8.1)$$

has a unique solution $u$ on $[0, \infty[$. The asymptotic behavior of this solution, as $t \to \infty$, is the subject of the present section. In order to study this asymptotic behavior we will assume that the conditions $(P_1)$–$(P_3)$ hold uniformly on $[0, \infty[$ i.e., that they hold for every $T > 0$ with constants $M$, $L$ and $\alpha$ which are independent of $T$. We will further assume:

$(P_4)$ The operators $A(t)A(s)^{-1}$ are uniformly bounded for $0 \leq s, t < \infty$ and there exists a closed operator $A(\infty)$ with domain $D$ such that

$$\lim_{t \to \infty} \|(A(t) - A(\infty))A(0)^{-1}\| = 0. \tag{8.2}$$

Under these assumptions the solutions of the homogeneous evolution equation decay exponentially to zero. This follows from:

**Theorem 8.1.** *Let* $\{A(t)\}_{t \geq 0}$ *satisfy the conditions* $(P_1)$–$(P_3)$ *uniformly on* $[0, \infty[$. *If* $\{A(t)\}_{t \geq 0}$ *satisfies also* $(P_4)$ *and* $U(t, s)$ *is the evolution system for* $\{A(t)\}$ *(see Theorem 6.1) then there exist constants* $C \geq 0$ *and* $\vartheta > 0$ *such that*

$$\|U(t, s)\| \leq Ce^{-\vartheta(t-s)} \qquad for \quad 0 \leq s \leq t. \tag{8.3}$$

PROOF. We note first that from the results of Section 2.5, the denseness of $D$ and the fact that $(P_2)$ holds for all $t \geq 0$ it follows that $-A(t)$ is the infinitesimal generator of an analytic semigroup $S_t(s)$, $s \geq 0$ for every $t \geq 0$ and that there are constants $C \geq 0$ and $\delta > 0$ (independent of $t$) such that;

$$\|S_t(s)\| \leq Ce^{-\delta s} \qquad for \quad s \geq 0 \tag{8.4}$$

$$\|A(t)S_t(s)\| \leq \frac{C}{s}e^{-\delta s} \qquad for \quad s > 0. \tag{8.5}$$

Set

$$\rho(\mu) = \sup\{\|(A(t) - A(s))A(r)^{-1}\| : 0 \leq \mu \leq s, t < \infty, 0 \leq r < \infty\}. \tag{8.6}$$

From $(P_3)$ and $(P_4)$ it follows that $\rho(\mu)$ is finite for $\mu \geq 0$ and $\rho(\mu) \to 0$ as $\mu \to \infty$. Combining (8.6) with the assumption $(P_3)$ we find

$$\|(A(t) - A(s))A(r)^{-1}\| \leq C\sqrt{\rho(\mu)}\,|t - s|^{\alpha/2} \qquad for \quad \mu \leq s, t \tag{8.7}$$

where throughout the rest of the proof $C$ will denote a generic constant. From (8.7) we deduce

$$\|R_1(t, s)\| = \|(A(s) - A(t))S_s(t - s)\|$$

$$\leq \|(A(s) - A(t))A(s)^{-1}\|\,\|A(s)S_s(t - s)\|$$

$$\leq C\Gamma\left(\frac{\alpha}{2}\right)\sqrt{\rho(\mu)}\,(t - s)^{\alpha/2-1}e^{-\delta(t-s)} \qquad for \quad \mu \leq s \leq t,$$

$$\tag{8.8}$$

and by induction on $m$ it follows easily that

$$\| R_m(t, s) \| = \left\| \int_s^t R_1(t, \tau) R_{m-1}(\tau, s)\, d\tau \right\|$$

$$\leq \frac{e^{-\delta(t-s)}}{t-s} \frac{\left( C\sqrt{\rho(\mu)}\, (t-s)^{\alpha/2} \right)^m}{\Gamma\left( \dfrac{m\alpha}{2} \right)}. \tag{8.9}$$

Since for $0 < \beta \leq 1$ there is a constant $C_\beta$ such that

$$\sum_{m=1}^{\infty} \frac{x^m}{\Gamma(m\beta)} \leq C_\beta x e^{2x^{1/\beta}} \qquad \text{for} \quad x \geq 0,$$

we have for $\mu \leq s \leq t$

$$\| R(t, s) \| \leq \sum_{m=1}^{\infty} \| R_m(t, s) \| \leq C\sqrt{\rho(\mu)}\, (t-s)^{\alpha/2 - 1}$$

$$\cdot \exp\left\{ - \left( \delta - C\rho(\mu)^{1/\alpha} \right)(t-s) \right\}$$

and thus, for every $0 < \vartheta' < \delta$ there is a $\mu_0 > 0$ such that for $t \geq s \geq \mu \geq \mu_0$

$$\| R(t, s) \| \leq C\sqrt{\rho(\mu)}\, (t-s)^{\alpha/2 - 1} \exp\left\{ -\vartheta'(t-s) \right\}. \tag{8.10}$$

Finally fixing $\vartheta$, $\vartheta < \vartheta' < \delta$, we find

$$\| U(t, s) \| \leq \| S_s(t-s) \| + \int_s^t \| S_\tau(t-\tau) \|\, \| R(\tau, s) \|\, d\tau$$

$$\leq C e^{-\delta(t-s)} + C\sqrt{\rho(\mu)}\, e^{-\vartheta'(t-s)}(t-s)^{\alpha/2} \leq C e^{-\vartheta(t-s)}$$

$$\text{for} \quad t \geq s \geq \mu \geq \mu_0. \tag{8.11}$$

By modifying, if necessary, the constant $C$, the inequality (8.11) holds for all $t \geq s \geq 0$ and the proof is complete.                                  $\square$

We turn now to the inhomogeneous initial value problem (8.1), and make the following additional assumption:

(F). The function $f:[0, \infty[ \to X$ satisfies a uniform Hölder condition on $[0, \infty[$, i.e., there are constants $C \geq 0$ and $0 < \gamma \leq 1$ such that

$$\| f(t) - f(s) \| \leq C |t-s|^\gamma \qquad \text{for} \quad 0 \leq s, t < \infty$$

and there is an element $f(\infty) \in X$ for which

$$\lim_{t \uparrow \infty} \| f(t) - f(\infty) \| = 0. \tag{8.12}$$

We have now,

**Theorem 8.2.** *Let* $\{ A(t) \}_{t \geq 0}$ *satisfy* $(P_1)$–$(P_3)$ *uniformly on* $[0, \infty[$. *Assume further that* $\{ A(t) \}_{t \geq 0}$ *satisfy* $(P_4)$ *and that* $f$ *satisfies* $(F)$. *If* $u$ *is the solution* *of the initial value problem* (1) *then there is an element* $u(\infty) \in X$, *independent of* $x \in X$, *such that*

$$\lim_{t \to \infty} u(t) = u(\infty), \tag{8.13}$$

$$u(\infty) \in D, A(\infty)u(\infty) = f(\infty) \tag{8.14}$$

*and*

$$\lim_{t \to \infty} u'(t) = 0. \tag{8.15}$$

PROOF. We start by showing that if $f(\infty) = 0$ then $u(t) \to 0$ as $t \to \infty$. From Theorem 7.1 it follows that

$$u(t) = U(t, 0)x + \int_0^t U(t, \sigma)f(\sigma) \, d\sigma.$$

Since by Theorem 8.1 $\| U(t, 0)x \| \le Ce^{-\vartheta t} \|x\|$ for some $\vartheta > 0$ it suffices to prove that $\int_0^t U(t, \sigma)f(\sigma) \, d\sigma \to 0$ as $t \to \infty$. Let $\| U(t, s) \| \le Ce^{-\vartheta(t-s)}$ and $\|f\|_\infty = \sup_{t \ge 0} \|f(t)\|$. Given $\varepsilon > 0$ choose $T > 0$ so that for $\sigma \ge T$, $\|f(\sigma)\| \le (\varepsilon/2)(\vartheta/C)$ and choose $T_1 > T$ such that for $t \ge T_1$, $e^{-\vartheta(t-T)} \le (\varepsilon/2)(\vartheta/C)(\|f\|_\infty + 1)^{-1}$. Then for $t > T_1$

$$\left\| \int_0^t U(t, \sigma)f(\sigma) \, d\sigma \right\| \le \left\| \int_0^T U(t, \sigma)f(\sigma) \, d\sigma \right\| + \left\| \int_T^t U(t, \sigma)f(\sigma) \, d\sigma \right\|$$

$$< \frac{\varepsilon}{2} + \frac{\varepsilon}{2} = \varepsilon$$

and therefore $u(t) \to 0$ as $t \to \infty$.

Next we show that the assumptions $(P_1)$–$(P_4)$ imply that $A(\infty)$ is invertible. Indeed, from $(P_4)$ it follows that

$$\|I - A(\infty)A(t)^{-1}\| \le \|(A(t) - A(\infty))A(0)^{-1}\| \, \|A(0)A(t)^{-1}\| \to 0$$

$$\text{as} \quad t \to \infty.$$

Therefore, for $t$ large enough $A(\infty)A(t)^{-1}$ is invertible. Denoting its inverse by $C(t)$, it is easy to see that $A(t)^{-1}C(t)$ is the inverse of $A(\infty)$.

We can now prove (8.13) and (8.14) as follows: Let $u(t)$ be the solution of (8.1) and set $u(\infty) = A(\infty)^{-1}f(\infty)$. If $v(t) = u(t) - u(\infty)$ then

$$\begin{cases} \dfrac{dv(t)}{dt} + A(t)v(t) = f(t) - A(t)u(\infty) = g(t) \\ v(0) = x - u(\infty). \end{cases} \tag{8.16}$$

But $g(t)$ is obviously Hölder continuous on $[0, \infty[$ and

$$\|g(t)\| \le \|f(t) - f(\infty)\| + \|(A(\infty) - A(t))A(\infty)^{-1}f(\infty)\|$$

$$\le \|f(t) - f(\infty)\| + \|(A(\infty) - A(t))A(0)^{-1}\| \, \|A(0)A(\infty)^{-1}f(\infty)\| \to 0 \tag{8.17}$$

as $t \to \infty$. Therefore by the first part of the proof $\|v(t)\| \to 0$ as $t \to \infty$ and hence $u(t) \to u(\infty)$ as $t \to \infty$ and the proof of (8.13) and (8.14) is complete.

To prove the last part of the claim, i.e., $u'(t) \to 0$ as $t \to \infty$ we need further estimates which we state and prove in the next lemma.

**Lemma 8.3.** *Let the conditions of Theorem 8.2 be satisfied. Then for $\mu \leq s < \tau, t$,*

$$\|A(t)S_t(t-s) - A(s)S_s(t-s)\| \leq C\sqrt{\rho(\mu)}\,(t-s)^{\alpha/2-1}e^{-\delta(t-s)}$$

$$(8.18)$$

*and*

$$\|R(t,s) - R(\tau,s)\| \leq C\sqrt{\rho(\mu)}\,(t-\tau)^{\beta}(\tau-s)^{\alpha/2-\beta-1}e^{-\vartheta(\tau-s)}$$

$$(8.19)$$

*where $0 < \vartheta < \delta$, $0 < \beta < \alpha/2$ and as before*

$$\rho(\mu) = \sup\{\|(A(t) - A(s))A(r)^{-1}\| : \mu \leq s, t < \infty, 0 \leq r < \infty\}.$$

PROOF OF LEMMA 8.3: The first estimate, (8.18), is obtained similarly to (6.18) by contour integration. Let $\Gamma = \{\lambda : |\arg \lambda| = \vartheta\}$ and $\Gamma_\delta = \{\lambda : \lambda - \delta \in \Gamma\}$ then

$$A(t)S_t(t-s) - A(s)S_s(t-s)$$

$$= \frac{1}{2\pi i}\int_{\Gamma_\delta} \lambda e^{-\lambda(t-s)}[R(\lambda : A(t)) - R(\lambda : A(s))]\,d\lambda$$

$$= \frac{1}{2\pi i}e^{-\delta(t-s)}\int_{\Gamma}(\lambda + \delta)e^{-\lambda(t-s)}$$

$$\cdot[R(\lambda + \delta : A(t)) - R(\lambda + \delta : A(s))]\,d\lambda. \qquad (8.20)$$

But,

$$\|R(\lambda + \delta : A(t)) - R(\lambda + \delta; A(s))\|$$

$$\leq \|R(\lambda + \delta : A(t))\|\,\|(A(t) - A(s))A(s)^{-1}\|$$

$$\cdot \|A(s)R(\lambda + \delta : A(s))\|$$

$$\leq C\|(A(t) - A(s))A(s)^{-1}\|\frac{1}{|\lambda + \delta|}$$

$$\leq C\sqrt{\rho(\mu)}\,(t-s)^{\alpha/2}\frac{1}{|\lambda + \delta|} \qquad (8.21)$$

where we used (8.7) and $(P_2)$. Using (8.21) to estimate (8.20) yields (8.18).

To prove (8.19) we note that if we use (8.7) to estimate $\|(A(t) - A(s))A(\tau)^{-1}\|$ and follow the proof of Lemma 6.4 we find that for $\mu \leq s < \tau \leq t$

$$\|R_1(t,s) - R_1(\tau,s)\| \leq C\sqrt{\rho(\mu)}\,(t-\tau)^{\beta}(\tau-s)^{\alpha/2-\beta-1}e^{-\delta(\tau-s)}$$

$$(8.22)$$

where $0 < \beta < \alpha/2$. Recalling that (see (8.8))

$$\|R_1(t, s)\| \le C\sqrt{\rho(\mu)}\,(t - s)^{\alpha/2 - 1}e^{-\delta(t-s)}$$

$$\le C_{\beta, \vartheta}\sqrt{\rho(\mu)}\,(t - s)^{\beta - 1}e^{-\vartheta(t-s)} \qquad (8.23)$$

and

$$\|R(t, s)\| \le C\sqrt{\rho(\mu)}\,(t - s)^{\alpha/2 - 1}e^{-\vartheta(t-s)} \qquad (8.24)$$

we find,

$$\left\|\int_\tau^t R_1(t, \sigma)R(\sigma, s)\,d\sigma\right\| \le C\rho(\mu)e^{-\vartheta(t-s)}(\tau - s)^{\alpha/2 - 1}\int_\tau^t (t - \sigma)^{\beta - 1}\,d\sigma$$

$$\le C\sqrt{\rho(\mu)}\,(t - \tau)^\beta(\tau - s)^{\alpha/2 - 1}e^{-\vartheta(t-s)}. \quad (8.25)$$

Estimating $\|R(t, s) - R(\tau, s)\|$ as in the proof of Corollary 6.5, using (8.22), (8.24) and (8.25) yields (8.19).                                                      □


We now return to the proof of Theorem 8.2. For every $t \ge s > 0$ the solution $u$ of the initial value problem (8.1) can be written as

$$u(t) = U(t, s)u(s) + \int_s^t U(t, \sigma)f(\sigma)\,d\sigma.$$

Since by (6.10)

$$\left\|\frac{\partial}{\partial t}U(t, s)u(s)\right\| \le \|A(t)U(t, s)\|\,\|u(s)\| \le \frac{C}{t - s}$$

it suffices to prove that the norm of

$$\frac{\partial}{\partial t}\int_s^t U(t, \sigma)f(\sigma)\,d\sigma = \frac{\partial}{\partial t}\int_s^t S_\sigma(t - \sigma)f(\sigma)\,d\sigma + \frac{\partial}{\partial t}\int_s^t W(t, \sigma)f(\sigma)\,d\sigma$$

$$(8.26)$$

can be made as small as we wish by choosing $s$ and $t > s$ large enough. To prove this claim we treat each term on the right-hand side of (8.26) separately.

In order to estimate the first term we denote

$$\delta(\mu) = \sup\{\|f(t) - f(s)\| : \mu \le s, t < \infty\}.$$

From the assumption (F) it follows that $\delta(\mu) \to 0$ as $\mu \to \infty$ and

$$\|f(t) - f(s)\| \le \sqrt{\delta(\mu)}\,|t - s|^{\gamma/2} \qquad \text{for} \quad \mu \le s, t < \infty. \quad (8.27)$$

A simple computation yields

$$\frac{\partial}{\partial t}\int_s^t S_\sigma(t-\sigma)f(\sigma)\,d\sigma = \int_s^t[A(t)S_t(t-\sigma) - A(\sigma)S_\sigma(t-\sigma)]f(\sigma)\,d\sigma$$

$$- \int_s^t A(t)S_t(t-\sigma)[f(\sigma) - f(t)]\,d\sigma$$

$$+ S_t(t-s)f(t) \qquad\qquad (8.28)$$

and therefore estimating each of the three terms on the right-hand side of (8.28), separately, using (8.18), (8.27) and (8.4) we get

$$\left\|\frac{\partial}{\partial t}\int_s^t S_\sigma(t-\sigma)f(\sigma)\,d\sigma\right\| \le C\|f\|_\infty\sqrt{\rho(\mu)} + C\sqrt{\delta(\mu)} + C\|f\|_\infty e^{-\delta(t-s)}.$$

$$(8.29)$$

To estimate the second term on the right-hand side of (8.26) we note that by the proof of Theorem 7.1 we have

$$\frac{\partial}{\partial t}\int_s^t W(t,\sigma)f(\sigma)\,d\sigma = \int_s^t\frac{\partial}{\partial t}W(t,\sigma)f(\sigma)\,d\sigma \qquad (8.30)$$

and from (6.35)

$$\frac{\partial W(t,s)}{\partial t} = \int_s^t[A(t)S_t(t-\tau) - A(\tau)S_\tau(t-\tau)]R(\tau,s)\,d\tau$$

$$+ \int_s^t A(t)S_t(t-\tau)[R(t,s) - R(\tau,s)]\,d\tau + S_t(t-s)R(t,s).$$

Estimating each one of the terms on the right-hand side, using (8.18), (8.24), (8.19), (8.5) and (8.4) yields

$$\left\|\frac{\partial W(t,s)}{\partial t}\right\| \le C\sqrt{\rho(\mu)}\,(t-s)^{\alpha/2-1}e^{-\vartheta(t-s)} \qquad (8.31)$$

and therefore,

$$\left\|\int_s^t\frac{\partial}{\partial t}W(t,\sigma)f(\sigma)\,d\sigma\right\| \le C\|f\|_\infty\sqrt{\rho(\mu)}. \qquad (8.32)$$

Combining (8.29) and (8.32) and noting that $\rho(\mu) \to 0$ and $\delta(\mu) \to 0$ as $\mu \to \infty$ yields that for any $\varepsilon > 0$ there is a $\mu$ such that if $t > s \ge \mu$

$$\left\|\frac{\partial}{\partial t}\int_s^t U(t,\sigma)f(\sigma)\,d\sigma\right\| < \varepsilon + C\|f\|_\infty e^{-\delta(t-s)}$$

which concludes the proof of Theorem 8.2.                                           □

Theorem 8.2 shows that if as $t \to \infty$, $A(t)$ converges to $A(\infty)$ and $f(t)$ to $f(\infty)$, then the solution $u(t)$ of the initial value problem (1) converges to a limit $u(\infty)$ as $t \to \infty$. In order to get more detailed information on the convergence of $u(t)$ to $u(\infty)$ more must be known on the convergence of

$A(t)$ to $A(\infty)$ and $f(t)$ to $f(\infty)$. We conclude this section with one result in this direction.

We will make the following assumptions:

$(A_n)$ The operator $A(t)$ has an expansion

$$A(t) = A_0 + \frac{1}{t}A_1 + \frac{1}{t^2}A_2 + \cdots + \frac{1}{t^n}A_n + \frac{1}{t^n}B_n(t) \qquad (8.33)$$

where $A_0$ is a densely defined closed linear operator for which the resolvent set $\rho(A_0)$ satisfies $\rho(A_0) \supset \{\lambda : \operatorname{Re}\lambda \leq 0\}$ and

$$\|R(\lambda : A_0)\| \leq \frac{M_0}{|\lambda| + 1} \qquad \text{for} \quad \lambda \in \rho(A_0). \qquad (8.34)$$

The operators $A_k$, $1 \leq k \leq n$ and $B_n(t)$ for $t \geq 0$ are closed linear operators satisfying $D(A_k) \supset D(A_0)$ and $D(B_n(t)) \supset D(A_0)$. Furthermore, the bounded linear operators $B_n(t)A_0^{-1}$ satisfy

$$\|B_n(t)A_0^{-1} - B_n(s)A_0^{-1}\| \leq C|t - s|^\rho \qquad (8.35)$$

for some $0 < \rho \leq 1$, $C > 0$ and

$$\lim_{t \to \infty} \|B_n(t)A_0^{-1}\| = 0 \qquad (8.36)$$

and

$(F_n)$ The function $f(t)$ has the following expansion

$$f(t) = f_0 + \frac{1}{t}f_1 + \frac{1}{t^2}f_2 + \cdots + \frac{1}{t^n}f_n + \frac{1}{t^n}\varphi_n(t) \qquad (8.37)$$

where $\varphi_n(t)$ is Hölder continuous in $t$ and

$$\lim_{t \to \infty} \|\varphi_n(t)\| = 0. \qquad (8.38)$$

We note that if $\{A(t)\}_{t \geq t_0}$ satisfies $(A_n)$ for some $n \geq 0$ it also satisfies $(A_k)$ with $0 \leq k \leq n$ where

$$B_k(t) = \sum_{l=k+1}^{n} t^{-l+k}A_l + t^{-n+k}B_n(t).$$

Furthermore if $A(t)$ satisfies $(A_n)$ with $n \geq 1$ so does $A(t) + (\alpha/t)I$ where $I$ is the identity operator. Finally if $f(t)$ satisfies $(F_n)$ it also satisfies $(F_k)$ with $0 \leq k \leq n$ with the appropriate definition of $\varphi_k(t)$.

We proceed by showing that the assumptions $(A_n)$ imply the existence of a $t_0 > 0$ such that the family $\{A(t)\}_{t \geq t_0}$ satisfies the necessary conditions for the existence of a unique solution $u(t)$ to the initial value problem

$$\begin{cases} \dfrac{du(t)}{dt} + A(t)u(t) = f(t) & \text{for } t > t_0 \\ u(t_0) = x \end{cases} \qquad (8.39)$$

where $f$ satisfies the condition (F). More precisely we have:

**Lemma 8.4.** *If* $\{A(t)\}_{t \geq 0}$ *satisfies* $(A_n)$ *with* $n \geq 0$ *then there is a* $t_0 > 0$ *such that* $\{A(t)\}_{t \geq t_0}$ *satisfies;*

(i) *For every* $t \geq t_0$, *the resolvent* $R(\lambda : A(t))$ *of* $A(t)$ *exists for all* Re $\lambda \leq 0$ *and*

$$\|R(\lambda : A(t))\| \leq \frac{M}{|\lambda| + 1} \qquad \text{for all} \quad \lambda \text{ with } \text{Re } \lambda \leq 0.$$

(ii) *There exist constants* $L$ *and* $0 < \alpha \leq 1$ *such that*

$$\|(A(t) - A(s))A(\tau)^{-1}\| \leq L|t - s|^\alpha \qquad \text{for} \quad t_0 \leq t, s, \tau.$$

(iii) *The operators* $\|A(t)A(s)^{-1}\|$ *are uniformly bounded for* $t_0 \leq s, t < \infty$ *and*

$$\lim_{t \to \infty} \|(A(t) - A_0)A_0^{-1}\| = 0. \qquad (8.40)$$

PROOF. (i) Set $Q(t) = A(t) - A_0$, from the closed graph theorem it follows that for every $t > 0$ and $\lambda \in \rho(A_0)$, $Q(t)R(\lambda : A_0)$ is a bounded linear operator. Furthermore, for $\lambda$ with Re $\lambda \leq 0$ we have

$$\|Q(t)R(\lambda : A_0)\| \leq \sum_{l=1}^{n} \gamma_l t^{-l} + \beta_n(t)t^{-n} \qquad (8.41)$$

where $\gamma_l = (M_0 + 1)\|A_l A_0^{-1}\|$ and $\beta_n(t) = (M_0 + 1)\|B_n(t)A_0^{-1}\|$. Therefore, there is a $t_0 > 0$ such that for $t \geq t_0$ and Re $\lambda \leq 0$, $\|Q(t)R(\lambda : A_0)\| < \frac{1}{2}$. Fix such a $t_0 > 0$, let $\lambda$ be such that Re $\lambda \leq 0$ and consider

$$\lambda I - A(t) = [I - Q(t)R(\lambda : A_0)](\lambda I - A_0). \qquad (8.42)$$

For $t \geq t_0$ the operator on the right-hand side of (8.42) is invertible and

$$\|R(\lambda : A(t))\| \leq \|R(\lambda : A_0)\| \, \|(I - Q(t)R(\lambda : A_0))^{-1}\| \leq \frac{2M_0}{|\lambda| + 1}$$

for all $\lambda$ with Re $\lambda \leq 0$. In particular it follows that for $t \geq t_0$, $A(t)^{-1}$ exists.

(ii) Using the Hölder continuity of $B_n(t)A_0^{-1}$ it follows easily that for $t \geq t_0 > 0$ the operator $A(t)A_0^{-1}$ is Hölder continuous with exponent $0 < \rho \leq 1$. For $\tau \geq t_0$, $\|Q(\tau)A_0^{-1}\| < \frac{1}{2}$ and consequently the operator $I + Q(\tau)A_0^{-1}$ is invertible and its inverse $A_0 A(\tau)^{-1}$ has norm less or equal to 2. Therefore,

$$\|(A(t) - A(s))A(\tau)^{-1}\| \leq \|(A(t) - A(s))A_0^{-1}\| \, \|A_0 A(\tau)^{-1}\|$$
$$\leq C|t - s|^\rho.$$

(iii) For $t, s \geq t_0$ we have

$$\|A(t)A(s)^{-1}\| \leq \|A(t)A_0^{-1}\| \|A_0A(s)^{-1}\| \leq 2\|A(t)A_0^{-1}\|$$
$$= 2\|I + Q(t)A_0^{-1}\| \leq 3.$$

Finally choosing $\lambda = 0$ in (8.41) and letting $t \to \infty$ yields (8.40).     □

Lemma 8.4 implies that if $A(t)$, $t \geq 0$, satisfies $(A_n)$ with $n \geq 0$ then $\{A(t)\}_{t \geq t_0}$ satisfies on $[t_0, \infty[$ the assumptions $(P_1)$–$(P_4)$ with $A(\infty) = A_0$. Moreover, it is easy to check that if $f$ satisfies $(F_n)$ with $n \geq 0$ then it satisfies the assumption $(F)$ with $f(\infty) = f_0$ and therefore under these assumptions the initial value problem (8.39) has a unique solution $u$ on $[t_0, \infty[$.

**Theorem 8.5.** *Let $A(t)$ satisfy the conditions $(A_n)$ with some $n > 0$ and let $f$ satisfy the condition $(F_n)$ with the same $n > 0$. If $u$ is the solution of the initial value problem (8.39) then for $t \geq t_0$,*

$$u(t) = u_0 + \frac{1}{t}u_1 + \frac{1}{t^2}u_2 + \cdots + \frac{1}{t^n}u_n + \frac{1}{t^n}v_n(t) \qquad (8.43)$$

*where $v_n(t) \to 0$ as $t \to \infty$ and*

$$A_0u_0 = f_0 \qquad (8.44)$$

$$A_0u_k - (k-1)u_{k-1} + \sum_{\nu=1}^{k} A_\nu u_{k-\nu} = f_k \qquad \text{for} \quad 1 \leq k \leq n. \qquad (8.45)$$

PROOF. For $n = 0$ Theorem 8.5 coincides, with the obvious changes of notations, with Theorem 8.2 and therefore it is true for $n = 0$. Assume that it is true for $(m - 1) < n$. Then the equations (8.43), (8.44) and (8.45) hold with $n$ replaced by $m - 1$. We will show that in this case the theorem is true also for $m$. Set

$$u(t) = u_0 + \frac{1}{t}u_1 + \cdots + \frac{1}{t^{m-1}}u_{m-1} + \frac{1}{t^m}w(t) \qquad (8.46)$$

where $u_k$, $0 \leq k \leq m - 1$ are determined consecutively by (8.45). Substituting (8.46) into the differential equation

$$\frac{du(t)}{dt} + A(t)u(t) = f(t) \qquad (8.47)$$

we get

$$\frac{1}{t^m}\left[\frac{dw}{dt} + \left(A(t) - \frac{m}{t}I\right)w\right]$$

$$= \frac{1}{t^m}\left[(m-1)u_{m-1} - \sum_{\nu=1}^{m} A_\nu u_{m-\nu} + f_m - B_m(t)u_0 + \varphi_m(t) + \frac{1}{t}g(t)\right]$$

$$(8.48)$$

where

$$B_m(t) = \sum_{l=m+1}^{n} t^{-l+m} A_l + t^{-n+m} B_n(t),$$

$$\varphi_m(t) = \sum_{l=m+1}^{n} t^{-l+m} f_l + t^{-n+m} \varphi_n(t)$$

and $g(t)$ is a finite sum of terms of the form $t^{-k}(A_l u_j + B_m(t)u_i)$ with $0 \le k, j \le m - 1$ and $0 \le i, l \le m$. It is easy to check that for $t \ge t_0 > 0$, $t^{-1}g(t)$ is Hölder continuous in $t$. Multiplying both sides of (8.48) by $t^m$ we find

$$\frac{dw}{dt} + \left( A(t) - \frac{m}{t}I \right)w = (m-1)u_{m-1} - \sum_{\nu=1}^{m} A_\nu u_{m-\nu} + f_m$$

$$+ \left[ \varphi_m(t) - B_m(t)u_0 + t^{-1}g(t) \right].$$

The term depending on $t$ on the right-hand side is Hölder continuous and tends to zero as $t \to \infty$. The operator $A(t) - (m/t)I$ clearly satisfies $(A_0)$ and therefore by our theorem with $n = 0$,

$$w(t) = u_m + v_m(t) \tag{8.49}$$

where

$$\lim_{t \to \infty} \|v_m(t)\| = 0.$$

Substituting (8.49) into (8.46) gives the desired result for $m$. The theorem follows by induction.                                                                                    $\square$

# CHAPTER 6

# Some Nonlinear Evolution Equations

## 6.1. Lipschitz Perturbations of Linear Evolution Equations

In this section we will study the following semilinear initial value problem:

$$\begin{cases} \dfrac{du(t)}{dt} + Au(t) = f(t, u(t)), & t > t_0 \\ u(t_0) = u_0 \end{cases} \tag{1.1}$$

where $-A$ is the infinitesimal generator of a $C_0$ semigroup $T(t)$, $t \geq 0$, on a Banach space $X$ and $f:[t_0, T] \times X \to X$ is continuous in $t$ and satisfies a Lipschitz condition in $u$.

Most of the results of this and the following sections, in which $A$ is assumed to be independent of $t$ can be easily extended to the case where $A$ depends on $t$ in a way that insures the existence of an evolution system $U(t, s)$, $0 \leq s \leq t \leq T$, for the family $\{A(t)\}_{t \in [0, T]}$. We will not deal with these extensions here and as a consequence the following sections (Section 6.1–6.3) are independent of the results of Chapter 5.

The initial value problem (6.1) does not necessarily have a solution of any kind. However, if it has a classical or strong solution (see Definition 4.2.1) then the argument given at the beginning of Section 4.2 shows that this solution $u$ satisfies the integral equation

$$u(t) = T(t - t_0)u_0 + \int_{t_0}^{t} T(t - s)f(s, u(s)) \, ds. \tag{1.2}$$

It is therefore natural to define,

**Definition 1.1.** A continuous solution $u$ of the integral equation (1.2) will be called a *mild solution* of the initial value problem (1.1).

We start with the following classical result which assures the existence and uniqueness of mild solutions of (1.1) for Lipschitz continuous functions $f$.

**Theorem 1.2.** *Let* $f : [t_0, T] \times X \to X$ *be continuous in* $t$ *on* $[t_0, T]$ *and uniformly Lipschitz continuous (with constant $L$) on $X$. If $-A$ is the infinitesimal generator of a $C_0$ semigroup $T(t)$, $t \geq 0$, on $X$ then for every $u_0 \in X$ the initial value problem* (1.1) *has a unique mild solution* $u \in C([t_0, T] : X)$. *Moreover, the mapping* $u_0 \to u$ *is Lipschitz continuous from* $X$ *into* $C([t_0, T] : X)$.

PROOF. For a given $u_0 \in X$ we define a mapping

$$F : C([t_0, T] : X) \to C([t_0, T] : X)$$

by

$$(Fu)(t) = T(t - t_0)u_0 + \int_{t_0}^{t} T(t - s)f(s, u(s)) \, ds, \qquad t_0 \leq t \leq T.$$

$$(1.3)$$

Denoting by $\|u\|_\infty$ the norm of $u$ as an element of $C([t_0, T] : X)$ it follows readily from the definition of $F$ that

$$\|(Fu)(t) - (Fv)(t)\| \leq ML(t - t_0)\|u - v\|_\infty \qquad (1.4)$$

where $M$ is a bound of $\|T(t)\|$ on $[t_0, T]$. Using (1.3), (1.4) and induction on $n$ it follows easily that

$$\|(F^n u)(t) - (F^n v)(t)\| \leq \frac{(ML(t - t_0))^n}{n!} \|u - v\|_\infty$$

whence

$$\|F^n u - F^n v\| \leq \frac{(MLT)^n}{n!} \|u - v\|_\infty. \qquad (1.5)$$

For $n$ large enough $(MLT)^n/n! < 1$ and by a well known extension of the contraction principle $F$ has a unique fixed point $u$ in $C([t_0, T] : X)$. This fixed point is the desired solution of the integral equation (1.2).

The uniqueness of $u$ and the Lipschitz continuity of the map $u_0 \to u$ are consequences of the following argument. Let $v$ be a mild solution of (1.1) on

$[t_0, T]$ with the initial value $v_0$. Then,

$$\|u(t) - v(t)\| \le \|T(t - t_0)u_0 - T(t - t_0)v_0\|$$
$$+ \int_{t_0}^{t} \|T(t - s)(f(s, u(s)) - f(s, v(s)))\| \, ds$$
$$\le M\|u_0 - v_0\| + ML\int_{t_0}^{t}\|u(s) - v(s)\| \, ds$$

which implies, by Gronwall's inequality, that

$$\|u(t) - v(t)\| \le Me^{ML(T - t_0)}\|u_0 - v_0\|$$

and therefore

$$\|u - v\|_\infty \le Me^{ML(T - t_0)}\|u_0 - v_0\|$$

which yields both the uniqueness of $u$ and the Lipschitz continuity of the map $u_0 \to u$.                                                                          □

It is not difficult to see that if $g \in C([t_0, T]: X)$ and in the proof of Theorem 1.2 we modify the definition of $F$ to

$$(Fu)(t) = g(t) + \int_{t_0}^{t} T(t - s)f(s, u(s)) \, ds$$

we obtain the following slightly more general result.

**Corollary 1.3.** *If $A$ and $f$ satisfy the conditions of Theorem 1.2 then for every $g \in C([t_0, T]; X)$ the integral equation*

$$w(t) = g(t) + \int_{t_0}^{t} T(t - s)f(s, w(s)) \, ds \tag{1.6}$$

*has a unique solution $w \in C([t_0, T]: X)$.*

The uniform Lipschitz condition of the function $f$ in Theorem 1.2 assures the existence of a global (i.e. defined on all of $[t_0, T]$) mild solution of (1.1). If we assume that $f$ satisfies only a local Lipschitz condition in $u$, uniformly in $t$ on bounded intervals, that is, for every $t' \ge 0$ and constant $c \ge 0$ there is a constant $L(c, t')$ such that

$$\|f(t, u) - f(t, v)\| \le L(c, t')\|u - v\| \tag{1.7}$$

holds for all $u, v \in X$ with $\|u\| \le c$, $\|v\| \le c$ and $t \in [0, t']$, then we have the following local version of Theorem 1.2.

**Theorem 1.4.** *Let $f: [0, \infty[ \times X \to X$ be continuous in $t$ for $t \ge 0$ and locally Lipschitz continuous in $u$, uniformly in $t$ on bounded intervals. If $-A$ is the infinitesimal generator of a $C_0$ semigroup $T(t)$ on $X$ then for every $u_0 \in X$*

*there is a $t_{max} \le \infty$ such that the initial value problem*

$$\begin{cases} \dfrac{du(t)}{dt} + Au(t) = f(t, u(t)), & t \ge 0 \\ u(0) = u_0 \end{cases} \qquad (1.8)$$

*has a unique mild solution $u$ on $[0, t_{max}[$. Moreover, if $t_{max} < \infty$ then*

$$\lim_{t \uparrow t_{max}} \|u(t)\| = \infty.$$

PROOF. We start by showing that for every $t_0 \ge 0$, $u_0 \in X$, the initial value problem (1.1) has, under the assumptions of our theorem, a unique mild solution $u$ on an interval $[t_0, t_1]$ whose length is bounded below by

$$\delta(t_0, \|u_0\|) = \min \left\{ 1, \frac{\|u_0\|}{K(t_0)L(K(t_0), t_0 + 1) + N(t_0)} \right\} \qquad (1.9)$$

where $L(c, t)$ is the local Lipschitz constant of $f$ as defined by (1.7), $M(t_0) = \sup\{\|T(t)\| : 0 \le t \le t_0 + 1\}$, $K(t_0) = 2\|u_0\|M(t_0)$ and $N(t_0) = \max\{\|f(t, 0)\| : 0 \le t \le t_0 + 1\}$. Indeed, let $t_1 = t_0 + \delta(t_0, \|u_0\|)$ where $\delta(t_0, \|u_0\|)$ is given by (1.9). The mapping $F$ defined by (1.3) maps the ball of radius $K(t_0)$ centered at 0 of $C([t_0, t_1] : X)$ into itself. This follows from the estimate

$\|(Fu)(t)\|$

$$\le M(t_0)\|u_0\| + \int_{t_0}^{t} \|T(t - s)\|(\|f(s, u(s)) - f(s, 0)\| + \|f(s, 0)\|) \, ds$$

$$\le M(t_0)\|u_0\| + M(t_0)K(t_0)L(K(t_0), t_0 + 1)(t - t_0)$$
$$\quad + M(t_0)N(t_0)(t - t_0)$$

$$\le M(t_0)\{\|u_0\| + K(t_0)L(K(t_0), t_0 + 1)(t - t_0) + N(t_0)(t - t_0)\}$$

$$\le 2M(t_0)\|u_0\| = K(t_0)$$

where the last inequality follows from the definition of $t_1$. In this ball, $F$ satisfies a uniform Lipschitz condition with constant $L = L(K(t_0), t_0 + 1)$ and thus as in the proof of Theorem 1.2 it possesses a unique fixed point $u$ in the ball. This fixed point is the desired solution of (1.1) on the interval $[t_0, t_1]$.

From what we have just proved it follows that if $u$ is a mild solution of (1.8) on the interval $[0, \tau]$ it can be extended to the interval $[0, \tau + \delta]$ with $\delta > 0$ by defining on $[\tau, \tau + \delta]$, $u(t) = w(t)$ where $w(t)$ is the solution of the integral equation

$$w(t) = T(t - \tau)u(\tau) + \int_{\tau}^{t} T(t - s)f(s, w(s)) \, ds, \qquad \tau \le t \le \tau + \delta.$$

$$(1.10)$$

Moreover, $\delta$ depends only on $\|u(\tau)\|$, $K(\tau)$ and $N(\tau)$.

Let $[0, t_{max}[$ be the maximal interval of existence of the mild solution $u$ of (1.8). If $t_{max} < \infty$ then $\lim_{t \to t_{max}} \|u(t)\| = \infty$ since otherwise there is a sequence $t_n \uparrow t_{max}$ such that $\|u(t_n)\| \leq C$ for all $n$. This would imply by what we have just proved that for each $t_n$, near enough to $t_{max}$, $u$ defined on $[0, t_n]$ can be extended to $[0, t_n + \delta]$ where $\delta > 0$ is independent of $t_n$ and hence $u$ can be extended beyond $t_{max}$ contradicting the definition of $t_{max}$.

To prove the uniqueness of the local mild solution $u$ of (1.8) we note that if $v$ is a mild solution of (1.8) then on every closed interval $[0, t_0]$ on which both $u$ and $v$ exist they coincide by the uniqueness argument given at the end of the proof of Theorem 1.2. Therefore, both $u$ and $v$ have the same $t_{max}$ and on $[0, t_{max}[, u \equiv v$.                                          $\square$

It is well known that in general, if $f$ just satisfies the conditions of Theorem 1.2 or Theorem 1.4 the mild solution of (1.1) need not be a classical solution or even a strong solution of (1.1). A sufficient condition for the mild solution of (1.1) to be a classical solution is given next.

**Theorem 1.5** (Regularity). *Let $-A$ be the infinitesimal generator of a $C_0$ semigroup $T(t)$ on $X$. If $f : [t_0, T] \times X \to X$ is continuously differentiable from $[t_0, T] \times X$ into $X$ then the mild solution of (1.1) with $u_0 \in D(A)$ is a classical solution of the initial value problem.*

PROOF. We note first that the continuous differentiability of $f$ from $[t_0, T]$ $\times X$ into $X$ implies that $f$ is continuous in $t$ and Lipschitz continuous in $u$, uniformly in $t$ on $[t_0, T]$. Therefore the initial value problem (1.1) possesses a unique mild solution $u$ on $[t_0, T]$ by Theorem 1.2. Next we show that this mild solution is continuously differentiable on $[t_0, T]$. To this end we set $B(s) = (\partial / \partial u) f(s, u)$ and

$$g(t) = T(t - t_0) f(t_0, u(t_0)) - AT(t - t_0) u_0$$

$$+ \int_{t_0}^{t} T(t - s) \frac{\partial}{\partial s} f(s, u(s))\, ds. \tag{1.11}$$

From our assumptions it follows that $g \in C([t_0, T] : X)$ and that the function $h(t, u) = B(t)u$ is continuous in $t$ from $[t_0, T]$ into $X$ and uniformly Lipschitz continuous in $u$ since $s \to B(s)$ is continuous from $[t_0, T]$ into $B(X)$. Let $w$ be the solution of the integral equation

$$w(t) = g(t) + \int_{t_0}^{t} T(t - s) B(s) w(s)\, ds. \tag{1.12}$$

The existence and uniqueness of $w \in C([t_0, T] : X)$ follows from Corollary 1.3. Moreover, from our assumptions we have

$$f(s, u(s + h)) - f(s, u(s)) = B(s)(u(s + h) - u(s)) + \omega_1(s, h) \tag{1.13}$$

and

$$f(s + h, u(s + h)) - f(s, u(s + h))$$
$$= (\partial/\partial s)f(s, u(s + h)) \cdot h + \omega_2(s, h)$$
(1.14)

where $h^{-1}\|\omega_i(s, h)\| \to 0$ as $h \to 0$ uniformly on $[t_0, T]$ for $i = 1, 2$. If $w_h(t) = h^{-1}(u(t + h) - u(t)) - w(t)$ then from the definition of $u$, (1.12), (1.13) and (1.14) we obtain

$$w_h(t) = \left[h^{-1}(T(t + h - t_0)u_0 - T(t - t_0)u_0) + AT(t - t_0)u_0\right]$$
$$+ \frac{1}{h}\int_{t_0}^t T(t - s)(\omega_1(s, h) + \omega_2(s, h))\, ds$$
$$+ \int_{t_0}^t T(t - s)\left(\frac{\partial}{\partial s}f(s, u(s + h)) - \frac{\partial}{\partial s}f(s, u(s))\right)\, ds$$
$$+ \left[\frac{1}{h}\int_{t_0}^{t_0 + h} T(t + h - s)f(s, u(s))\, ds - T(t - t_0)f(t_0, u(t_0))\right]$$
$$+ \int_{t_0}^t T(t - s)B(s)w_h(s)\, ds.$$
(1.15)

It is not difficult to see that the norm of each one of the four first terms on the right-hand side of (1.15) tends to zero as $h \to 0$. Therefore we have

$$\|w_h(t)\| \le \varepsilon(h) + M\int_{t_0}^t \|w_h(s)\|\, ds$$
(1.16)

where $M = \max\{\|T(t - s)\|\, \|B(s)\| : t_0 \le s \le T\}$ and $\varepsilon(h) \to 0$ as $h \to 0$. From (1.16) it follows by Gronwall's inequality that $\|w_h(t)\| \le \varepsilon(h)e^{(T - t_0)M}$ and therefore $\|w_h(t)\| \to 0$ as $h \to 0$. This implies that $u(t)$ is differentiable on $[t_0, T]$ and that its derivative is $w(t)$. Since $w \in C([t_0, T] : X)$, $u$ is continuously differentiable on $[t_0, T]$.

Finally, to show that $u$ is the classical solution of (1.1) we note that from the continuous differentiability of $u$ and the assumptions on the differentiability of $f$ it follows that $s \to f(s, u(s))$ is continuously differentiable on $[t_0, T]$. From Corollary 4.2.5 it then follows that

$$v(t) = T(t - t_0)u_0 + \int_{t_0}^t T(t - s)f(s, u(s))\, ds$$
(1.17)

is the classical solution of the initial value problem

$$\begin{cases} \dfrac{dv(t)}{dt} + Av(t) = f(t, u(t)) \\ v(t_0) = u_0 . \end{cases}$$
(1.18)

But, by definition, $u$ is a mild solution of (1.18) and by the uniqueness of mild solutions of (1.18) it follows that $u = v$ on $[t_0, T]$. Thus, $u$ is a classical solution of the initial value problem (1.1).                                    $\square$

In general if $f:[t_0, T] \times X \to X$ is just Lipschitz continuous in both variables i.e.,

$$\|f(t_1, x_1) - f(t_2, x_2)\| \le C(|t_1 - t_2| + \|x_1 - x_2\|), \qquad t_1, t_2 \in [t_0, T]$$

$$(1.19)$$

the mild solution of (1.1) need not be a strong solution of the initial value problem. However, if $X$ is reflexive, the Lipschitz continuity of $f$ suffices to assure that the mild solution $u$ with initial data $u_0 \in D(A)$ is a strong solution. Indeed we have:

**Theorem 1.6.** *Let* $-A$ *be the infinitesimal generator of a* $C_0$ *semigroup* $T(t)$ *on a reflexive Banach space* $X$. *If* $f:[t_0, T] \times X \to X$ *is Lipschitz continuous in both variables,* $u_0 \in D(A)$ *and* $u$ *is the mild solution of the initial value problem* (1.1) *then* $u$ *is the strong solution of this initial value problem.*

PROOF. Let $\|T(t)\| \le M$ and $\|f(t, u(t))\| \le N$ for $t_0 \le t \le T$ and let $f$ satisfy (1.19). For $0 < h < t - t_0$ we have

$$u(t + h) - u(t) = T(t + h - t_0)u_0 - T(t - t_0)u_0$$
$$+ \int_{t_0}^{t_0 + h} T(t + h - s)f(s, u(s))\, ds$$
$$+ \int_{t_0}^{t} T(t - s)[f(s + h, u(s + h)) - f(s, u(s))]\, ds$$

and therefore,

$$\|u(t + h) - u(t)\| \le hM\|Au_0\| + hMN$$
$$+ MC\int_{t_0}^{t}(h + \|u(s + h) - u(s)\|)\, ds$$
$$\le C_1 h + MC\int_{t_0}^{t}\|u(s + h) - u(s)\|\, ds$$

which by Gronwall's inequality implies

$$\|u(t + h) - u(t)\| \le C_1 e^{TMC} h \qquad (1.20)$$

and $u$ is Lipschitz continuous.

The Lipschitz continuity of $u$ combined with the Lipschitz continuity of $f$ imply that $t \to f(t, u(t))$ is Lipschitz continuous on $[t_0, T]$. From Corollary 4.2.11 it then follows that the initial value problem

$$\begin{cases} \dfrac{dv}{dt} + Av = f(t, u(t)) \\ v(t_0) = u_0 \end{cases} \qquad (1.21)$$

has a unique strong solution $v$ on $[t_0, T]$ satisfying

$$v(t) = T(t - t_0)u_0 + \int_{t_0}^{t} T(t - s)f(s, u(s))\, ds = u(t)$$

and so $u$ is a strong solution of (1.1).                    □

We conclude this section with an application of Theorem 1.2 which provides us with a classical solution of the initial value problem (1.1). Let $-A$ be the infinitesimal generator of the $C_0$ semigroup $T(t)$ on $X$. We endow the domain $D(A)$ of $A$ with the graph norm, that is, for $x \in D(A)$ we define $|x|_A = \|x\| + \|Ax\|$. It is not difficult to show that $D(A)$ with the norm $| \cdot |_A$ is a Banach space which we denote by $Y$. The completeness of $Y$ is a direct consequence of the closedness of $A$. Clearly $Y \subset X$ and since $T(t): D(A) \to D(A)$, $T(t)$, $t \geq 0$ is a semigroup on $Y$ which is easily seen to be a $C_0$ semigroup on $Y$.

**Theorem 1.7.** *Let $f:[t_0, T] \times Y \to Y$ be uniformly Lipschitz in $Y$ and for each $y \in Y$ let $f(t, y)$ be continuous from $[t_0, T]$ into $Y$. If $u_0 \in D(A)$ then the initial value problem* (1.1) *has a unique classical solution on $[t_0, T]$.*

PROOF. We apply Theorem 1.2 in $Y$ and obtain a function $u \in C([t_0, T]: Y)$ satisfying in $Y$ and a fortiori in $X$,

$$u(t) = T(t - t_0)u_0 + \int_{t_0}^{t} T(t - s)f(s, u(s)) \, ds. \qquad (1.22)$$

Let $g(s) = f(s, u(s))$. From our assumptions it follows that $g(s) \in D(A)$ for $s \in [t_0, T]$ and that $s \to Ag(s)$ is continuous in $X$. Therefore it follows from Corollary 4.2.6 that the initial value problem

$$\begin{cases} \dfrac{dv}{dt} + Av = g(t) \\ v(t_0) = u_0 \end{cases} \qquad (1.23)$$

has a unique classical solution $v$ on $[t_0, T]$. This solution is then clearly also a mild solution of (1.23) and therefore

$$v(t) = T(t - t_0)u_0 + \int_{t_0}^{t} T(t - s)g(s) \, ds$$

$$= T(t - t_0)u_0 + \int_{t_0}^{t} T(t - s)f(s, u(s)) \, ds = u(t)$$

and $u$ is a classical solution of (1, 1) on $[t_0, T]$.  □

If in the previous theorem we assume only that $f:[t_0, T] \times Y \to Y$ is locally Lipschitz continuous in $Y$ uniformly in $[t_0, T]$ we obtain, using Theorem 1.4, that for every $u_0 \in D(A)$ the initial value problem possesses a classical solution on a maximal interval $[t_0, t_{\max}[$ and if $t_{\max} < T$ then

$$\lim_{t \uparrow t_{\max}} (\|u(t)\| + \|Au(t)\|) = \infty. \qquad (1.24)$$

We note that in this situation it may well be that $\|u(t)\|$ is bounded on

$[t_0, t_{\max}[$ and only $\|Au(t)\| \to \infty$ as $t \uparrow t_{\max}$. This is indeed the case in many applications to partial differential equations.

## 6.2.  Semilinear Equations with Compact Semigroups

We continue our study of the semilinear initial value problem

$$\begin{cases} \dfrac{du(t)}{dt} + Au(t) = f(t, u(t)), & t > 0 \\ u(0) = u_0 \end{cases} \tag{2.1}$$

In the previous section we have proved the existence of mild solutions (Definition 1.1) of the initial value problem (2.1) under the assumptions that $-A$ is the infinitesimal generator of a $C_0$ semigroup of operators while $f(t, x)$ is continuous in both its variables and uniformly locally Lipschitz continuous in $x$. If the Lipschitz continuity of $f$ in $x$ is dropped then, as is well known, the existence of a mild solution of (2.1) is no more guaranteed even if $A \equiv 0$. In order to assure the existence of a mild solution of (2.1) in this case, we have to impose further conditions on the operator $A$. Our main assumption in this section will be that $-A$ is the infinitesimal generator of a compact $C_0$ semigroup (Definition 2.3.1). We note in passing that in applications generators of compact semigroups occur often in the case where $-A$ has a compact resolvent and generates an analytic semigroup $T(t)$, $t \geq 0$. Indeed, in this case,

$$\|T(t_1) - T(t_2)\| \leq \frac{C}{t_1}|t_2 - t_1| \qquad \text{for} \quad 0 < t_1 < t_2$$

by Theorem 2.5.2(d), so $T(t)$ is continuous in the uniform operator topology for $t > 0$ and hence by Theorem 2.3.3 it is also compact for $t > 0$.

The main result of this section is the following local existence theorem.

**Theorem 2.1.** *Let $X$ be a Banach space and $U \subset X$ be open. Let $-A$ be the infinitesimal generator of a compact semigroup $T(t)$, $t \geq 0$. If $0 < a \leq \infty$ and $f:[0, a[ \times U \to X$ is continuous then for every $u_0 \in U$ there exists a $t_1 = t_1(u_0)$, $0 < t_1 < a$ such that the initial value problem (2.1) has a mild solution $u \in C([0, t_1] : U)$.*

PROOF. Since we are interested here only in local solutions, we may assume that $a < \infty$. Let $\|T(t)\| \leq M$ for $0 \leq t \leq a$ and let $t' > 0$, $\rho > 0$ be such that $B_\rho(u_0) = \{v : \|v - u_0\| \leq \rho\} \subset U$ and $\|f(s, v)\| \leq N$ for $0 \leq s \leq t'$

and $v \in B_\rho(u_0)$. Choose $t'' > 0$ such that

$$\|T(t)u_0 - u_0\| < \rho/2 \qquad \text{for} \quad 0 \le t \le t''$$

and let

$$t_1 = \min\left(t', t'', a, \frac{\rho}{2MN}\right).$$

Set $Y = C([0, t_1]: X)$ and $Y_0 = \{u : u \in Y, u(0) = u_0, u(t) \in B_\rho(u_0)$ for $0 \le t \le t_1\}$. $Y_0$ is clearly a bounded closed convex subset of $Y$. We define a mapping $F: Y \to Y_0$ by

$$(Fu)(t) = T(t)u_0 + \int_0^t T(t-s)f(s, u(s))\, ds. \qquad (2.3)$$

Since

$$\|(Fu)(t) - u_0\| \le \|T(t)u_0 - u_0\| + \left\|\int_0^t T(t-s)f(s, u(s))\, ds\right\|$$

$$\le \rho/2 + t_1 MN \le \rho,$$

$F$ maps $Y_0$ into $Y_0$. Also, from the continuity of $f$ on $[0, a[\times U$ it follows easily that $F$ is a continuous map of $Y_0$ into $Y_0$. Moreover, $F$ maps $Y_0$ into a precompact subset of $Y_0$. To prove this, we first show that for every fixed $t$, $0 \le t \le t_1$, the set $Y_0(t) = \{(Fu)(t): u \in Y_0\}$ is precompact in $X$. This is clear for $t = 0$ since $Y_0(0) = \{u_0\}$. Let $t > 0$ be fixed. For $0 < \varepsilon < t$ set

$$(F_\varepsilon u)(t) = T(t)u_0 + \int_0^{t-\varepsilon} T(t-s)f(s, u(s))\, ds$$

$$= T(t)u_0 + T(\varepsilon)\int_0^{t-\varepsilon} T(t-s-\varepsilon)f(s, u(s))\, ds.$$

Since $T(t)$ is compact for every $t > 0$, the set $Y_\varepsilon(t) = \{(F_\varepsilon u)(t): u \in Y_0\}$ is precompact in $X$ for every $\varepsilon$, $0 < \varepsilon < t$. Furthermore, for $u \in Y_0$ we have

$$\|(Fu)(t) - (F_\varepsilon u)(t)\| \le \int_{t-\varepsilon}^t \|T(t-s)f(s, u(s))\|ds \le \varepsilon MN$$

which implies that $Y_0(t)$ is totally bounded, i.e, precompact in $X$. We continue and show that

$$F(Y_0) = \tilde{Y} = \{Fu : u \in Y_0\} \qquad (2.4)$$

is an equicontinuous family of functions. For $t_2 > t_1 > 0$ we have

$$\|(Fu)(t_1) - (Fu)(t_2)\| \le \|(T(t_1) - (T(t_2))u_0\|$$

$$+ N\int_0^{t_1}\|T(t_2 - s) - T(t_1 - s)\|ds + (t_2 - t_1)MN.$$

$$(2.5)$$

The right-hand side of (2.5) is independent of $u \in Y_0$ and tends to zero as $t_2 - t_1 \to 0$ as a consequence of the continuity of $T(t)$ in the uniform operator topology for $t > 0$ which in turn follows from the compactness of $T(t)$, $t > 0$ (Theorem 2.3.2).

It is also clear that $\tilde{Y}$ is bounded in $Y$. The desired precompactness of $\tilde{Y} = F(Y_0)$ is now a consequence of Arzela-Ascoli's theorem. Finally, it follows from Schauder's fixed point theorem that $F$ has a fixed point in $Y_0$ and any fixed point of $F$ is a mild solution of (2.1) on $[0, t_1]$ satisfying $u(t) \in U$ for $0 \leq t \leq t_1$.                                       □

We turn now to global existence. Here further assumptions must be made since global existence fails quite commonly. We start with the following result.

**Theorem 2.2.** *Let* $-A$ *be the infinitesimal generator of a compact semigroup* $T(t)$, $t \geq 0$ *on* $X$. *If* $f:[0, \infty[ \times X \to X$ *is continuous and maps bounded sets in* $[0, \infty[ \times X$ *into bounded sets in* $X$ *then for every* $u_0 \in X$ *the initial value problem* (2.1) *has a mild solution* $u$ *on a maximal interval of existence* $[0, t_{max}[$. *If* $t_{max} < \infty$ *then*

$$\lim_{t \uparrow t_{max}} \|u(t)\| = \infty. \tag{2.6}$$

PROOF. First we note that a mild solution $u$ of (2.1) defined on a closed interval $[0, t_1]$ can be extended to a larger interval $[0, t_1 + \delta]$, $\delta > 0$, by defining $u(t + t_1) = w(t)$ where $w(t)$ is a mild solution of

$$\begin{cases} \dfrac{dw(t)}{dt} + Aw(t) = f(t + t_1, w(t)) \\ w(0) = u(t_1) \end{cases} \tag{2.7}$$

the existence of which on an interval of positive length $\delta > 0$ is assured by Theorem 2.1. Let $[0, t_{max}[$ be the maximal interval to which the mild solution $u$ of (2.1) can be extended. We will show that if $t_{max} < \infty$ then $\|u(t)\| \to \infty$ as $t \uparrow t_{max}$. To do so we will first prove that $t_{max} < \infty$ implies $\lim_{t \uparrow t_{max}} \|u(t)\| = \infty$. Indeed, if $t_{max} < \infty$ and $\lim_{t \uparrow t_{max}} \|u(t)\| < \infty$ we can assume that $\|T(t)\| \leq M$ and $\|u(t)\| \leq K$ for $0 \leq t < t_{max}$ where $M$ and $K$ are constants. By our assumption on the function $f$ we also have a constant $N$ such that $\|f(t, u(t))\| \leq N$ for $0 \leq t < t_{max}$. Now if $0 < \rho < t < t' < t_{max}$ then

$$\|u(t') - u(t)\| \leq \|T(t')u_0 - T(t)u_0\|$$

$$+ \left\| \left( \int_0^{t-\rho} + \int_{t-\rho}^{t} \right) (T(t' - s) - T(t - s)) f(s, u(s)) \, ds \right\|$$

$$+ \left\| \int_t^{t'} T(t' - s) f(s, u(s)) \, ds \right\|$$

$$\leq \|T(t')u_0 - T(t)u_0\| + N \int_0^{t-\rho} \|T(t' - s) - T(t - s)\| \, ds$$

$$+ 2MN\rho + (t' - t)MN. \tag{2.8}$$

Since $t > \rho > 0$ is arbitrary and since $T(t)$ is continuous in the uniform operator topology for $t \geq \rho > 0$, the right-hand side of (2.8) tends to zero as $t, t'$ tend to $t_{max}$. Therefore $\lim_{t \uparrow t_{max}} u(t) = u(t_{max})$ exists and by the first part of the proof the solution $u$ can be extended beyond $t_{max}$, contradicting the maximality of $t_{max}$. Therefore the assumption that $t_{max} < \infty$ implies that $\overline{\lim}_{t \uparrow t_{max}} \|u(t)\| = \infty$. To conclude the proof we will show that $\lim_{t \uparrow t_{max}} \|u(t)\| = \infty$. If this is false then there is a sequence $t_n \uparrow t_{max}$ and a constant $K$ such that $\|u(t_n)\| \leq K$ for all $n$. Let $\|T(t)\| \leq M$ for $0 \leq t \leq t_{max}$ and let $N = \sup\{\|f(t, x)\| : 0 \leq t \leq t_{max}, \|x\| \leq M(K + 1)\}$. Since $t \rightarrow \|u(t)\|$ is continuous and $\overline{\lim}_{t \uparrow t_{max}} \|u(t)\| = \infty$ we can find a sequence $\langle h_n \rangle$ with the following properties: $h_n \rightarrow 0$ as $n \rightarrow \infty$, $\|u(t)\| \leq M(K + 1)$ for $t_n \leq t \leq t_n + h_n$ and $\|u(t_n + h_n)\| = M(K + 1)$. But then we have

$$M(K + 1) = \|u(t_n + h_n)\| \leq \|T(h_n)u(t_n)\|$$

$$+ \int_{t_n}^{t_n + h_n} \|T(t_n + h_n - s)f(s, u(s))\| ds$$

$$\leq MK + h_n NM$$

which is absurd as $h_n \rightarrow 0$. Therefore we have $\lim_{t \uparrow t_{max}} \|u(t)\| = \infty$ and the proof is complete.                                                                        $\square$

We conclude this section with two useful conditions for the existence of global mild solutions of the initial value problem (2.1) under the assumptions of Theorem 2.2.

**Corollary 2.3.** *Let* $-A$ *be the infinitesimal generator of a compact* $C_0$ *semigroup,* $T(t)$, $t \geq 0$ *on* $X$. *Let* $f: [0, \infty[ \times X \rightarrow X$ *be continuous and map bounded sets in* $[0, \infty[ \times X$ *into bounded sets in* $X$. *Then for every* $u_0 \in X$ *the initial value problem (2.1) has a global solution* $u \in C([0, \infty[ \times X)$ *if either one of the following conditions is satisfied:*

(i) *There exists a continuous function* $k_0(s):[0, \infty[ \rightarrow ]0, \infty[$ *such that* $\|u(t)\|$ $\leq k_0(t)$ *for every* $t$ *in the interval of existence of* $u$.
(ii) *There exist two locally integrable functions* $k_1(s)$ *and* $k_2(s)$ *such that*

$$\|f(s, x)\| \leq k_1(s)\|x\| + k_2(s) \quad for \quad 0 \leq s < \infty, x \in X. \quad (2.9)$$

PROOF. Part (i) is a trivial consequence of Theorem 2.2. To prove (ii) we reduce it to (i) as follows: Assume that the solution $u$ exists on the interval $[0, t[$. Set $\|T(t)\| \leq Me^{\omega t}$ and

$$\psi(t) = M\|u_0\| + \int_0^t Me^{-\omega s}k_2(s) \, ds.$$

The function $\psi$ thus defined is obviously continuous on $[0, \infty[$ and we have

$$\|u(t)\|e^{-\omega t} \le e^{-\omega t}\|T(t)u_0\| + e^{-\omega t}\int_0^t \|T(t-s)f(s, u(s))\|\,ds$$

$$\le \psi(t) + \int_0^t Mk_1(s)\|u(s)\|e^{-\omega s}\,ds \qquad (2.10)$$

and by Gronwall's inequality

$$\|u(t)\|e^{-\omega t} \le \psi(t) + M\int_0^t k_1(s)\psi(s)\exp\left\{M\int_s^t k_1(r)\,dr\right\}ds$$

which implies the boundedness of $\|u(t)\|$ by a continuous function.         $\square$

## 6.3. Semilinear Equations with Analytic Semigroups

As we have noted briefly at the beginning of section 6.2, the initial value problem

$$\begin{cases} \dfrac{du(t)}{dt} + Au(t) = f(t, u(t)), & t > t_0 \\ u(t_0) = x_0 \end{cases} \qquad (3.1)$$

occurs often in the applications with an operator $-A$ which is the infinitesimal generator of an analytic semigroup on a Banach space $X$. In this case if $R(\lambda : -A)$ is compact for some $\lambda \in \rho(-A)$ and $f(t, x)$ is continuous in $[t_0, T] \times X$ the problem has a (possibly non unique) mild local solution by Theorem 2.1. If we assume further, as we will do in the sequel, that $f$ is regular with respect to $-A$ in some sense, we will be able to obtain unique local strong solutions of the initial value problem (3.1).

Throughout this section we will assume that $-A$ is the infinitesimal generator of an analytic semigroup $T(t)$ on the Banach space $X$. For convenience we will also assume that $T(t)$ is bounded, that is $\|T(t)\| \le M$ for $t \ge 0$, and that $0 \in \rho(-A)$, i.e. $-A$ is invertible.

We note that if $-A$ is the infinitesimal generator of an analytic semigroup then $-A - \alpha I$ is invertible and generates a bounded analytic semigroup for $\alpha > 0$ large enough. This enables one to reduce the general case where $-A$ is the infinitesimal generator of an analytic semigroup to the case where the semigroup is bounded and $-A$ is invertible.

From our assumptions on $A$ and the results of Section 2.2.6 it follows that $A^\alpha$ can be defined for $0 \le \alpha \le 1$ and $A^\alpha$ is a closed linear invertible operator with domain $D(A^\alpha)$ dense in $X$. The closedness of $A^\alpha$ implies that $D(A^\alpha)$ endowed with the graph norm of $A^\alpha$, i.e. the norm $\||x\|| = \|x\| + \|A^\alpha x\|$, is a Banach space. Since $A^\alpha$ is invertible its graph norm $\|| \cdot \||$ is equivalent to the norm $\|x\|_\alpha = \|A^\alpha x\|$. Thus, $D(A^\alpha)$ equipped with the norm $\| \|_\alpha$ is a Banach space which we denote by $X_\alpha$. From this definition it is clear that

$0 < \alpha < \beta$ implies $X_\alpha \supset X_\beta$ and that the imbedding of $X_\beta$ in $X_\alpha$ is continuous.

Our main assumption concerning the function $f$ in (3.1) will be,

**Assumption (F).** *Let $U$ be an open subset of $\mathbb{R}^+ \times X_\alpha$. The function $f: U \to X$ satisfies the assumption $(F)$ if for every $(t, x) \in U$ there is a neighborhood $V \subset U$ and constants $L \geq 0, 0 < \vartheta \leq 1$ such that*

$$\|f(t_1, x_1) - f(t_2, x_2)\| \leq L(|t_1 - t_2|^\vartheta + \|x_1 - x_2\|_\alpha) \qquad (3.2)$$

*for all $(t_i, x_i) \in V$.*

We can now state and prove the main existence result of this section.

**Theorem 3.1.** *Let $-A$ be the infinitesimal generator of an analytic semigroup $T(t)$ satisfying $\|T(t)\| \leq M$ and assume further that $0 \in \rho(-A)$. If, $0 < \alpha < 1$ and $f$ satisfies the assumption $(F)$ then for every initial data $(t_0, x_0) \in U$ the initial value problem (3.1) has a unique local solution $u \in C([t_0, t_1[: X) \cap C^1(]t_0, t_1[: X)$ where $t_1 = t_1(t_0, x_0) > t_0$.*

PROOF. From our assumptions on the operator $A$ it follows (Theorem 2.6.13) that

$$\|A^\alpha T(t)\| \leq C_\alpha t^{-\alpha} \qquad \text{for} \quad t > 0. \qquad (3.3)$$

For the rest of the proof, we fix $(t_0, x_0) \in U$ and choose $t_1' > t_0, \delta > 0$ such that the estimate (3.2) with some fixed constants $L$ and $\vartheta$ holds in the set $V = \{(t, x): t_0 \leq t \leq t_1', \|x - x_0\|_\alpha \leq \delta\}$. Let,

$$B = \max_{t_0 \leq t \leq t_1'} \|f(t, x_0)\| \qquad (3.4)$$

and choose $t_1$ such that

$$\|T(t - t_0)A^\alpha x_0 - A^\alpha x_0\| < \delta/2 \qquad \text{for} \quad t_0 \leq t < t_1 \qquad (3.5)$$

and

$$0 < t_1 - t_0 < \min\left\{ t_1' - t_0, \left[ \frac{\delta}{2}(1 - \alpha)C_\alpha^{-1}(B + \delta L)^{-1} \right]^{1/1 - \alpha} \right\}.$$

$$(3.6)$$

Let $Y$ be the Banach space $C([t_0, t_1]: X)$ with the usual supremum norm which we denote by $\| \cdot \|_Y$. On $Y$ we define a mapping $F$ by

$$Fy(t) = T(t - t_0)A^\alpha x_0 + \int_{t_0}^t A^\alpha T(t - s)f(s, A^{-\alpha}y(s)) \, ds. \qquad (3.7)$$

Clearly, $F: Y \to Y$ and for every $y \in Y$, $Fy(t_0) = A^\alpha x_0$. Let $S$ be the nonempty closed and bounded subset of $Y$ defined by

$$S = \{ y: y \in Y, y(t_0) = A^\alpha x_0, \|y(t) - A^\alpha x_0\| \leq \delta \} \qquad (3.8)$$

For $y \in S$ we have,

$$\|Fy(t) - A^\alpha x_0\| \leq \|T(t - t_0)A^\alpha x_0 - A^\alpha x_0\|$$

$$+ \left\| \int_{t_0}^{t} A^\alpha T(t - s)[f(s, A^{-\alpha}y(s)) - f(s, x_0)]\, ds \right\|$$

$$+ \left\| \int_{t_0}^{t} A^\alpha T(t - s)f(s, x_0)\, ds \right\|$$

$$\leq \frac{\delta}{2} + C_\alpha(L\delta + B) \int_{t_0}^{t} (t - s)^{-\alpha}\, ds$$

$$= \frac{\delta}{2} + C_\alpha(1 - \alpha)^{-1}(L\delta + B)(t_1 - t_0)^{1-\alpha} \leq \delta$$

where we used (3.2), (3.3), (3.6) and (3.8). Therefore $F: S \to S$. Furthermore, if $y_1, y_2 \in S$ then

$$\|Fy_1(t) - Fy_2(t)\|$$

$$\leq \int_{t_0}^{t} \|A^\alpha T(t - s)\| \, \|f(s, A^{-\alpha}y_1(s)) - f(s, A^{-\alpha}y_2(s))\|\, ds$$

$$\leq LC_\alpha(1 - \alpha)^{-1}(t_1 - t_0)^{1-\alpha}\|y_1 - y_2\|_Y \leq \tfrac{1}{2}\|y_1 - y_2\|_Y$$

which implies

$$\|Fy_1 - Fy_2\|_Y \leq \tfrac{1}{2}\|y_1 - y_2\|_Y \qquad \text{for} \quad y_1, y_2 \in S. \tag{3.9}$$

By the contraction mapping theorem the mapping $F$ has a unique fixed point $y \in S$. This fixed point satisfies the integral equation

$$y(t) = T(t - t_0)A^\alpha x_0 + \int_{t_0}^{t} A^\alpha T(t - s)f(s, A^{-\alpha}y(s))\, ds$$

$$\text{for} \quad t_0 \leq t \leq t_1. \tag{3.10}$$

From (3.2) and the continuity of $y$ it follows that $t \to f(t, A^{-\alpha}y(t))$ is continuous on $[t_0, t_1]$ and a fortiori bounded on this interval. Let

$$\|f(t, A^{-\alpha}y(t))\| \leq N \qquad \text{for} \quad t_0 \leq t \leq t_1. \tag{3.11}$$

Next we want to show that $t \to f(t, A^{-\alpha}y(t))$ is locally Hölder continuous on $]t_0, t_1]$. To this end we show first that the solution $y$ of (3.10) is locally Hölder continuous on $]t_0, t_1]$.

We note that for every $\beta$ satisfying $0 < \beta < 1 - \alpha$ and every $0 < h < 1$ we have by Theorem 2.6.13.

$$\|(T(h) - I)A^\alpha T(t - s)\| \leq C_\beta h^\beta \|A^{\alpha+\beta}T(t - s)\| \leq Ch^\beta(t - s)^{-(\alpha+\beta)}. \tag{3.12}$$

If $t_0 < t < t + h \le t_1$, then

$$\|y(t + h) - y(t)\| \le \|(T(h) - I)A^\alpha T(t - t_0)x_0\|$$

$$+ \int_{t_0}^t \|(T(h) - I)A^\alpha T(t - s)f(s, A^{-\alpha}y(s))\| ds$$

$$+ \int_t^{t+h} \|A^\alpha T(t + h - s)f(s, A^{-\alpha}y(s))\| ds$$

$$= I_1 + I_2 + I_3. \tag{3.13}$$

Using (3.11) and (3.12) we estimate each of the terms of (3.13) separately.

$$I_1 \le C(t - t_0)^{-(\alpha + \beta)} h^\beta \|x_0\| \le M_1 h^\beta \tag{3.14}$$

$$I_2 \le CNh^\beta \int_{t_0}^t (t - s)^{-(\alpha + \beta)} ds \le M_2 h^\beta \tag{3.15}$$

$$I_3 \le NC_\alpha \int_t^{t+h} (t + h - s)^{-\alpha} = \frac{NC_\alpha}{1 - \alpha} h^{1-\alpha} \le M_3 h^\beta. \tag{3.16}$$

Note that $M_2$ and $M_3$ can be chosen to be independent of $t \in [t_0, t_1]$ while $M_1$ depends on $t$ and blows up at $t \downarrow t_0$. Combining (3.13) with these estimates it follows that for every $t_0' > t_0$ there is a constant $C$ such that

$$\|y(t) - y(s)\| \le C|t - s|^\beta \quad \text{for} \quad t_0 < t_0' \le t, s \le t_1 \tag{3.17}$$

and therefore $y$ is locally Hölder continuous on $]t_0, t]$. The local Hölder continuity of $t \to f(t, A^{-\alpha}y(t))$ follows now from

$$\|f(s, A^{-\alpha}y(s)) - f(t, A^{-\alpha}y(t))\| \le L(|t - s|^\vartheta + \|y(t) - y(s)\|)$$

$$\le C_1(|t - s|^\vartheta + |t - s|^\beta). \tag{3.18}$$

Let $y$ be the solution of (3.10) and consider the inhomogeneous initial value problem

$$\begin{cases} \dfrac{du(t)}{dt} + Au(t) = f(t, A^{-\alpha}y(t)) \\ u(t_0) = x_0. \end{cases} \tag{3.19}$$

By Corollary 4.3.3. this problem has a unique solution $u \in C^1(]t_0, t_1]: X)$. The solution of (3.19) is given by

$$u(t) = T(t - t_0)x_0 + \int_{t_0}^t T(t - s)f(s, A^{-\alpha}y(s)) ds. \tag{3.20}$$

For $t > t_0$ each term of (3.20) is in $D(A)$ and a fortiori in $D(A^\alpha)$. Operating on both sides of (3.20) with $A^\alpha$ we find

$$A^\alpha u(t) = T(t - t_0)A^\alpha x_0 + \int_{t_0}^t A^\alpha T(t - s)f(s, A^{-\alpha}y(s)) ds. \tag{3.21}$$

But by (3.10) the right-hand side of (3.21) equals $y(t)$ and therefore $u(t) = A^{-\alpha}y(t)$ and by (3.20), $u$ is a $C^1(]t_0, t_1]: X)$ solution of (3.1). The uniqueness of $u$ follows readily from the uniqueness of the solutions of (3.10) and (3.19) and the proof is complete. ☐

Theorem 3.1 states that under suitable conditions we have a continuously differentiable solution of the initial value problem (3.1) on the interval $]t_0, t_1]$. More is actually true. The derivative $u'$ of the solution is locally Hölder continuous on $]t_0, t_1]$. This is a consequence of the following regularity result.

**Corollary 3.2.** *Let $A$ and $f$ satisfy the assumptions of Theorem 3.1 and assume further that $f$ satisfies (3.2) for every $(t, x) \in U$ (i.e. the constants $\vartheta$ and $L$ are uniform in $U$). If $u$ is the solution of the initial value problem (3.1) on $[t_0, t_1]$ then $du/dt$ is locally Hölder continuous on $]t_0, t_1]$ with exponent $v = min(\vartheta, \beta)$ for any $\beta$ satisfying $0 < \beta < 1 - \alpha$.*

PROOF. Let $0 < \beta < 1 - \alpha$. In the proof of Theorem 3.2 we showed that if $u$ is the solution of the initial value problem (3.1) then for every $t_0 < t_0' < t_1$, $f(t, u(t))$ is Hölder continuous on $[t_0', t_1]$ with exponent $v = min(\vartheta, \beta)$. From Theorem 4.3.5 it then follows that for every $t_0'' > t_0'$, $du/dt$ is Hölder continuous on $[t_0'', t_1]$ with the same exponent $v$. ☐

We conclude this section with a result on the existence of global solutions of (3.1).

**Theorem 3.3.** *Let $0 \in \rho(-A)$ and let $-A$ be the infinitesimal generator of an analytic semigroup $T(t)$ satisfying $\|T(t)\| \leq M$ for $t \geq 0$. Let $f:[t_0, \infty[ \times X_\alpha \to X$ satisfy $(F)$. If there is a continuous nondecreasing real valued function $k(t)$ such that*

$$\|f(t, x)\| \leq k(t)(1 + \|x\|_\alpha) \qquad \text{for} \quad t \geq t_0, x \in X_\alpha \qquad (3.22)$$

*then for every $x_0 \in X_\alpha$ the initial value problem (3.1) has a unique solution $u$ which exists for all $t \geq t_0$.*

PROOF. Applying Theorem 3.1 we can continue the solution of (3.1) as long as $\|u(t)\|_\alpha$ stays bounded. It is therefore sufficient to show that if $u$ exists on $[0, T[$ then $\|u(t)\|_\alpha$ is bounded as $t \uparrow T$. Since

$$A^\alpha u(t) = A^\alpha T(t - t_0) x_0 + \int_{t_0}^t A^\alpha T(t - s) f(s, u(s)) \, ds$$

it follows that

$$\|u(t)\|_\alpha \leq M\|A^\alpha x_0\| + \frac{k(T) C_\alpha T^{1-\alpha}}{1 - \alpha} + k(T) C_\alpha \int_{t_0}^t (t - s)^{-\alpha} \|u(s)\|_\alpha \, ds$$

which implies by Lemma 5.6.7 that $\|u(t)\|_\alpha \leq C$ on $[0, T[$ and the proof is complete. ☐

## 6.4. A Quasilinear Equation of Evolution

In this section we will discuss the Cauchy problem for the *quasilinear* initial value problem

$$\begin{cases} \dfrac{du(t)}{dt} + A(t, u)u = 0 & \text{for } 0 \le t \le T \\ u(0) = u_0 \end{cases} \tag{4.1}$$

in a Banach space $X$.

The initial value problem (4.1) differs from the semilinear initial value problems that were treated in the previous sections by the fact that here the linear operator $A(t, u)$ appearing in the problem depends explicitly on the solution $u$ of the problem, while in the semilinear case the nonlinear operator was the sum of a fixed linear operator (independent of the solution $u$) and a nonlinear "function" of $u$.

In general the study of quasilinear initial value problems is quite complicated. For the sake of simplicity we will restrict ourselves in this section to a rather simple framework starting with mild solutions of the initial value problem (4.1). We begin by indicating briefly the general idea behind the definition and the existence proof of such mild solutions.

Let $u \in C([0, T] : X)$ and consider the linear initial value problem

$$\begin{cases} \dfrac{dv}{dt} + A(t, u)v = 0 & \text{for } 0 \le t \le T \\ v(0) = u_0 \end{cases} \tag{4.2}$$

If this problem has a unique mild solution $v \in C([0, T] : X)$, for every given $u \in C([0, T] : X)$, then it defines a mapping $u \to v = F(u)$ of $C([0, T] : X)$ into itself. The fixed points of this mapping are defined to be mild solution of (4.1).

To prove the existence of a local mild solution of (4.1) we will show that under suitable conditions, there exists always a $T', 0 < T' \le T$ such that the restriction of the mapping $F$ to $C([0, T'] : X)$ is a contraction which maps some ball of $C([0, T'] : X)$ into itself. The contraction mapping principle will then imply the existence of a unique fixed point $u$ of $F$ in this ball and $u$ is then, by definition, the desired mild solution of (4.1).

In order to carry out the program as indicated above, we will need some preliminaries. We start with the existence of mild solutions of the linear initial value problem (4.2). To this end we modify the assumptions $(H_1)$–$(H_3)$ of section 5.3 so that they depend on an additional parameter.

**Definition 4.1.** Let $B$ be a subset of $X$ and for every $0 \le t \le T$ and $b \in B$ let $A(t, b)$ be the infinitesimal generator of a $C_0$ semigroup $S_{t, b}(s), s \ge 0$, on $X$. The family of operators $\{A(t; b)\}, (t, b) \in [0, T] \times B$, is *stable* if there are constants $M \geq 1$ and $\omega$ such that

$$\rho(A(t, b)) \supset ]\omega, \infty[ \quad \text{for} \quad (t, b) \in [0, T] \times B \quad (4.3)$$

and

$$\left\| \prod_{j=1}^{k} R(\lambda : A(t_j, b_j)) \right\| \leq M(\lambda - \omega)^{-k} \quad \text{for} \quad \lambda > \omega \quad (4.4)$$

and every finite sequences $0 \leq t_1 \leq t_2 \leq \cdots \leq t_k \leq T$, $b_j \in B$, $1 \leq j \leq k$.

It is not difficult to show (see proof of Theorem 5.2.2) that the stability of $\{A(t, b)\}$, $(t, b) \in [0, T] \times B$ implies that

$$\left\| \prod_{j=1}^{k} S_{t_j, b_j}(s_j) \right\| \leq M \exp \left\{ \omega \sum_{j=1}^{k} s_j \right\} \quad \text{for} \quad s_j \geq 0 \quad (4.5)$$

and any finite sequences $0 \leq t_1 \leq t_2 \leq \cdots \leq t_k \leq T$, $b_j \in B$, $1 \leq j \leq k$.

Let $X$ and $Y$ be Banach spaces such that $Y$ is densely and continuously imbedded in $X$. Let $B \subset X$ be a subset of $X$ such that for every $(t, b) \in [0, T] \times B$, $A(t, b)$ is the infinitesimal generator of a $C_0$ semigroup $S_{t, b}(s)$, $s \geq 0$, on $X$. We make the following assumptions:

$(\tilde{H}_1)$ The family $\{A(t, b)\}$, $(t, b) \in [0, T] \times B$ is stable.
$(\tilde{H}_2)$ $Y$ is $A(t, b)$-admissible for $(t, b) \in [0, T] \times B$ and the family $\{\tilde{A}(t, b)\}$, $(t, b) \in [0, T] \times B$ of parts $\tilde{A}(t, b)$ of $A(t, b)$ in $Y$, is stable in $Y$.
$(\tilde{H}_3)$ For $(t, b) \in [0, T] \times B$, $D(A(t, b)) \supset Y$, $A(t, b)$ is a bounded linear operator from $Y$ to $X$ and $t \to A(t, b)$ is continuous in the $B(Y, X)$ norm $\| \, \|_{Y \to X}$ for every $b \in B$.
$(\tilde{H}_4)$ There is a constant $L$ such that

$$\|A(t, b_1) - A(t, b_2)\|_{Y \to X} \leq L \|b_1 - b_2\| \quad (4.6)$$

holds for every $b_1, b_2 \in B$ and $0 \leq t \leq T$.

**Lemma 4.2.** *Let $B \subset X$ and let $u \in C([0, T] : X)$ have values in $B$. If $\{A(t, b)\}$, $(t, b) \in [0, T] \times B$ is a family of operators satisfying the assumptions $(\tilde{H}_1)$–$(\tilde{H}_4)$ then $\{A(t, u(t))\}_{t \in [0, T]}$ is a family of operators satisfying the assumptions $(H_1)$–$(H_3)$ of Theorem 5.3.1.*

PROOF. From $(\tilde{H}_1)$ and $(\tilde{H}_2)$ it follows readily that $\{A(t, u(t))\}_{t \in [0, T]}$ satisfies $(H_1)$ and $(H_2)$. Moreover it is clear from $(\tilde{H}_3)$ that for $t \in [0, T]$ $D(A(t, u(t))) \supset Y$ and that $A(t, u(t))$ is a bounded linear operator from $Y$ to $X$. It remains only to show that $t \to A(t, u(t))$ is continuous in the $B(Y, X)$ norm. But by $(\tilde{H}_4)$ we have

$$\|A(t_1, u(t_1)) - A(t_2, u(t_2))\|_{Y \to X}$$
$$\leq \|A(t_1, u(t_1)) - A(t_2, u(t_1))\|_{Y \to X} + C \|u(t_1) - u(t_2)\|. \quad (4.7)$$

Since $u(t)$ is continuous in $X$, the continuity of $t \to A(t, b)$ together with (4.7) imply the continuity of $t \to A(t, u(t))$ in the $B(Y, X)$ norm.    $\square$

As a consequence of Lemma 4.2 and Theorem 5.3.1 we now have:

**Theorem 4.3.** *Let $B \subset X$ and let $\{A(t, b)\}, (t, b) \in [0, T] \times B$ be a family of operators satisfying the conditions $(\tilde{H}_1)$–$(\tilde{H}_4)$. If $u \in C([0, T]: X)$ has values in $B$ then there is a unique evolution system $U_u(t, s), 0 \leq s \leq t \leq T$, in $X$ satisfying*

$$\|U_u(t, s)\| \leq Me^{\omega(t-s)} \qquad \text{for} \quad 0 \leq s \leq t \leq T \qquad (4.8)$$

$$\frac{\partial^+}{\partial t} U_u(t, s)w\bigg|_{t=s} = A(s, u(s))w \qquad \text{for} \quad w \in Y, 0 \leq s \leq T \quad (4.9)$$

$$\frac{\partial}{\partial s} U_u(t, s)w = -U_u(t, s)A(s, u(s))w$$

$$\text{for} \quad w \in Y, 0 \leq s \leq t \leq T. \quad (4.10)$$

For every function $u \in C([0, T]: X)$ with values in $B$ and $u_0 \in X$ the function $v(t) = U_u(t, 0)u_0$ is defined to be the mild solution of the initial value problem (4.2). From Theorem 4.3 it therefore follows that if the family $\{A(t, b)\}, (t, b) \in [0, T] \times B$, satisfies the conditions $(\tilde{H}_1)$–$(\tilde{H}_4)$ then for every $u_0 \in X$ and $u \in C([0, T]: X)$ with values in $B$ the initial value problem (4.2) possesses a unique mild solution $v$ given by

$$v(t) = U_u(t, 0)u_0 \qquad (4.12)$$

In the sequel we will need also the following continuous dependence result.

**Lemma 4.4.** *Let $B \subset X$ and let $\{A(t, b)\}, (t, b) \in [0, T] \times B$, satisfy the conditions $(\tilde{H}_1)$–$(\tilde{H}_4)$. There is a constant $C_1$ such that for every $u, v \in C([0, T]: X)$ with values in $B$ and every $w \in Y$ we have*

$$\|U_u(t, s)w - U_v(t, s)w\| \leq C_1\|w\|_Y \int_s^t \|u(\tau) - v(\tau)\| d\tau. \quad (4.13)$$

PROOF. As in the proof of Theorem 5.3.1 we obtain for every $w \in Y$ the estimate,

$$\|U_u(t, s)w - U_v(t, s)w\| \leq C\|w\|_Y \int_s^t \|A(\tau, u(\tau)) - A(\tau, v(\tau))\|_{Y \to X} d\tau$$

$$(4.14)$$

where $C$ depends only on the stability constants of $\{A(t, b)\}$ and $\{\tilde{A}(t, b)\}$. Combining (4.14) with $(\tilde{H}_4)$ yields (4.13). $\qquad \square$

We turn now to the existence of local mild solutions of the initial value problem (4.1). In the first result the initial value $u_0$ will be assumed to be in $Y$ and $B$ will be a ball of radius $r$ in $X$ centered at $u_0$.

**Theorem 4.5.** *Let $u_0 \in Y$ and let $B = \{x : \|x - u_0\| \le r\}, r > 0$. If $\{A(t, b)\}$, $(t, b) \in [0, T] \times B$ satisfy the assumptions $(\tilde{H}_1)$–$(\tilde{H}_4)$ then there is a $T', 0 < T' \le T$ such that the initial value problem*

$$\begin{cases} \dfrac{du}{dt} + A(t, u)u = 0, & 0 \le t \le T' \\ u(0) = u_0 \end{cases} \tag{4.15}$$

*has a unique mild solution $u \in C([0, T'] : X)$ with $u(t) \in B$ for $0 \le t \le T'$.*

PROOF. We note first that the constant function $u(t) \equiv u_0$ satisfies the assumptions of Theorem 4.3 and there is therefore an evolution system $U_{u_0}(t, s), 0 \le s \le t \le T$ associated to $u_0$. Let $0 < t_1 \le T$ be such that

$$\max_{0 \le t \le t_1} \| U_{u_0}(t, 0)u_0 - u_0 \| < \frac{r}{2}$$

and choose

$$T' = \min \left\{ t_1, \tfrac{1}{2}(C\|u_0\|_Y + 1)^{-1} \right\} \tag{4.16}$$

where $C$ is the constant appearing in Lemma 4.4. On the closed subset $\mathcal{S}$ of $C([0, T'] : X)$ defined by

$$\mathcal{S} = \{u : u \in C([0, T'] : X), u(0) = u_0, \|u(t) - u_0\| \le r \text{ for } 0 \le t \le T\} \tag{4.17}$$

we consider the mapping

$$Fu(t) = U_u(t, 0)u_0 \qquad \text{for} \quad 0 \le t \le T'. \tag{4.18}$$

By our assumptions and Theorem 4.3 it is clear that $F$ is well defined on $\mathcal{S}$ and that its range is in $C([0, T'] : X)$. We claim that $F : \mathcal{S} \to \mathcal{S}$. Indeed, for $u \in \mathcal{S}$ we clearly have $Fu(0) = u_0$ and by Lemma 4.4 and (4.16),

$$\| Fu(t) - u_0 \| \le \| U_u(t, 0)u_0 - U_{u_0}(t, 0)u_0 \| + \| U_{u_0}(t, 0)u_0 - u_0 \|$$

$$\le Cr\|u_0\|_Y T' + \frac{r}{2} \le r.$$

Moreover, if $u_1, u_2 \in \mathcal{S}$ then by Lemma 4.4

$$\| Fu_1(t) - Fu_2(t) \| = \| U_{u_1}(t, 0)u_0 - U_{u_2}(t, 0)u_0 \|$$

$$\le C\|u_0\|_Y \int_0^t \|u_1(\tau) - u_2(\tau)\| d\tau$$

$$\le C\|u_0\|_Y T' \|u_1 - u_2\|_\infty \le \tfrac{1}{2} \|u_1 - u_2\|_\infty$$

where $\| \ \|_\infty$ is the usual supremum norm in $C([0, T'] : X)$. From the last inequality it follows readily that

$$\| Fu_1 - Fu_2 \|_\infty \le \tfrac{1}{2} \|u_1 - u_2\|_\infty \tag{4.19}$$

so that $F$ is a contraction. From the contraction mapping theorem it follows that $F$ has a unique fixed point $u \in \mathcal{S}$ which is the desired mild solution of (4.15) on $[0, T']$.                                                                    □

A different version of Theorem 4.5 which is often very useful in the applications of the theory to partial differential equations, is obtained by restricting the set $B$ that appears in the conditions $(\tilde{H}_1)-(\tilde{H}_4)$ to a ball in $Y$ rather than a ball in $X$ as we have assumed above. The price that we have to pay for this relaxation of the conditions are the following further assumptions,

$(\tilde{H}_5)$ For every $u \in C([0,T]:X)$ satisfying $u(t) \in B$ for $0 \le t \le T$, we have

$$U_u(t,s)Y \subset Y, \qquad 0 \le s \le t \le T \qquad (4.20)$$

and $U_u(t,s)$ is strongly continuous in $Y$ for $0 \le s \le t \le T$.

$(\tilde{H}_6)$ Closed convex bounded subsets of $Y$ are also closed in $X$.

We note that the condition $(\tilde{H}_6)$ is always satisfied if $X$ and $Y$ are reflexive Banach spaces. We now have:

**Theorem 4.6.** *Let* $u_0 \in Y$ *and let* $B = \{y: \|y - u_0\|_Y \le r\}$, $r > 0$. *Let* $\{A(t,b)\}$, $(t,b) \in [0,T] \times B$ *be a family of linear operators satisfying the assumptions* $(\tilde{H}_1)-(\tilde{H}_6)$. *If* $A(t,b)u_0 \in Y$ *and*

$$\|A(t,b)u_0\|_Y \le k \qquad \text{for} \quad (t,b) \in [0,T] \times B \qquad (4.21)$$

*then there exists a* $T'$, $0 < T' \le T$ *such that the initial value problem* (4.15) *has a unique classical solution* $u \in C([0,T']:B) \cap C^1([0,T']:X)$.

PROOF. We start by showing the existence of a unique mild solution of (4.15). We note first that from the construction of $U_u(t,s)$ and $(\tilde{H}_5)$ it follows that

$$\|U_u(t,s)\|_Y \le C_1 \qquad \text{for} \quad 0 \le s \le t \le T \qquad (4.22)$$

and every $u \in C([0,T']:X)$ with values in $B$. After choosing

$$T' = \min\left\{T, \frac{r}{kC_1}, \frac{1}{2}(C\|u_0\|_Y + 1)^{-1}\right\} \qquad (4.23)$$

where $C$ is the constant appearing in Lemma 4.4, we consider the subset $\mathbb{S}$ of $C([0,T']:X)$ defined by

$$\mathbb{S} = \{u: u \in C([0,T']:X), u(0) = u_0, u(t) \in B \qquad \text{for} \quad 0 \le t \le T'\}.$$

From $(\tilde{H}_6)$ it follows that $\mathbb{S}$ is a closed convex subset of $C([0,T']:X)$. Next we define on $\mathbb{S}$ the mapping

$$Fu(t) = U_u(t,0)u_0 \qquad 0 \le t \le T' \qquad (4.24)$$

and show that $F: \mathbb{S} \to \mathbb{S}$. Clearly $Fu(0) = u_0$ and $Fu(t) \in C([0,T']:X)$. From $(\tilde{H}_5)$ it follows that $Fu(t) \in Y$ for $0 \le t \le T'$ and it remains to show that $\|Fu(t) - u_0\|_Y \le r$ for $0 \le t \le T'$. Integrating (4.10) in $X$ from $s$ to $t$ we find

$$U_u(t,0)u_0 - u_0 = \int_0^t U_u(t,\tau)A(\tau,u(\tau))u_0\,d\tau. \qquad (4.25)$$

Estimating (4.25) in $Y$ and using (4.21), (4.22) and (4.23) yields

$$\|Fu(t) - u_0\|_Y = \|U_u(t, 0)u_0 - u_0\|_Y \le C_1 kT \le r.$$

Therefore $F: \mathbb{S} \to \mathbb{S}$. Now, exactly as in the proof of Theorem 4.5, we also have for any $u_1, u_2 \in \mathbb{S}$

$$\|Fu_1 - Fu_2\|_\infty \le \tfrac{1}{2}\|u_1 - u_2\|_\infty$$

where $\|\ \|_\infty$ is the supremum norm in $C([0, T']: X)$. Thus by the contraction mapping theorem $F$ has a unique fixed point $u \in \mathbb{S}$ which is the mild solution of (4.15) of $[0, T']$. But $u(t) = U_u(t, 0)u_0$ and therefore by $(\tilde{H}_5)$ and Theorem 5.4.3, $u$ is the unique $Y$-valued solution of the linear evolution equation

$$\begin{cases} \dfrac{dv}{dt} + A(t, u)v = 0 \\ v(0) = u_0 \end{cases} \tag{4.26}$$

and thus $u$ is a classical solution of (4.15) and $u \in C([0, T']: Y) \cap C^1([0, T']: X)$. The uniqueness of $u$ is obvious and the proof is complete. $\square$

# Applications to Partial Differential Equations—Linear Equations

## 7.1. Introduction

The theory of semigroups of linear operators has applications in many branches of analysis. Such applications to Harmonic analysis, approxima-tion theory, ergodic theory and many other subjects can be found in the general texts that are mentioned at the beginning of the bibliographical remarks.

In the present and following chapter we will restrict our attention to applications which are related to the solution of initial value problems for partial differential equations. The different sections of these two chapters are essentially independent of each other and each one of them describes a special application. Basic results from the general theory of partial differen-tial equations which will be used will be stated, without proof, when needed.

In the applications of the abstract theory, it is usually shown that a given differential operator $A$ is the infinitesimal generator of a $C_0$ semigroup in a certain concrete Banach space of functions $X$. This provides us with the existence and uniqueness of a solution of the initial value problem

$$\begin{cases} \dfrac{\partial u(t, x)}{\partial t} = Au(t, x) \\ u(0, x) = u_0(x) \end{cases} \tag{1.1}$$

in the sense of the Banach space $X$. The solution $u$ thus obtained may actually be a classical solution of the initial value problem (1.1). If this is the case, it is usually proved by showing that $u$ is more regular than the regularity provided by the abstract theory. A common tool in such regular-ity proofs is the Sobolev's imbedding theorem, a version of which will be stated at the end of this section.

We turn now to the description of the main concrete Banach spaces that will be used in the sequel. In doing so we will use the following notations; $x = (x_1, x_2, \ldots, x_n)$ is a variable point in the $n$-dimensional Euclidean space $\mathbb{R}^n$. For any two such points $x = (x_1, \ldots, x_n), y = (y_1, \ldots, y_n)$ we set $x \cdot y = \sum_{i=1}^n x_i y_i$ and $|x|^2 = x \cdot x$.

An $n$-tuple of nonnegative integers $\alpha = (\alpha_1, \alpha_2, \ldots, \alpha_n)$ is called a multi-index and we define

$$|\alpha| = \sum_{i=1}^n \alpha_i$$

and

$$x^\alpha = x_1^{\alpha_1} x_2^{\alpha_2} \cdots x_n^{\alpha_n} \quad \text{for} \quad x = (x_1, x_2, \ldots, x_n).$$

Denoting $D_k = \partial/\partial x_k$ and $D = (D_1, D_2, \ldots, D_n)$ we have

$$D^\alpha = D_1^{\alpha_1} D_2^{\alpha_2} \cdots D_n^{\alpha_n} = \frac{\partial^{\alpha_1}}{\partial x_1^{\alpha_1}} \frac{\partial^{\alpha_2}}{\partial x_2^{\alpha_2}} \cdots \frac{\partial^{\alpha_n}}{\partial x_n^{\alpha_n}}.$$

Let $\Omega$ be a fixed domain in $\mathbb{R}^n$ with boundary $\partial\Omega$ and closure $\overline{\Omega}$. We will usually assume that $\partial\Omega$ is smooth. This will mean that $\partial\Omega$ is of the class $C^k$ for some suitable $k \geq 1$. Recall that $\partial\Omega$ is of the class $C^k$ if for each point $x \in \partial\Omega$ there is a ball $B$ with center at $x$ such that $\partial\Omega \cap B$ can be represented in the form $x_i = \varphi(x_1, \ldots, x_{i-1}, x_{i+1}, \ldots, x_n)$ for some $i$ with $\varphi$ $k$-times continuously differentiable.

By $C^m(\Omega)$ $(C^m(\overline{\Omega}))$ we denote the set of all $m$-times continuously differentiable real-valued (or sometimes complex-valued) functions in $\Omega$ $(\overline{\Omega})$. $C_0^m(\Omega)$ will denote the subspace of $C^m(\Omega)$ consisting of those functions which have compact support in $\Omega$.

For $u \in C^m(\Omega)$ and $1 \leq p < \infty$ we define

$$\|u\|_{m,p} = \left( \int_\Omega \sum_{|\alpha| \leq m} |D^\alpha u|^p \, dx \right)^{1/p}. \tag{1.2}$$

If $p = 2$ and $u, v \in C^m(\Omega)$ we also define

$$(u, v)_m = \int_\Omega \sum_{|\alpha| \leq m} D^\alpha u \overline{D^\alpha v} \, dx. \tag{1.3}$$

Denoting by $\tilde{C}_p^m(\Omega)$ the subset of $C^m(\Omega)$ consisting of those functions $u$ for which $\|u\|_{m,p} < \infty$ we define $W^{m,p}(\Omega)$ and $W_0^{m,p}(\Omega)$ to be the completions in the norm $\| \cdot \|_{m,p}$ of $\tilde{C}_p^m(\Omega)$ and $C_0^m(\Omega)$ respectively.

It is well known that $W^{m,p}(\Omega)$ and $W_0^{m,p}(\Omega)$ are Banach spaces and obviously $W_0^{m,p}(\Omega) \subset W^{m,p}(\Omega)$. For $p = 2$ we denote $W^{m,2}(\Omega) = H^m(\Omega)$ and $W_0^{m,2}(\Omega) = H_0^m(\Omega)$. The spaces $H^m(\Omega)$ and $H_0^m(\Omega)$ are Hilbert spaces with the scalar product $(\ ,\ )_m$ given by (1.3).

The spaces $W^{m,p}(\Omega)$ defined above, consist of functions $u \in L^p(\Omega)$ whose derivatives $D^\alpha u$, in the sense of distributions, of order $k \leq m$ are in $L^p(\Omega)$.

If $\Omega$ is a bounded domain then the Hölder inequality implies

$$W^{m,p}(\Omega) \subset W^{m,r}(\Omega) \qquad \text{for} \quad 1 \le r \le p \qquad (1.4)$$

and the imbedding is continuous. Furthermore we have:

**Theorem 1.1.** *Let $\Omega$ be a bounded domain in $\mathbb{R}^n$ with a smooth boundary $\partial\Omega$ (e.g. $\partial\Omega$ is of the class $C^1$) and let $1 \le r, p < \infty$. If $j, m$ are integers such that $0 \le j < m$ and*

$$\frac{1}{p} > \frac{1}{r} + \frac{j}{n} - \frac{m}{n} \qquad (1.5)$$

*then $W^{m,r}(\Omega) \subset W^{j,p}(\Omega)$ and the imbedding is compact.*

Some relations between the space $W^{m,p}(\Omega)$ and the spaces of continuously differentiable functions $C^k(\Omega)$ and integrable functions $L^r(\Omega)$, $r \ge 1$ are given next.

**Theorem 1.2** (Sobolev). *Let $\Omega$ be a bounded domain in $\mathbb{R}^n$ with a smooth boundary $\partial\Omega$ (e.g. $\partial\Omega$ is of class $C^m$) then,*

$$W^{k,p}(\Omega) \subset L^{np/(n-kp)}(\Omega) \qquad \text{for} \quad kp < n \qquad (1.6)$$

*and*

$$W^{k,p}(\Omega) \subset C^m(\bar{\Omega}) \qquad \text{for} \quad 0 \le m < k - \frac{n}{p}. \qquad (1.7)$$

*Moreover, there exist constants $C_1$ and $C_2$ such that for any $u \in W^{m,p}(\Omega)$*

$$\|u\|_{0,np/(n-kp)} \le C_1 \|u\|_{k,p} \qquad \text{for} \quad kp < n \qquad (1.8)$$

*and*

$$\sup\{|D^\alpha u(x)| : |\alpha| \le m, x \in \bar{\Omega}\} \le C_2 \|u\|_{k,p} \qquad \text{for} \quad 0 \le m < k - \frac{n}{p}. \qquad (1.9)$$

The space $W_0^{k,p}(\Omega)$ is the subspace of elements of $W^{k,p}(\Omega)$ which vanish in some generalized sense on $\partial\Omega$. It can be shown that if $\partial\Omega$ is of class $C^k$ and $u \in C^{k-1}(\bar{\Omega}) \cap W_0^{k,p}(\Omega)$ then $u$ and its first $k-1$ normal derivatives vanish on $\partial\Omega$ and conversely, if $u \in C^k(\bar{\Omega})$ and its $k-1$ first normal derivatives vanish on $\partial\Omega$ then $u \in W_0^{k,p}(\Omega)$.

# 7.2. Parabolic Equations—$L^2$ Theory

Let $\Omega$ be a bounded domain in $\mathbb{R}^n$ with smooth boundary $\partial\Omega$. Consider the differential operator of order $2m$,

$$A(x, D) = \sum_{|\alpha| \le 2m} a_\alpha(x) D^\alpha \qquad (2.1)$$

where the coefficients $a_\alpha(x)$ are sufficiently smooth complex-valued functions of $x$ in $\bar{\Omega}$. The principal part $A'(x, D)$ of $A(x, D)$ is the operator

$$A'(x, D) = \sum_{|\alpha| = 2m} a_\alpha(x) D^\alpha. \tag{2.2}$$

**Definition 2.1.** The operator $A(x, D)$ is strongly elliptic if there exists a constant $c > 0$ such that

$$\mathrm{Re}\, (-1)^m A'(x, \xi) \geq c|\xi|^{2m} \tag{2.3}$$

for all $x \in \bar{\Omega}$ and $\xi \in \mathbb{R}^n$.

For strongly elliptic operators we have the following important estimate.

**Theorem 2.2** (Gårding's inequality). *If $A(x, D)$ is a strongly elliptic operator of order $2m$ then there exist constants $c_0 > 0$ and $\lambda_0 \geq 0$ such that for every $u \in H^{2m}(\Omega) \cap H_0^m(\Omega)$ we have*

$$\mathrm{Re}\, (Au, u)_0 \geq c_0 \|u\|_{m,2}^2 - \lambda_0 \|u\|_{0,2}^2. \tag{2.4}$$

The proof of Gårding's inequality is usually based on the definition of strong ellipticity and the use of the Fourier transform. For certain simple cases it can be obtained through integration by parts. Consider for example the operator $-\Delta$ given by

$$\Delta u = \sum_{i=1}^n \frac{\partial^2 u}{\partial x_i^2}.$$

$-\Delta$ is clearly strongly elliptic and for every $u \in C_0^\infty(\Omega)$ we have

$$(-\Delta u, u)_0 = -\int_\Omega u\, \Delta u\, dx = \int_\Omega \nabla u \cdot \nabla u\, dx = \|u\|_{1,2}^2 - \|u\|_{0,2}^2 \tag{2.5}$$

where the second equality follows from integration by parts while the last equality is a direct consequence of the definition of the norms $\|\ \|_{m,2}$. A simple limiting argument shows that (2.5) stays true for every $u \in H^2(\Omega) \cap H_0^1(\Omega)$ and (2.4) holds for $-\Delta$.

Let $A(x, D)$ be a strongly elliptic operator of order $2m$ with smooth coefficients $a_\alpha(x)$ in $\bar{\Omega}$. Using integration by parts $(Au + \lambda u, v)_0$ can be extended to a continuous sesquilinear form on $H_0^m(\Omega) \times H_0^m(\Omega)$ for every complex $\lambda$. If $\mathrm{Re}\,\lambda \geq \lambda_0$ then it follows from Gårding's inequality (2.4) that this form is coercive. We may therefore apply the classical Lax-Milgram lemma and derive the existence of a unique weak solution $u \in H_0^m(\Omega)$ of the boundary value problem

$$A(x, D)u + \lambda u = f \tag{2.6}$$

for every $f \in L^2(\Omega)$ and $\mathrm{Re}\,\lambda \geq \lambda_0$. It can then be shown, but this is not easy, that the weak solution of (2.6) actually satisfies $u \in H^{2m}(\Omega)$ and therefore we have,

**Theorem 2.3.** *Let $A(x, D)$ be strongly elliptic of order $2m$. For every $\lambda$ satisfying* Re $\lambda \geq \lambda_0$ *and every $f \in L^2(\Omega)$ there exists a unique $u \in H^{2m}(\Omega)$ $\cap H_0^m(\Omega)$ satisfying the equation*

$$A(x, D)u + \lambda u = f.$$

With a given strongly elliptic operator $A(x, D)$ on a bounded domain $\Omega \subset \mathbb{R}^n$ we associate an unbounded linear operator $A$ acting in the Hilbert space $H = L^2(\Omega)$. This is done as follows:

**Definition 2.4.** Let $A(x, D) = \sum_{|\alpha| \leq 2m} a_\alpha(x) D^\alpha$ be strongly elliptic in $\Omega$ and set $D(A) = H^{2m}(\Omega) \cap H_0^m(\Omega)$. For every $u \in D(A)$ define

$$Au = A(x, D)u.$$

With this definition we have:

**Theorem 2.5.** *Let $H = L^2(\Omega)$ and let $A$ be the operator defined above. For every $\lambda$ satisfying* Re $\lambda \geq \lambda_0$ *the operator $-A_\lambda = -(A + \lambda I)$ is the infinitesimal generator of a $C_0$ semigroup of contractions on $H = L^2(\Omega)$.*

PROOF. Clearly $D(A_\lambda) = D(A) \supset C_0^\infty(\Omega)$. Since $C_0^\infty(\Omega)$ is dense in $H$ it follows that $D(A_\lambda)$ is dense in $H$. Also, from Gårding's inequality we have,

$$\text{Re}\,(-A_\lambda u, u)_0 \leq -c_0\|u\|_{m,2}^2 + (\lambda_0 - \text{Re}\,\lambda)\|u\|_{0,2}^2.$$

Since Re $\lambda \geq \lambda_0$, Re$(-A_\lambda u, u)_0 \leq 0$ and $-A_\lambda$ is dissipative. Finally if Re $\lambda \geq \lambda_0$ then the range of $\mu I + A_\lambda$ is all of $H$ for every $\mu > 0$. This is a direct consequence of Theorem 2.3. From Theorem 1.4.3 it follows now that $-A_\lambda$ is the infinitesimal generator of a $C_0$ semigroup of contractions on $H = L^2(\Omega)$. $\square$

An immediate consequence of Theorem 2.5 is:

**Corollary 2.6.** *Let $A(x, D)$ be a strongly elliptic operator of order $2m$ on a bounded domain $\Omega$ with smooth boundary $\partial\Omega$ in $\mathbb{R}^n$. For every $u_0 \in H^{2m}(\Omega)$ $\cap H_0^m(\Omega)$ the initial value problem*

$$\begin{cases} \dfrac{\partial u(t, x)}{\partial t} + A(x, D)u(t, x) = 0 & \text{in } \Omega \\ u(0, x) = u_0(x) \end{cases} \tag{2.7}$$

*has a unique solution $u(t, x) \in C^1([0, \infty[: H^{2m}(\Omega) \cap H_0^m(\Omega))$.*

Theorem 2.5 implies that if $A(x, D)$ is a strongly elliptic operator then $-A$, defined by definition 2.4, is the infinitesimal generator of a $C_0$

semigroup on $H = L^2(\Omega)$. Actually more is true in this case. Indeed, we have:

**Theorem 2.7.** *If $A(x, D)$ is a strongly elliptic operator of order $2m$ then the operator $-A$ (given by definition 2.4) is the infinitesimal generator of an analytic semigroup of operators on $H = L^2(\Omega)$.*

PROOF. Let $A_{\lambda_0} = A + \lambda_0 I$. From Gårding's inequality we have

$$\mathrm{Re}\left(A_{\lambda_0}u, u\right)_0 \geq c_0 \|u\|_{m,2}^2. \tag{2.8}$$

A simple integration by parts yields for every $u \in D(A_{\lambda_0})$

$$\left|\mathrm{Im}\left(A_{\lambda_0}u, u\right)_0\right| \leq \left|\left(A_{\lambda_0}u, u\right)_0\right| \leq b\|u\|_{m,2}^2 \tag{2.9}$$

for some constant $b > 0$. From (2.8) and (2.9) it follows that the numerical range $S(A_{\lambda_0})$ of $A_{\lambda_0}$ satisfies

$$S\left(A_{\lambda_0}\right) \subset S_{\vartheta_1} = \{\lambda : -\vartheta_1 < \arg\lambda < \vartheta_1\} \tag{2.10}$$

where $\vartheta_1 = \arctan(b/c_0) < \pi/2$. Choosing $\vartheta$ such that $\vartheta_1 < \vartheta < \pi/2$ and denoting $\Sigma_\vartheta = \{\lambda : |\arg\lambda| > \vartheta\}$ there exists a constant $C_\vartheta$ such that

$$d\left(\lambda : S\left(A_{\lambda_0}\right)\right) \geq C_\vartheta |\lambda| \qquad \text{for all } \lambda \in \Sigma_\vartheta \tag{2.11}$$

where $d(\lambda : S)$ denotes the distance between $\lambda$ and the set $S \subset \mathbb{C}$. From Theorem 2.3 it follows that all real $\mu$, $\mu < 0$ are in the resolvent set of $A_{\lambda_0}$ and therefore $\Sigma_\vartheta$ is contained in a component of the complement of $\overline{S\left(A_{\lambda_0}\right)}$ which has a nonempty intersection with $\rho(A_{\lambda_0})$. Theorem 1.3.9 then implies that $\rho(A_{\lambda_0}) \supset \Sigma_\vartheta$ and that for every $\lambda \in \Sigma_\vartheta$,

$$\left\|R\left(\lambda : A_{\lambda_0}\right)\right\| \leq d\left(\lambda : \overline{S\left(A_{\lambda_0}\right)}\right)^{-1} \leq \frac{1}{C_\vartheta|\lambda|} \tag{2.12}$$

and therefore $-A_{\lambda_0}$ is the infinitesimal generator of an analytic semigroup by Theorem 2.5.2 (c). This implies finally (see e.g., Corollary 3.2.2) that $-A$ is the infinitesimal generator of an analytic semigroup of operators on $L^2(\Omega)$.          □

As a direct consequence of Theorem 2.7 and Corollary 4.3.3 we have:

**Corollary 2.8.** *Let $A(x, D)$ be a strongly elliptic operator of order $2m$ in a bounded domain $\Omega \subset \mathbb{R}^n$ and let $f(t, x) \in L^2(\Omega)$ for every $t \geq 0$. If*

$$\int_\Omega |f(t, x) - f(s, x)|^2 \, dx \leq K|t - s|^{2\vartheta} \tag{2.13}$$

*then for every $u_0(x) \in L^2(\Omega)$ the initial value problem*

$$\begin{cases} \dfrac{\partial u}{\partial t} + A(x, D)u = f(t, x) & \text{in} \quad \Omega \times \mathbb{R}^+ \\ u(0, x) = u_0 & \text{in} \quad \Omega \end{cases} \tag{2.14}$$

*has a unique solution* $u(t, x) \in C^1(]0, \infty[ : H^{2m}(\Omega) \cap H_0^m(\Omega))$.

**Remark 2.9.** It is worthwhile to note that if the operator $A$ has constant coefficients, Theorems 2.5 and 2.7 remain true for the domain $\Omega = \mathbb{R}^n$. The proofs of this particular case can be carried out easily using the Fourier transform.

# 7.3. Parabolic Equations—$L^p$ Theory

Let $\Omega$ be a bounded domain with smooth boundary in $\mathbb{R}^n$. In the previous section we considered semigroups defined on the Hilbert space $L^2(\Omega)$. It is often useful to replace the Hilbert space $L^2(\Omega)$ by the Banach space $L^p(\Omega)$, $1 \le p \le \infty$. This is usually important if one wishes to obtain optimal regularity results. In the present section we will discuss the theory of semigroups associated with strongly elliptic differential operators in $L^p(\Omega)$. During most of the section we will restrict ourselves to the values $1 < p < \infty$. Some comments on the cases $p = 1$ and $p = \infty$ will be made at the end of the section.

Let $1 < p < \infty$ and let $\Omega$ be a bounded domain with smooth boundary $\partial\Omega$ in $\mathbb{R}^n$. Let

$$A(x, D)u = \sum_{|\alpha| \le 2m} a_\alpha(x) D^\alpha u \tag{3.1}$$

be a strongly elliptic differential operator in $\Omega$. The operator

$$A^*(x, D)u = \sum_{|\alpha| \le 2m} (-1)^{|\alpha|} D^\alpha\overline{(a_\alpha(x)u)} \tag{3.2}$$

is called the formal adjoint of $A(x, D)$. From the definition of strong ellipticity it is clear that if $A(x, D)$ is strongly elliptic so is $A^*(x, D)$. The coefficients $a_\alpha(x)$ of $A(x, D)$ are tacitly assumed to be smooth enough, e.g. $a_\alpha(x) \in C^{2m}(\overline{\Omega})$ or $a_\alpha(x) \in C^\infty(\overline{\Omega})$. Many of the results however, hold under the weaker assumptions that $a_\alpha(x) \in L^\infty(\Omega)$ for $0 \le |\alpha| < 2m$ and $a_\alpha(x) \in C(\overline{\Omega})$ for $|\alpha| = 2m$. For strongly elliptic differential operators the following fundamental a-priori estimates have been established.

**Theorem 3.1.** *Let $A$ be a strongly elliptic operator of order $2m$ on a bounded domain $\Omega$ with smooth boundary $\partial\Omega$ in $\mathbb{R}^n$ and let $1 < p < \infty$. There exists a* *constant C such that*

$$\|u\|_{2m,p} \leq C(\|Au\|_{0,p} + \|u\|_{0,p}) \tag{3.3}$$

*for every $u \in W^{2m,p}(\Omega) \cap W_0^{m,p}(\Omega)$.*

Using this a-priori estimate together with an argument of S. Agmon one proves the following theorem.

**Theorem 3.2.** *Let A be a strongly elliptic operator of order $2m$ on a bounded domain $\Omega$ with smooth boundary $\partial\Omega$ in $\mathbb{R}^n$ and let $1 < p < \infty$. There exist constants $C > 0$, $R \geq 0$ and $0 < \vartheta < \pi/2$ such that*

$$\|u\|_{0,p} \leq \frac{C}{|\lambda|} \|(\lambda I + A)u\|_{0,p} \tag{3.4}$$

*for every $u \in W^{2m,p}(\Omega) \cap W_0^{m,p}(\Omega)$ and $\lambda \in \mathbb{C}$ satisfying $|\lambda| \geq R$ and $\vartheta - \pi < \arg\lambda < \pi - \vartheta$.*

With a strongly elliptic operator $A(x, D)$ we associate a linear (unbounded) operator $A_p$ in $L^p(\Omega)$ as follows:

**Definition 3.3.** *Let $A = A(x, D)$ be a strongly elliptic operator of order $2m$ on a bounded domain $\Omega$ in $\mathbb{R}^n$ and let $1 < p < \infty$. Set*

$$D(A_p) = W^{2m,p}(\Omega) \cap W_0^{m,p}(\Omega) \tag{3.5}$$

*and*

$$A_p u = A(x, D)u \qquad \text{for} \quad u \in D(A_p). \tag{3.6}$$

The domain $D(A_p)$ of $A_p$ contains $C_0^\infty(\Omega)$ and it is therefore dense in $L^p(\Omega)$. Moreover, from Theorem 3.1 it follows readily that $A_p$ is a closed operator in $L^p(\Omega)$.

With this definition we have:

**Lemma 3.4.** *Let $A(x, D)$ be a strongly elliptic operator of order $2m$ on $\Omega$ and let $A_p$, $1 < p < \infty$, be the operator associated with it by Definition 3.3. The operator $A_q^*$, $q = p/(p - 1)$ associated by Definition 3.3 with the formal adjoint $A^*(x, D)$ of $A(x, D)$ on $L^q(\Omega)$ is the adjoint operator of $A_p$.*

From Theorems 3.1 and 3.2 we deduce,

**Theorem 3.5.** *Let $A(x, D)$ be a strongly elliptic operator of order $2m$ on a bounded domain $\Omega$ with smooth boundary $\partial\Omega$ in $\mathbb{R}^n$ and let $1 < p < \infty$. If $A_p$*

*is the operator associated with A by Definition 3.3 then* $-A_p$ *is the infinitesimal generator of an analytic semigroup on* $L^p(\Omega)$.

PROOF. We have already noted that $D(A_p)$ is dense in $L^p(\Omega)$ and that $A_p$ is a closed operator in $L^p(\Omega)$. From Theorem 3.2 it follows that for every

$$\lambda \in \Sigma_{\vartheta} = \{\mu : \vartheta - \pi < \arg \mu < \pi - \vartheta, |\mu| \geq R\} \tag{3.7}$$

the operator $\lambda I + A_p$ is injective and has closed range. Similarly, it follows from Theorem 3.2, applied to the operator $A_q^*$ on $L^q(\Omega)$, that there are constants $R' \geq 0$ and $0 < \vartheta' < \pi/2$ such that for every $\lambda \in \Sigma_{\vartheta'} = \{\mu : \vartheta' - \pi < \arg \mu < \pi - \vartheta', |\mu| \geq R'\}$ $\bar{\lambda} I + A_q^*$ is injective. Let $\vartheta_1 = \min(\vartheta, \vartheta')$ and $R_1 = \max(R, R')$ then for every

$$\lambda \in \Sigma_{\vartheta_1} = \{\mu : \vartheta_1 - \pi < \arg \mu < \pi - \vartheta_1, |\mu| \geq R_1\}$$

$\lambda I + A_p$ is bijective. Indeed, we have already seen that it is injective so it remains only to show that it is surjective. Let $\lambda \in \Sigma_{\vartheta_1}$. If $v \in L^q(\Omega)$ satisfies $\langle (\lambda I + A_p)u, v \rangle = 0$ for all $u \in D(A_p)$ then it follows from Lemma 3.4 that $v \in D(A_q^*)$ and that $\langle u, (\bar{\lambda} I + A_q^*)v \rangle = 0$ for all $u \in D(A_p)$. Since $D(A_p)$ is dense in $L^p(\Omega)$, $(\bar{\lambda} I + A_q^*)v = 0$ and the injectivity of $\bar{\lambda} I + A_q^*$ implies $v = 0$. Thus for $\lambda \in \Sigma_{\vartheta_1}$, $\lambda I + A_p$ is invertible and from (3.4) it follows that

$$\|(\lambda I + A_p)^{-1}\| \leq \frac{C}{|\lambda|} \qquad \text{for all} \quad \lambda \in \Sigma_{\vartheta_1}$$

which, by Theorem 2.5.2 (c), implies that $-A_p$ is the infinitesimal generator of an analytic semigroup on $L^p(\Omega)$.                                                    $\square$

Theorem 3.5 is based on the deep a-priori $L^p$ estimate (3.3). For second order strongly elliptic operators with real coefficients written in divergence form, Theorem 3.5 can be proved directly without the use of Theorems 3.1 and 3.2. This will be done next.

Let $\Omega$ be a bounded domain with smooth boundary $\partial \Omega$ in $\mathbb{R}^n$ and let $A(x, D)$ be the symmetric second order differential operator given by

$$A(x, D)u = - \sum_{k,l=1}^n \frac{\partial}{\partial x_k}\left(a_{k,l}(x)\frac{\partial u}{\partial x_l}\right). \tag{3.8}$$

We assume that the coefficients $a_{k,l}(x) = a_{l,k}(x)$ are real valued and continuously differentiable in $\bar{\Omega}$ and that $A(x, D)$ is strongly elliptic, i.e.

that there is a constant $C_0 > 0$ such that

$$\sum_{k,l=1}^{n} a_{k,l}(x)\xi_k\xi_l \geq C_0 \sum_{k=1}^{n} \xi_k^2 = C_0|\xi|^2 \tag{3.9}$$

for all real $\xi_k$, $1 \leq k \leq n$.

With the second order symmetric differential operator $A(x, D)$ given by (3.8) we associate the operator $A_p$ on $L^p(\Omega)$, $1 < p < \infty$, by Definition 3.3. We then have:

**Theorem 3.6.** *Let $1 < p < \infty$, the operator $-A_p$ is the infinitesimal generator of an analytic semigroup of contractions on $L^p(\Omega)$.*

PROOF. Let $1 < p < \infty$ be fixed and let $q = p/(p-1)$. We denote the pairing between $L^p(\Omega)$ and $L^q(\Omega)$ by $\langle \ , \ \rangle$. If $u \in D(A_p)$ then the function $u^* = |u|^{p-2}\bar{u}$ is in $L^q(\Omega)$ and $\langle u, u^* \rangle = \|u\|_{0,p}^p$. Let $2 \leq p < \infty$, integration by parts yields

$$\langle A_p u, u^* \rangle = -\int_{\Omega} \sum_{k,l=1}^{n} \frac{\partial}{\partial x_k}\left(a_{k,l}\frac{\partial u}{\partial x_l}\right)\bar{u}|u|^{p-2}\,dx$$

$$= \int_{\Omega} \sum_{k,l=1}^{n} a_{k,l}\frac{\partial u}{\partial x_l}\frac{\partial}{\partial x_k}\left(\bar{u}|u|^{p-2}\right)\,dx$$

$$= \int_{\Omega} \sum_{k,l=1}^{n} a_{k,l}\left(|u|^{p-2}\frac{\partial u}{\partial x_l}\frac{\partial \bar{u}}{\partial x_k} + \bar{u}\frac{\partial u}{\partial x_l}\frac{\partial |u|^{p-2}}{\partial x_k}\right)\,dx.$$

But,

$$\frac{\partial}{\partial x_k}|u|^{p-2} = \frac{1}{2}(p-2)|u|^{p-4}\left(\bar{u}\frac{\partial u}{\partial x_k} + u\frac{\partial \bar{u}}{\partial x_k}\right).$$

Denoting, $|u|^{(p-4)/2}\bar{u}(\partial u/\partial x_k) = \alpha_k + i\beta_k$ we find after a simple computation that

$$\langle A_p u, u^* \rangle = \int_{\Omega} \sum_{k,l=1}^{n} a_{k,l}((p-1)\alpha_k\alpha_l + \beta_k\beta_l + i(p-2)\alpha_k\beta_l)\,dx. \tag{3.10}$$

Let $|\alpha_{k,l}(x)| \leq M$ for $1 \leq k, l \leq n$ and $x \in \bar{\Omega}$ and set

$$|\alpha|^2 = \sum_{k=1}^{n} \int_{\Omega} \alpha_k^2\,dx \qquad |\beta|^2 = \sum_{k=1}^{n} \int_{\Omega} \beta_k^2\,dx$$

then it follows easily from (3.9) and (3.10) that

$$\mathrm{Re}\,\langle A_p u, u^* \rangle \geq C_0((p-1)|\alpha|^2 + |\beta|^2) \geq 0 \tag{3.11}$$

and

$$\frac{|\mathrm{Im}\,\langle A_p u, u^* \rangle|}{|\mathrm{Re}\,\langle A_p u, u^* \rangle|} \leq \frac{|p-2|M\left(\frac{\rho}{2}|\alpha|^2 + \frac{1}{2\rho}|\beta|^2\right)}{C_0((p-1)|\alpha|^2 + |\beta|^2)} \tag{3.12}$$

for every $\rho > 0$. Choosing $\rho = \sqrt{p - 1}$ in (3.12) yields

$$\frac{|\operatorname{Im} \langle A_p u, u^* \rangle|}{|\operatorname{Re} \langle A_p u, u^* \rangle|} \leq \frac{M|p - 2|}{2 C_0 \sqrt{p - 1}}. \tag{3.13}$$

From (3.11) it follows readily that for every $\lambda > 0$ and $u \in D(A_p)$ we have

$$\lambda \|u\|_{0, p} \leq \|(\lambda I + A_p) u\|_{0, p} \tag{3.14}$$

and therefore $\lambda I + A_p$ is injective and has closed range for every $\lambda > 0$. Since (3.14) holds for every $2 \leq p < \infty$ it follows that for $\lambda > 0$, $\lambda I + A_p$ is also surjective. Indeed, if $v \in L^q(\Omega)$ satisfies $\langle (\lambda I + A_p) u, v \rangle = 0$ for all $u \in D(A_p)$ then, since $A(x, D)$ is formally self adjoint, it follows from Lemma 3.4 that $v \in D(A_q)$, $q = p/(p - 1)$, and that $\langle u, (\bar{\lambda} I + A_q) v \rangle = 0$ for every $u \in D(A_p)$. Since $D(A_p)$ is dense in $L^p(\Omega)$, $(\bar{\lambda} I + A_q) v = 0$ and (3.14), with $p$ replaced by $q$, implies $v = 0$. Thus, $\lambda I + A_p$ is bijective for $\lambda > 0$ and as a consequence of (3.14) we have

$$\|(\lambda I + A_p)^{-1}\|_{0, p} \leq \frac{1}{\lambda} \qquad \text{for} \quad \lambda > 0. \tag{3.15}$$

The Hille-Yosida theorem (Theorem 1.3.1) now implies that $-A_p$ is the infinitesimal generator of a contraction semigroup on $L^p(\Omega)$ for every $2 \leq p < \infty$. Finally, to prove that the semigroup generated by $-A_p$ is analytic we observe that by (3.11) and (3.13) the numerical range $S(-A_p)$ of $-A_p$ is contained in the set $S_{\vartheta_1} = \{\lambda : |\arg \lambda| > \pi - \vartheta_1\}$ where $\vartheta_1 = \arctan(M|p - 2|/2 C_0 \sqrt{p - 1})$, $0 < \vartheta_1 < \pi/2$. Choosing $\vartheta_1 < \vartheta < \pi/2$ and denoting

$$\Sigma_\vartheta = \{\lambda : |\arg \lambda| < \pi - \vartheta\} \tag{3.16}$$

it follows that there is a constant $C_\vartheta > 0$ for which

$$d\left(\lambda : \overline{S(-A_p)}\right) \geq C_\vartheta |\lambda| \qquad \text{for} \quad \lambda \in \Sigma_\vartheta.$$

Since $\lambda > 0$ is in the resolvent set $\rho(-A_p)$ of $-A_p$ by the first part of the proof, it follows from Theorem 1.3.9 that $\rho(-A_p) \supset \Sigma_\vartheta$ and that

$$\|(\lambda I + A_p)^{-1}\|_{0, p} \leq \frac{1}{C_\vartheta |\lambda|} \qquad \text{for} \quad \lambda \in \Sigma_\vartheta \tag{3.17}$$

whence by Theorem 2.5.2 (c), $-A_p$ is the infinitesimal generator of an analytic semigroup on $L^p(\Omega)$ for $2 \leq p < \infty$. The case $1 < p < 2$ is obtained from the previous case $2 \leq p < \infty$ by a duality argument. $\qquad \square$

We turn now to the cases $p = 1$ and $p = \infty$ and start with $p = \infty$. Recall that the norm in $L^\infty(\Omega)$ is defined by

$$\|u\|_{0,\infty} = \operatorname{ess\,sup} \{|u(x)| : x \in \Omega\}.$$

Let $A(x, D)$ be the uniformly elliptic operator of order $2m$ given by (3.1) and defined on a bounded domain $\Omega \subset \mathbb{R}^n$ with smooth boundary $\partial\Omega$. We associate with $A(x, D)$ an operator $A_\infty$ on $L^\infty(\Omega)$ as follows:

$$D(A_\infty) = \{u : u \in W^{2m,p}(\Omega) \qquad \text{for all} \quad p > n,$$

$$A(x, D)u \in L^\infty(\Omega), D^\beta u = 0 \qquad \text{on} \quad \partial\Omega \text{ for } 0 \le |\beta| < m\} \qquad (3.18)$$

and

$$A_\infty u = A(x, D)u \qquad \text{for} \quad u \in D(A_\infty). \qquad (3.19)$$

We note first that from Sobolev's theorem (Theorem 1.2) it follows that $D(A_\infty) \subset C^{2m-1}(\bar{\Omega})$ and therefore, since by our assumptions $\partial\Omega$ is smooth the conditions $D^\beta u = 0$ for $0 \le |\beta| < m$ on $\partial\Omega$ make sense. Moreover, from the regularity of the boundary and the definition of $D(A_\infty)$ it follows that $D(A_\infty) \subset W^{2m,p}(\Omega) \cap W_0^{m,p}(\Omega) = D(A_p)$ for every $p > n$. But $-A_\infty$ is not the infinitesimal generator of a $C_0$ semigroup of operators on $L^\infty(\Omega)$. The reason for this is that $D(A_\infty)$ is never dense in $L^\infty(\Omega)$. Indeed, we have noted above that $D(A_\infty) \subset C(\bar{\Omega})$ and therefore also $\overline{D(A_\infty)} \subset C(\bar{\Omega})$, where $\overline{D(A_\infty)}$ is the closure of $D(A_\infty)$ in the $\| \ \|_{0,\infty}$ norm. Since $C(\bar{\Omega})$ is not dense in $L^\infty(\Omega)$, $D(A_\infty)$ cannot be dense in $L^\infty(\Omega)$.

To overcome this difficulty we restrict ourselves to spaces of continuous functions on $\bar{\Omega}$. We define

$$D(A_c) = \{u : u \in D(A_\infty), A(x, D)u \in C(\bar{\Omega}), A(x, D)u = 0 \text{ on } \partial\Omega\}$$

$$\qquad (3.20)$$

and

$$A_c u = A(x, D)u \qquad \text{for} \quad u \in D(A_c). \qquad (3.21)$$

The operator $A_c$ thus defined is considered as an operator on the space:

$$C = \{u : u \in C(\bar{\Omega}), u = 0 \text{ on } \partial\Omega\} \qquad (3.22)$$

and we have:

**Theorem 3.7.** *The operator* $-A_c$ *is the infinitesimal generator of an analytic semigroup on* $C$.

The proof of Theorem 3.7 is based on a-priori estimates in the norms $\| \ \|_{k,\infty}$, similar to the estimates (3.3). Since we have no such a-priori estimates for the case $p = 1$, the results for this case will be derived in a different way, which will exploit a duality between continuous functions and $L^1$ functions. We start with a lemma.

**Lemma 3.8.** *Let $\Omega$ be a bounded domain in $\mathbb{R}^n$. For $u \in L^1(\Omega)$ we have*

$$\|u\|_{0,1} = \sup \left\{ \int_\Omega u(x)\varphi(x)\,dx : \varphi \in C_0^\infty(\Omega), \|\varphi\|_{0,\infty} \le 1 \right\}. \quad (3.23)$$

PROOF. Since for every $\varphi \in C_0^\infty(\Omega)$ satisfying $\|\varphi\|_{0,\infty} \le 1$ we have

$$\left| \int_\Omega u\varphi\,dx \right| \le \|\varphi\|_{0,\infty} \|u\|_{0,1} \le \|u\|_{0,1}$$

the sup on the right-hand side of (3.23) is clearly less or equal to $\|u\|_{0,1}$. Since $C_0^\infty(\Omega)$ is dense in $L^1(\Omega)$ it suffices to prove the result for $u \in C_0^\infty(\Omega)$. Let $p_n(z) \in C^\infty(\mathbb{C})$ be such that $p_n(0) = 0$, $|p_n(z)| \le 1$ and $p_n(z) = \bar{z}/|z|$ for $|z| \ge 1/n$. Then $p_n(u(x)) \in C_0^\infty(\Omega)$ and $\|p_n(u(x))\|_{0,\infty} \le 1$. Also,

$$\lim_{n\to\infty} \int_\Omega u(x)p_n(u(x))\,dx = \int_\Omega |u(x)|\,dx = \|u\|_{0,1}$$

and thus the sup on the right-hand side of (3.23) is larger or equal to $\|u\|_{0,1}$. $\square$

We turn now to the definition of the operator $A$, associated with the strongly elliptic operator $A(x, D)$ given by (3.1), on the space $L^1(\Omega)$.

**Definition 3.9.** Let $A(x, D)$ be the strongly elliptic operator of order $2m$ on the bounded domain $\Omega \subset \mathbb{R}^n$ with smooth boundary $\partial\Omega$ given by (3.1). Set,

$$D(A_1) = \{u : u \in W^{2m-1,1}(\Omega) \cap W_0^{m,1}(\Omega), A(x, D)u \in L^1(\Omega)\} \quad (3.24)$$

where $A(x, D)u$ is understood in the sense of distributions. For $u \in D(A_1)$, $A_1$ is defined by

$$A_1 u = A(x, D)u. \quad (3.25)$$

**Theorem 3.10.** *The operator $-A_1$ is the infinitesimal generator of an analytic semigroup on $L^1(\Omega)$.*

PROOF. Let

$$A(x, D)u = \sum_{|\alpha| \le 2m} a_\alpha(x) D^\alpha u$$

and

$$\tilde{A}(x, D)u = \sum_{|\alpha| \le 2m} (-1)^{|\alpha|} D^\alpha(a_\alpha(x)u).$$

Let $\tilde{A}_c$ be the operator associated with $\tilde{A}(x, D)$ on the space $C$ (given by

(3.22)). Since $\tilde{A}(x, D)$ is strongly elliptic together with $A(x, D)$ it follows from Theorem 3.7 that $-\tilde{A}_c$ is the infinitesimal generator of an analytic semigroup on $C$. Theorem 2.5.2 then implies that there are constants $M > 0$, $R \geq 0$ and $0 < \vartheta < \pi/2$ such that

$$\|(\lambda I + \tilde{A}_c)^{-1}\|_{0,\infty} \leq M|\lambda|^{-1} \tag{3.26}$$

for every $\lambda \in \Sigma_\vartheta = \{\mu : |\arg \mu| > \vartheta, |\mu| \geq R\}$. Now, let $u \in D(A_1)$. From Lemma 3.8 it follows that

$$\|u\|_{0,1} = \sup \left\{ \int_\Omega u\varphi \, dx : \varphi \in C_0^\infty(\Omega), \|\varphi\|_{0,\infty} \leq 1 \right\}. \tag{3.27}$$

Since $C_0^\infty(\Omega)$ is contained in the range of $\lambda I + \tilde{A}_c$ for every $\lambda \in \Sigma_\vartheta$ it follows from (3.26) and (3.27) that

$$\|u\|_{0,1} = \sup \left\{ \int_\Omega u(\lambda I + \tilde{A}_c)v \, dx : v \in D(\tilde{A}_c), \|v\|_{0,\infty} \leq M|\lambda|^{-1} \right\}$$

which implies that for every $v \in D(\tilde{A}_c)$, $\|v\|_{0,\infty} \leq M|\lambda|^{-1}$ we have

$$\|u\|_{0,1} \leq \left| \int_\Omega u(\lambda I + \tilde{A}_c)v \, dx \right| = \left| \int_\Omega (\lambda I + A_1)uv \, dx \right|$$

$$\leq \|(\lambda I + A_1)u\|_{0,1}\|v\|_{0,\infty} \leq M|\lambda|^{-1}\|(\lambda I + A_1)u\|_{0,1}.$$

Thus for every $\lambda \in \Sigma_\vartheta$, $\lambda I + A_1$ is injective and has closed range. Moreover, since $D(A_2) \subset D(A_1)$ the range of $\lambda I + A_1$ contains $L^2(\Omega)$, which is dense in $L^1(\Omega)$, and therefore the range of $\lambda I + A_1$ is all of $L^1(\Omega)$ and

$$\|(\lambda I + A_1)^{-1}\|_{0,1} \leq M|\lambda|^{-1}$$

for every $\lambda \in \Sigma_\vartheta$. From Theorem 2.5.2 it follows therefore that $-A_1$ is the infinitesimal generator of an analytic semigroup on $L^1(\Omega)$. $\qquad \square$

## 7.4. The Wave Equation

In this section we consider the initial value problem for the wave equation in $\mathbb{R}^n$ i.e., the initial value problem

$$\begin{cases} \dfrac{\partial^2 u}{\partial t^2} = \Delta u, & \text{for } x \in \mathbb{R}^n, t > 0 \\ u(0, x) = u_1(x), \dfrac{\partial u}{\partial t}(0, x) = u_2(x), & \text{for } x \in \mathbb{R}^n. \end{cases} \tag{4.1}$$

This problem is equivalent to the first order system:

$$\begin{cases} \dfrac{\partial}{\partial t}\begin{pmatrix} u_1 \\ u_2 \end{pmatrix} = \begin{pmatrix} 0 & I \\ \Delta & 0 \end{pmatrix}\begin{pmatrix} u_1 \\ u_2 \end{pmatrix} & \text{for } x \in \mathbb{R}^n, t > 0 \end{cases} \tag{4.2}$$

and

$$\left\{ \begin{pmatrix} u_1(0, x) \\ u_2(0, x) \end{pmatrix} = \begin{pmatrix} u_1(x) \\ u_2(x) \end{pmatrix} \right. \qquad \text{for} \quad x \in \mathbb{R}^n$$

In order to use the theory of semigroups we are interested in showing that the operator $\begin{pmatrix} 0 & I \\ \Delta & 0 \end{pmatrix}$ is the infinitesimal generator of a $C_0$ semigroup of operators in some appropriately chosen Banach space of functions. It turns out that the right space is the Hilbert space $H^1(\mathbb{R}^n) \times L^2(\mathbb{R}^n)$.

The spaces $H^k(\mathbb{R}^n)$ have been defined in Section 7.1. For the special case where $\Omega = \mathbb{R}^n$ they can be characterized in the following useful way: Let $f \in L^2(\mathbb{R}^n)$ and let

$$\hat{f}(\xi) = (2\pi)^{-n/2} \int_{\mathbb{R}^n} e^{-ix \cdot \xi} f(x) \, dx \qquad (4.3)$$

be its Fourier transform. The function $f \in H^k(\mathbb{R}^n)$ if and only if $(1 + |\xi|^2)^{k/2} \hat{f}(\xi) \in L^2(\mathbb{R}^n)$. This characterization is a simple consequence of Parseval's identity and the elementary properties of the Fourier Transform.

Given a vector $U = [u_1, u_2] \in C_0^\infty(\mathbb{R}^n) \times C_0^\infty(\mathbb{R}^n)$ we define the norm

$$\||U|\| = \||[u_1, u_2]|\| = \left( \int_{\mathbb{R}^n} (|u_1|^2 + |\nabla u_1|^2 + |u_2|^2) \, dx \right)^{1/2}. \qquad (4.4)$$

It is easy to check that the completion of $C_0^\infty(\mathbb{R}^n) \times C_0^\infty(\mathbb{R}^n)$ with respect to the norm $\|| \cdot |\|$ is the Hilbert space $H = H^1(\mathbb{R}^n) \times L^2(\mathbb{R}^n)$. In this Hilbert space we define the operator $A$ associated with the differential operator $\begin{pmatrix} 0 & I \\ \Delta & 0 \end{pmatrix}$ as follows;

**Definition 4.1.** Let

$$D(A) = H^2(\mathbb{R}^n) \times H^1(\mathbb{R}^n) \qquad (4.5)$$

and for $U = [u_1, u_2] \in D(A)$ let

$$AU = A[u_1, u_2] = [u_2, \Delta u_1]. \qquad (4.6)$$

To prove that the operator $A$ defined by (4.5), (4.6) is the infinitesimal generator of a $C_0$ group of operators on $H$ we will need the following simple preliminaries.

**Lemma 4.2.** If $\nu > 0$ and $f \in H^k(\mathbb{R}^n)$, $k \geq 0$, then there is a unique function $u \in H^{k+2}(\mathbb{R}^n)$ satisfying

$$u - \nu \Delta u = f. \qquad (4.7)$$

PROOF. Let $\hat{f}(\xi)$ be the Fourier transform of $f$ and let $\bar{u}(\xi) = (1 + \nu|\xi|^2)^{-1} \hat{f}(\xi)$. Since $f \in H^k(\mathbb{R}^n)$, $(1 + |\xi|^2)^{k/2} \hat{f}(\xi) \in L^2(\mathbb{R}^n)$ and therefore $(1 + |\xi|^2)^{(k+2)/2} \tilde{u}(\xi) \in L^2(\mathbf{R}^n)$. If $u$ is defined by

$$u(x) = (2\pi)^{-n/2} \int_{\mathbf{R}^n} e^{ix \cdot \xi} \tilde{u}(\xi) \, d\xi$$

then $u \in H^{k+2}(\mathbf{R}^n)$ and $u$ is a solution of (4.7). The uniqueness of the solution $u$ of (4.7) follows from the fact that if $w \in H^{k+2}(\mathbf{R}^n)$ satisfies $w - \nu\,\Delta w = 0$ then $\hat{w} = 0$ and therefore $w = 0$.                    $\square$

**Lemma 4.3.** *For every $F = [f_1, f_2] \in C_0^\infty(\mathbf{R}^n) \times C_0^\infty(\mathbf{R}^n)$ and real $\lambda \neq 0$ the equation*

$$U - \lambda A U = F \tag{4.8}$$

*has a unique solution $U = [u_1, u_2] \in H^k(\mathbf{R}^n) \times H^{k-2}(\mathbf{R}^n)$ for every $k \geq 2$. Moreover,*

$$||| U ||| \leq (1 - |\lambda|^{-1}) ||| F |||      \text{for}   0 < |\lambda| < 1. \tag{4.9}$$

PROOF. Let $\lambda \neq 0$ be real and let $w_1, w_2$ be solutions of

$$w_i - \lambda^2 \, \Delta w_i = f_i    i = 1, 2. \tag{4.10}$$

From Lemma 4.2 it is clear that such solutions exist and that $w_i \in H^k(\mathbf{R}^n)$ for every $k \geq 0$. Set $u_1 = w_1 + \lambda w_2$, $u_2 = w_2 + \lambda\,\Delta w_1$. It is easy to check that $U = [u_1, u_2]$ is a solution of (4.8) and therefore $u_1 - \lambda u_2 = f_1$ and $u_2 - \lambda\,\Delta u_1 = f_2$. Moreover, $U \in H^k(\mathbf{R}^n) \times H^{k-2}(\mathbf{R}^n)$ for every $k \geq 2$. Denoting $(\,,\,)_0$ the scalar product in $L^2(\mathbf{R}^n)$ we have

$$\begin{aligned}
||| F |||^2 &= (f_1 - \Delta f_1, f_1)_0 + (f_2, f_2)_0 \\
&= (u_1 - \lambda u_2 - \Delta u_1 + \lambda\,\Delta u_2, u_1 - \lambda u_2)_0 \\
&\quad + (u_2 - \lambda\,\Delta u_1, u_2 - \lambda\,\Delta u_1)_0 \\
&\geq (u_1 - \Delta u_1, u_1)_0 + \|u_2\|_{0,2}^2 - 2|\lambda|\,\mathrm{Re}\,(u_1, u_2)_0 \\
&\geq (1 - |\lambda|) ||| U |||^2.
\end{aligned}$$

Therefore if $0 < |\lambda| < 1$,

$$||| F |||^2 \geq (1 - |\lambda|)^2 ||| U |||^2. \tag{4.11}$$

$\square$

Lemma 4.3 shows that the range of the operator $I - \lambda A$ contains $C_0^\infty(\mathbf{R}^n) \times C_0^\infty(\mathbf{R}^n)$ for all real $\lambda$ satisfying $0 < |\lambda| < 1$. Since the operator $A$ defined by Definition 4.1 is closed, the range of $I - \lambda A$ is all of $H = H^1(\mathbf{R}^n) \times L^2(\mathbf{R}^n)$ and we have

**Corollary 4.4.** *For every $F \in H^1(\mathbf{R}^n) \times L^2(\mathbf{R}^n)$ and real $\lambda$ satisfying $0 < |\lambda| < 1$ the equation*

$$U - \lambda A U = F \tag{4.12}$$

*has a unique solution $U \in H^2(\mathbb{R}^n) \times H^1(\mathbb{R}^n)$ and*

$$\||U\|| \leq (1 - |\lambda|)^{-1} \||F\||. \tag{4.13}$$

**Theorem 4.5.** *The operator $A$, defined in Definition 4.1 is the infinitesimal generator of a $C_0$ group on $H = H^1(\mathbb{R}^n) \times L^2(\mathbb{R}^n)$, satisfying*

$$\|T(t)\| \leq e^{|t|}. \tag{4.14}$$

PROOF. The domain of $A$, $H^2(\mathbb{R}^n) \times H^1(\mathbb{R}^n)$ is clearly dense in $H$. From Corollary 4.4 it follows that $(\mu I - A)^{-1}$ exists for $|\mu| > 1$ and satisfies

$$\left\|(\mu I - A)^{-1}\right\| \leq \frac{1}{|\mu| - 1} \qquad \text{for} \quad |\mu| > 1. \tag{4.15}$$

From Theorem 1.6.3 it follows that $A$ is the infinitesimal generator of a group $T(t)$ satisfying (4.14). $\qquad\square$

**Corollary 4.6.** *For every $f_1 \in H^2(\mathbb{R}^n)$, $f_2 \in H^1(\mathbb{R}^n)$ there exists a unique $u(t, x) \in C^1([0, \infty[; H^2(\mathbb{R}^n))$ satisfying the initial value problem*

$$\begin{cases} \dfrac{\partial^2 u}{\partial t^2} = \Delta u \\ u(0, x) = f_1(x) \\ u_t'(0, x) = f_2(x). \end{cases} \tag{4.16}$$

PROOF. Let $T(t)$ be the semigroup generated by $A$ and set

$$[u_1(t, x), u_2(t, x)] = T(t)[f_1(x), f_2(x)]$$

then

$$\frac{\partial}{\partial t}[u_1, u_2] = A[u_1, u_2] = [u_2, \Delta u_1]$$

and $u_1$ is the desired solution. $\qquad\square$

We conclude this section by showing that if the initial values $f_1$, $f_2$ in the initial value problem (4.16) are smooth so is the solution. To this end we note that Sobolev's theorem (Theorem 1.2) can be extended to the special unbounded domain $\Omega = \mathbb{R}^n$ as follows:

**Theorem 4.7.** *For $0 \leq m < k - n/2$ we have*

$$H^k(\mathbb{R}^n) \subset C^m(\mathbb{R}^n). \tag{4.17}$$

PROOF. Let $v \in C_0^\infty(\mathbb{R}^n)$ then, as is well known, $\xi^\alpha \hat{v}(\xi) \in L^2(\mathbb{R}^n)$ for every $\alpha$ and

$$D^\alpha v(x) = (2\pi)^{-(n/2)} \int_{\mathbb{R}^n} i^{|\alpha|} \xi^\alpha e^{ix \cdot \xi} \hat{v}(\xi) \, d\xi.$$

Estimating $D^\alpha v(x)$, by the Cauchy-Schwartz inequality, we find for every $N > n/2$,

$$|D^\alpha v(x)|^2 \leq (2\pi)^{-n} \int_{\mathbb{R}^n} (1 + |\xi|^2)^{-N} d\xi \int_{\mathbb{R}^n} |\xi|^{2|\alpha|} (1 + |\xi|^2)^N |\hat{v}(\xi)|^2 d\xi$$

$$\leq C_1 \int_{\mathbb{R}^n} (1 + |\xi|^2)^{N+|\alpha|} |\hat{v}(\xi)|^2 d\xi \leq C_2 \|v\|_{N+|\alpha|,2}^2 \qquad (4.18)$$

where $C_1$ and $C_2$ are constants depending on $N$ and $|\alpha|$. Let $u \in H^k(\mathbb{R}^n)$ and let $u_n \in C_0^\infty(\mathbb{R}^n)$ be such that $u_n \to u$ in $H^k(\mathbb{R}^n)$. Then, from (4.18) it follows that $D^\alpha u_n \to D^\alpha u$ uniformly in $\mathbb{R}^n$ for all $\alpha$ satisfying $|\alpha| \leq m < k - n/2$ and therefore $u \in C^m(\mathbb{R}^n)$ as desired.          $\square$

Consider now the initial value problem (4.16) with $f_1, f_2 \in C_0^\infty(\mathbb{R}^n)$. Clearly, $[f_1, f_2] \in D(A^k)$ for every $k \geq 1$, where $A$ is the operator defined in Definition 4.1. Therefore $[u_1, u_2] = T(t)[f_1, f_2] \in D(A^k)$ for every $k \geq 1$ and in particular $\Delta^k u_1 \in L^2(\mathbb{R}^n)$ for all $k \geq 0$. This implies that $u_1 \in H^k(\mathbb{R}^n)$ for every $k \geq 0$ and from Theorem 4.8 it follows that $u_1$, the solution of (4.16), satisfies $u_1(t, x) \in C^\infty(\mathbb{R}^n)$ for every $t \geq 0$. With a little more effort one can show that actually, $u_1(t, x) \in C^\infty(\mathbb{R} \times \mathbb{R}^n)$ and is a classical smooth solution of (4.16), but we will not do this here.

## 7.5. A Schrödinger Equation

The Schrödinger equation is given by

$$\frac{1}{i} \frac{\partial u}{\partial t} = \Delta u - Vu \qquad (5.1)$$

where the function $V$ is called the potential. We will consider this equation in the Hilbert space $H = L^2(\mathbb{R}^n)$. We start with the definition of the operator $A_0$ associated with the differential operator $i\Delta$.

**Definition 5.1.** Let $D(A_0) = H^2(\mathbb{R}^n)$ where the space $H^2(\mathbb{R}^n)$ is defined in Section 7.1. For $u \in D(A_0)$ let

$$A_0 u = i \Delta u \qquad (5.2)$$

**Lemma 5.2.** *The operator $iA_0$ is self adjoint in $L^2(\mathbb{R}^n)$.*

PROOF. Integration by parts yields

$$(-\Delta u, v)_0 = -\int_{\mathbb{R}^n} \Delta u \cdot \bar{v} \, dx = -\int_{\mathbb{R}^n} u \cdot \overline{\Delta v} \, dx = (u, -\Delta v)_0$$

and therefore $iA_0 = -\Delta$ is symmetric. To show that it is self adjoint it suffices to show that for every $\lambda$ with $\text{Im } \lambda \neq 0$ the range of $\lambda I - iA_0$ is dense in $L^2(\mathbb{R}^n)$. But, if $f \in C_0^\infty(\mathbb{R}^n)$ then, using the Fourier transform, it follows that

$$u(x) = (2\pi)^{-(n/2)} \int_{\mathbb{R}^n} \frac{\hat{f}(\xi)e^{ix\cdot\xi}}{\lambda + |\xi|^2} d\xi \qquad (5.3)$$

is in $D(A_0) = H^2(\mathbb{R}^n)$ and it is the solution of $(\lambda I - iA_0)u = f$. The range of $\lambda I - iA_0$ contains therefore $C_0^\infty(\mathbb{R}^n)$ and is thus dense in $L^2(\mathbb{R}^n)$.  $\square$

From Stone's theorem (Theorem 1.10.8) we now have:

**Corollary 5.3.** *$A_0$ is the infinitesimal generator of a group of unitary operators on $L^2(\mathbb{R}^n)$.*

Next we treat the potential $V$. To this end we define an operator $V$ in $L^2(\mathbb{R}^n)$ by,
$$D(V) = \{u : u \in L^2(\mathbb{R}^n), V \cdot u \in L^2(\mathbb{R}^n)\}$$
and for $u \in D(V)$, $Vu = V(x)u(x)$.

**Lemma 5.4.** *Let $V(x) \in L^p(\mathbb{R}^n)$. If $p > n/2$ and $p \geq 2$ then for every $\varepsilon > 0$ there exists a constant $C(\varepsilon)$ such that*

$$\|Vu\| \leq \varepsilon \|\Delta u\| + C(\varepsilon)\|u\| \qquad for \quad u \in H^2(\mathbb{R}^n) \qquad (5.4)$$

*where the norm $\|\cdot\|$ denotes the $L^2$ norm in $\mathbb{R}^n$.*

PROOF. If $u \in H^2(\mathbb{R}^n)$ then $(1 + |\xi|^2)\hat{u}(\xi) \in L^2(\mathbb{R}^n)$ and since $p > n/2$ we also have $(1 + |\xi|^2)^{-1} \in L^p(\mathbb{R}^n)$. Using Hölder's inequality and Parseval's identity we have for $q = 2p/(2 + p)$

$$\|\hat{u}\|_{0,q} = \left( \int_{\mathbb{R}^n} |\hat{u}(\xi)|^q d\xi \right)^{1/q}$$

$$= \left( \int_{\mathbb{R}^n} (1 + |\xi|^2)^{-q}(1 + |\xi|^2)^q |\hat{u}(\xi)|^q d\xi \right)^{1/q}$$

$$\leq \left( \int_{\mathbb{R}^n} (1 + |\xi|^2)^{-p} d\xi \right)^{1/p} \left( \int_{\mathbb{R}^n} (1 + |\xi|^2)^2 |\hat{u}(\xi)|^2 d\xi \right)^{1/2}$$

$$\leq C_p(\|\Delta u\| + \|u\|).$$

Since $p \geq 2$, $1 \leq q \leq 2$ and therefore by the classical theorem of Hausdorff and Young we have $\|u\|_{0,r} \leq \|\hat{u}\|_{0,q}$ where $1/r + 1/q = 1$. Thus,

$$\|u\|_{0,r} \leq C_p(\|\Delta u\| + \|u\|). \qquad (5.5)$$

Replacing the function $u(x)$ in (5.5) by $u(\rho x)$, $\rho > 0$ and choosing an appropriate $\rho$ we can make the coefficient of $\|\Delta u\|$ as small as we wish. Given $\varepsilon > 0$ we choose it so that

$$\|u\|_{0,r}\|V\|_{0,p} \leq \varepsilon \|\Delta u\| + C(\varepsilon)\|u\|. \qquad (5.6)$$

Finally, using Hölder's inequality again we have

$$\|Vu\|^2 = \int_{\mathbf{R}^n} V^2 u^2 \, dx \le \left( \int_{\mathbf{R}^n} |V|^p \, dx \right)^{2/p} \left( \int_{\mathbf{R}^n} |u|^r \, dx \right)^{2/r}$$

and therefore by (5.6),

$$\|Vu\| \le \|V\|_{0,p} \|u\|_{0,r} \le \varepsilon \|\Delta u\| + C(\varepsilon)\|u\|$$

as desired. $\qquad\qquad\square$

**Theorem 5.5.** *Let $V(x)$ be real, $V(x) \in L^p(\mathbf{R}^n)$. If $p > n/2$, $p \ge 2$ then $A_0 - iV$ is the infinitesimal generator of a group of unitary operators on $L^2(\mathbf{R}^n)$.*

PROOF. We have already seen that the operator $iA_0$ is self adjoint (Lemma 5.2) and in particular $\pm A_0$ is $m$-dissipative. Since $V$ is real the operator $V$ is symmetric and therefore $iA_0 + V$ is a symmetric operator. To prove that it is self adjoint we have to show that the range of $I \pm (A_0 - iV)$ is all of $L^2(\mathbf{R}^n)$. This follows readily from the fact that $\pm(A_0 - iV)$ is $m$-dissipative which in turn follows from the $m$-dissipativity of $\pm A_0$, the estimate

$$\|Vu\| \le \varepsilon \|A_0 u\| + C(\varepsilon)\|u\| \qquad \text{for} \quad u \in D(A_0)$$

and the perturbation Theorem 3.3.2. Thus, $A_0 - iV$ is self adjoint and by Stone's theorem it is the infinitesimal generator of a group of unitary operators on $L^2(\mathbf{R}^n)$. $\qquad\square$

**Remark 5.6.** Adding to $V$ in Theorem 5.5 any real $V_0$ such that $V_0 \in L^\infty(\mathbf{R}^n)$ will not change the conclusion of the theorem. This follows from the fact that $\pm V_0$ is symmetric and bounded and therefore $iA_0 + V + V_0$ is again a self adjoint operator. The fact that the range of $I \pm (A_0 - iV - iV_0)$ is all of $L^2(\mathbf{R}^n)$ follows from the same fact for $I \pm (A_0 - iV)$ and Theorem 3.1.1.

## 7.6. A Parabolic Evolution Equation

In the previous sections we have applied the theory of semigroups to obtain existence and uniqueness results for solutions of initial value problems for partial differential operators. All these applications dealt with partial differential operators which were independent of the $t$-variable. Once these operators depend on $t$, the problem ceases to be autonomous and we have to use the theory of evolution systems, as developed in Chapter 5, to obtain similar results.

The use of the theory of evolution systems is technically more complicated than the use of the semigroup theory. Therefore we will restrict ourselves here only to one example of such an application which extends some of the results of Section 7.3 to the non autonomous situation.

Let $1 < p < \infty$ and let $\Omega$ be a bounded domain with smooth boundary $\partial\Omega$ in $\mathbb{R}^n$. Consider the initial value problem

$$\begin{cases} \dfrac{\partial u}{\partial t} + A(t, x, D)u = f(t, x) & \text{in } \Omega \times [0, T] \\ D^\alpha u(t, x) = 0, |\alpha| < m, & \text{on } \partial\Omega \times [0, T] \\ u(0, x) = u_0(x) & \text{in } \Omega \end{cases} \tag{6.1}$$

where

$$A(t, x, D) = \sum_{|\alpha| \leq 2m} a_\alpha(t, x) D^\alpha \tag{6.2}$$

with the notations introduced in Section 7.1. We will make the following assumptions;

$(H_1)$ The operators $A(t, x, D)$, $t \geq 0$, are uniformly strongly elliptic in $\Omega$ i.e., there is a constant $c > 0$ such that

$$(-1)^m \operatorname{Re} \sum_{|\alpha|=2m} a_\alpha(t, x)\xi^\alpha \geq c|\xi|^{2m} \tag{6.3}$$

for every $x \in \overline{\Omega}$, $0 \leq t \leq T$ and $\xi \in \mathbb{R}^n$.

$(H_2)$ The coefficients $a_\alpha(t, x)$ are smooth functions of the variables $x$ in $\overline{\Omega}$ for every $0 \leq t \leq T$ and satisfy for some constants $C_1 > 0$ and $0 < \beta \leq 1$

$$|a_\alpha(t, x) - a_\alpha(s, x)| \leq C_1|t - s|^\beta \tag{6.4}$$

for $x \in \overline{\Omega}$, $0 \leq s, t \leq T$ and $|\alpha| \leq 2m$.

With the family $A(t, x, D)$, $t \in [0, T]$, of strongly elliptic operators, we associate a family of linear operators $A_p(t)$, $t \in [0, T]$, in $L^p(\Omega)$, $1 < p < \infty$. This is done as follows:

$$D\big(A_p(t)\big) = D = W^{2m, p}(\Omega) \cap W_0^{m, p}(\Omega)$$

and

$$A_p(t)u = A(t, x, D)u \quad \text{for } u \in D.$$

If $u_0 \in L^p(\Omega)$ and $f(t, x) \in L^p(\Omega)$ for every $0 \leq t \leq T$ then a classical solution $u$ of the (abstract) initial value problem

$$\begin{cases} \dfrac{du}{dt} + A_p(t)u = f \\ u(0) = u_0 \end{cases} \tag{6.5}$$

in $L^p(\Omega)$ is defined to be a generalized solution of the initial value problem (6.1). Recall that such a generalized solution $u$, if it exists, satisfies by its definition; $u(t, x) \in W^{2m, p}(\Omega) \cap W_0^{m, p}(\Omega)$ for every $t > 0$, $du/dt$ exists, in the sense of $L^p(\Omega)$ and is continuous on $]0, T]$, $u$ itself is continuous on $[0, T]$ and satisfies (6.5) in $L^p(\Omega)$.

The main result of this section is the existence and uniqueness of generalized solutions of (6.1) under the assumptions $(H_1),(H_2)$ and the Hölder continuity of the function $f$. We start with the following technical lemma.

**Lemma 6.1.** *Under the assumptions* $(H_1),(H_2)$ *there is a constant* $k \geq 0$ *such that the family of operators* $\{A_p(t) + kI\}_{t \in [0, T]}$ *satisfies the conditions* $(P_1)$-$(P_3)$ *of Section 5.6.*

PROOF. From the definition of the operators $A_p(t)$ given above it follows readily that for every real $k$ the domain of $D(A_p(t) + kI) = D(A_p(t)) = D$ is independent of $t$ and therefore, for any choice of $k \geq 0$, the family $\{A_p(t) + kI\}_{t \in [0, T]}$ satisfies the condition $(P_1)$.

Since the constant $C$ in the a-priori estimate stated in Theorem 3.1 (equation (3.3)) depends only on $\Omega, n, m, p$ and the ellipticity constant $c$, we have

$$\|u\|_{2m, p} \leq C\big(\|A_p(t)u\|_{0, p} + \|u\|_{0, p}\big) \tag{6.6}$$

for every $u \in D$. The a-priori estimate (6.6) implies, via the argument of S. Agmon, that

$$\|u\|_{0, p} \leq \frac{M_1}{|\lambda|} \|(\lambda I + A_p(t))u\|_{0, p} \tag{6.7}$$

for $u \in D$ and $\lambda$ satisfying Re $\lambda \geq 0$ and $|\lambda| \geq R$ for some constant $R \geq 0$. Choosing $k > R$, (6.7) implies that

$$\|u\|_{0, p} \leq \frac{M_1}{|\lambda + k|} \|(\lambda I + (A_p(t) + kI))u\|_{0, p}$$

$$\leq \frac{M}{|\lambda| + 1} \|(\lambda I + A_p(t) + kI)u\|_{0, p} \tag{6.8}$$

holds for $u \in D$ and $\lambda$ satisfying Re $\lambda \geq 0$. Using Lemma 3.1, as in the proof of Theorem 3.5, it can be shown that for Re $\lambda \geq 0$, $0 \leq t \leq T$ the operator $\lambda I + (A_p(t) + kI)$ is surjective and hence (6.8) implies

$$\|R(\lambda : A_p(t) + kI)u\|_{0, p} \leq \frac{M}{1 + |\lambda|} \|u\|_{0, p} \tag{6.9}$$

for $u \in L^p(\Omega)$ and $\lambda$ satisfying Re $\lambda \leq 0$. Therefore, fixing a $k > R$, as we will now do, implies that the family $\{A_p(t) + kI\}_{t \in [0, T]}$ satisfies $(P_2)$.

Finally, for $u \in L^p(\Omega)$ and $w = (A_p(\tau) + kI)^{-1}u$ we have $w \in D$ and

$$\|(A_p(t) + kI)w - (A_p(s) + kI)w\|_{0, p}$$

$$= \left\| \sum_{|\alpha| \leq 2m} (a_\alpha(t, x) - a_\alpha(s, x))D^\alpha w \right\|_{0, p}$$

$$\leq C_1|t - s|^\beta \sum_{|\alpha| \leq 2m} \|D^\alpha w\|_{0, p} \leq C_2|t - s|^\beta \|w\|_{2m, p}. \tag{6.10}$$

From (6.7) and (6.9) it follows that

$$\|w\|_{2m,p} \le C\Big(\|A_p(\tau)(A_p(\tau) + kI)^{-1}u\|_{0,p} + \|(A_p(\tau) + kI)^{-1}u\|_{0,p}\Big)$$

$$\le C(1 + kM + M)\|u\|_{0,p}. \tag{6.11}$$

Combining (6.10) and (6.11) yields

$$\Big\|\big((A_p(t) + kI) - (A_p(s) + kI)\big)(A_p(\tau) + kI)^{-1}u\Big\|_{0,p}$$

$$\le C_3|t - s|^{\beta}\|u\|_{0,p} \tag{6.12}$$

for every $u \in L^p(\Omega)$ and the family $\{A_p(t) + kI\}_{t \in [0, T]}$ satisfies also the condition $(P_3)$ of Section 5.6.                                                                      □

From Lemma 6.1 and Theorem 5.7.1 we now deduce our main result.

**Theorem 6.2.** *Let the family $A(t, x, D)$, $0 \le t \le T$, satisfy the conditions $(H_1)$ and $(H_2)$ and let $f(t, x) \in L^p(\Omega)$ for $0 \le t \le T$ satisfy*

$$\Big(\int_{\Omega}|f(t, x) - f(s, x)|^p\, dx\Big)^{1/p} \le C|t - s|^{\gamma} \tag{6.13}$$

*for some constants $C > 0$ and $0 \le \gamma < 1$. Then for every $u_0(x) \in L^p(\Omega)$ the evolution equation (6.1) possesses a unique generalized solution.*

PROOF. We note first that if $f$ satisfies (6.13) so does $e^{-kt}f$ for every real $k$. From Lemma 6.1 it follows that there are values of $k \ge 0$ such that the family $\{A_p(t) + kI\}_{t \in [0, T]}$ satisfies the assumptions $(P_1)$–$(P_3)$ of Section 5.6. We choose and fix such a $k$.

Given $u_0(x) \in L^p(\Omega)$, it follows from Theorem 5.7.1 that the initial value problem

$$\frac{dv}{dt} + (A_p(t) + kI)v = e^{-kt}f, \qquad v(0) = u_0 \tag{6.14}$$

has a unique (classical) solution $v$. A simple computation shows that the function $u = e^{kt}v$ is a solution of the initial value problem

$$\frac{du}{dt} + A_p(t)u = f, \qquad u(0) = u_0 \tag{6.15}$$

and therefore (by definition) it is a generalized solution of the initial value problem (6.1).

The uniqueness of this generalized solution follows from the uniqueness of the solution $v$ of (6.14) combined with the fact that $u$ is a solution of (6.15) if and only if $v = e^{-kt}u$ is a solution of (6.14).                                             □

**Remark 6.3.** It can be shown that if the boundary $\partial\Omega$ of $\Omega$ is smooth enough and the coefficients $a_\alpha(t, x)$ and $f(t, x)$ are smooth enough then the generalized solution of (6.1) is a classical solution of this initial value problem. For example, if all the data is $C^\infty$ i.e., the boundary $\partial\Omega$ is of class $C^\infty$, the coefficients $a_\alpha(t, x)$ and $f(t, x)$ are in $C^\infty([0, T] \times \bar{\Omega})$ then the generalized solution $u$ is in $C^\infty(]0, T] \times \bar{\Omega})$.

# CHAPTER 8

# Applications to Partial Differential Equations—Nonlinear Equations

## 8.1. A Nonlinear Schrödinger Equation

In this section we consider a simple application of the results of Section 6.1 to the initial value problem for the following nonlinear Schrödinger equation in $\mathbb{R}^2$

$$\begin{cases} \dfrac{1}{i} \dfrac{\partial u}{\partial t} - \Delta u + k|u|^2 u = 0 & \text{in } ]0, \infty[ \times \mathbb{R}^2 \\ u(x,0) = u_0(x) & \text{in } \mathbb{R}^2 \end{cases} \tag{1.1}$$

where $u$ is a complex valued function and $k$ a real constant. The space in which this problem will be considered is $L^2(\mathbb{R}^2)$. Defining the linear operator $A_0$ by $D(A_0) = H^2(\mathbb{R}^2)$ and $A_0 u = -i \Delta u$ for $u \in D(A_0)$, the initial value problem (1.1) can be rewritten as

$$\begin{cases} \dfrac{du}{dt} + A_0 u + F(u) = 0 & \text{for } t > 0 \\ u(0) = u_0 \end{cases} \tag{1.2}$$

where $F(u) = ik|u|^2 u$.

From Corollary 7.5.3 it follows that the operator $-A_0$ is the infinitesimal generator of a $C_0$ group of unitary operators $S(t)$, $-\infty < t < \infty$, on $L^2(\mathbb{R}^2)$. A simple application of the Fourier transform gives the following explicit formula for $S(t)$;

$$(S(t)u)(x) = \frac{1}{4\pi it} \int_{\mathbb{R}^2} \exp\left\{ i \frac{|x-y|^2}{4t} \right\} u(y) \, dy. \tag{1.3}$$

Moreover, we have

**Lemma 1.1.** *Let $S(t)$, $t \geq 0$ be the semigroup given by (1.3). If $2 \leq p \leq \infty$ and $1/q + 1/p = 1$ then $S(t)$ can be extended in a unique way to a bounded operator from $L^q(\mathbb{R}^2)$ into $L^p(\mathbb{R}^2)$ and*

$$\|S(t)u\|_{0,p} \leq (4\pi t)^{-(2/q-1)}\|u\|_{0,q}. \tag{1.4}$$

PROOF. Since $S(t)$ is a unitary operator on $L^2(\mathbb{R}^2)$ we have $\|S(t)u\|_{0,2} = \|u\|_{0,2}$ for $u \in L^2(\mathbb{R}^2)$. On the other hand it is clear from (1.3) that $S(t): L^1(\mathbb{R}^2) \to L^\infty(\mathbb{R}^2)$ and that for $t > 0$, $\|S(t)u\|_{0,\infty} \leq (4\pi t)^{-1}\|u\|_{0,1}$. The Riesz convexity theorem implies in this situation that $S(t)$ can be extended uniquely to an operator from $L^q(\mathbb{R}^2)$ into $L^p(\mathbb{R}^2)$ and that (1.4) holds. □

In order to prove the existence of a local solution of the initial value problem (1.2) for every $u_0 \in H^2(\mathbb{R}^2)$ we will use Theorem 6.1.7 and the remarks following it. To do so we note first that the graph norm of the operator $A_0$ in $L^2(\mathbb{R}^2)$ i.e., the norm $\|\|u\|\| = \|u\|_{0,2} + \|A_0 u\|_{0,2}$ for $u \in D(A_0)$ is equivalent to the norm $\| \cdot \|_{2,2}$ in $H^2(\mathbb{R}^2)$. Therefore $D(A_0)$ equipped with the graph norm is the space $H^2(\mathbb{R}^2)$. Next we prove the needed properties of the nonlinear operator $F$.

**Lemma 1.2.** *The nonlinear mapping $F(u) = ik|u|^2 u$ maps $H^2(\mathbb{R}^2)$ into itself and satisfies for $u, v \in H^2(\mathbb{R}^2)$,*

$$\|F(u)\|_{2,2} \leq C\|u\|_{0,\infty}^2 \|u\|_{2,2} \tag{1.5}$$

$$\|F(u) - F(v)\|_{2,2} \leq C(\|u\|_{2,2}^2 + \|v\|_{2,2}^2)\|u - v\|_{2,2}. \tag{1.6}$$

PROOF. From Sobolev's theorem in $\mathbb{R}^2$ (see Theorem 7.4.7) it follows that $H^2(\mathbb{R}^2) \subset L^\infty(\mathbb{R}^2)$ and that there is a constant $C$ such that

$$\|u\|_{0,\infty} \leq C\|u\|_{2,2} \quad \text{for} \quad u \in H^2(\mathbb{R}^2). \tag{1.7}$$

Denoting by $D$ any first order differential operator we have for every $u \in H^2(\mathbb{R}^2)$

$$|D^2(|u|^2 u)| \leq C(|u|^2|D^2 u| + |u| |Du|^2)$$

and therefore

$$\||u|^2 u\|_{2,2} \leq C(\|u\|_{0,\infty}^2 \|u\|_{2,2} + \|u\|_{0,\infty}\|u\|_{1,4}^2). \tag{1.8}$$

From Gagliardo-Nirenberg inequalities we have

$$\|u\|_{1,4} \leq C\|u\|_{0,\infty}^{1/2}\|u\|_{2,2}^{1/2} \tag{1.9}$$

and combining (1.8) and (1.9) we obtain (1.5).

The inequality (1.6) is proved similarly using Leibnitz's formula for the derivatives of products and the estimates (1.7) and (1.9). □

Denoting $D(A_0)$ equipped with the graph norm of $A_0$ by $Y$ it follows from Lemma 1.2 that $F: Y \to Y$ and that it is locally Lipschitz continuous in $Y$. Therefore the remark following Theorem 6.1.7 implies:

**Lemma 1.3.** *For every $u_0 \in H^2(\mathbb{R}^2)$ there exists a unique solution $u$ of the initial value problem (1.2) defined for $t \in [0, T_{\max}[$ such that*

$$u \in C^1([0, T_{\max}[: L^2(\mathbb{R}^2)) \cap C([0, T_{\max}[: H^2(\mathbb{R}^2))$$

*with the property that either $T_{\max} = \infty$ or $T_{\max} < \infty$ and $\lim_{t \to T_{\max}} \|u(t)\|_{2,2} = \infty$.*

From Lemma 1.3 it follows that the initial value problem (1.2) has a unique local solution. To prove that this local solution is a global solution it suffices, by Lemma 1.3, to prove that for every $T > 0$ if $u$ is a solution of (1.2) on $[0, T[$ then $\|u(t)\|_{2,2} \leq C(T)$ for $0 \leq t < T$ and some constant $C(T)$. That this is indeed so in our case, at least if $k \geq 0$, is proved next.

**Lemma 1.4.** *Let $u_0 \in H^2(\mathbb{R}^2)$ and let $u$ be the solution of the initial value problem (1.2) on $[0, T[$. If $k \geq 0$ then $\|u(t)\|_{2,2}$ is bounded on $[0, T[$.*

PROOF. We will first show that $\|u(t)\|_{1,2}$ is bounded on $[0, T[$. To this end we multiply the equation

$$\frac{1}{i}\frac{\partial u}{\partial t} - \Delta u + k|u|^2 u = 0 \tag{1.10}$$

by $\bar{u}$ and integrate over $\mathbb{R}^2$. Then, taking the imaginary part of the result gives $d/dt\|u\|_{0,2}^2 = 0$ and therefore

$$\|u(t)\|_{0,2} = \|u_0\|_{0,2} \qquad \text{for} \quad 0 \leq t < T. \tag{1.11}$$

Next we multiply (1.10) by $\partial \bar{u}/\partial t$, integrate over $\mathbb{R}^2$ and consider the real part of the result. This leads to

$$\frac{1}{2}\int_{\mathbb{R}^2}|\nabla u(t, x)|^2\, dx + \frac{k}{4}\int_{\mathbb{R}^2}|u(t, x)|^4\, dx$$

$$= \frac{1}{2}\int_{\mathbb{R}^2}|\nabla u_0(x)|^2\, dx + \frac{k}{4}\int_{\mathbb{R}^2}|u_0(x)|^4\, dx. \tag{1.12}$$

Therefore, since $k \geq 0$, $\|u\|_{1,2}$ is bounded on $[0, T[$.

To prove that $\|u(t)\|_{2,2}$ is bounded on $[0, T[$ we note first that from Sobolev's theorem it follows that $H^1(\mathbb{R}^2) \subset L^p(\mathbb{R}^2)$ for $p > 2$ and that

$$\|v\|_{0,p} \leq C\|v\|_{1,2} \qquad \text{for} \quad v \in H^1(\mathbb{R}^2). \tag{1.13}$$

Therefore if $u$ is the solution of (1.2) on $[0, T[$ it follows from the boundedness of $\|u(t)\|_{1,2}$ on $[0, T[$ and (1.13) that

$$\|u(t)\|_{0,p} \leq C \qquad \text{for} \quad p > 2, \quad 0 \leq t < T. \tag{1.14}$$

Since $u$ is the solution of (1.2) it is also the solution of the integral equation

$$u(t) = S(t)u_0 - \int_0^t S(t - s)F(u(s))\, ds. \qquad (1.15)$$

Denoting by $D$ any first order derivative we have

$$Du(t) = S(t)Du_0 - \int_0^t S(t - s)DF(u(s))\, ds. \qquad (1.16)$$

We fix now $p > 2$ and let $q = p/(p - 1)$ and $r = 4p/(p - 2)$. Then denoting by $C$ a generic constant and using Lemma 1.1, (1.16) and the Hölder inequality we find

$$\|Du(t)\|_{0,p} \leq \|S(t)Du_0\|_{0,p} + C\int_0^t (t - s)^{1-2/q}\big\||u(s)|^2|Du(s)|\big\|_{0,q}\, ds$$

$$\leq C\|u_0\|_{2,2} + C\int_0^t (t - s)^{1-2/q}\|u(s)\|_{0,r}^2\|Du(s)\|_{0,2}\, ds$$

$$\leq C\|u_0\|_{2,2} + C\int_0^t (t - s)^{1-2/q}\, ds \leq C(t)$$

where in the last inequality we used the facts that $r > 2$ and therefore $\|u(s)\|_{0,r} \leq C$ by (1.14) and that $\|Du(s)\|_{0,2} \leq C\|u(s)\|_{1,2} \leq C$. Therefore, $\|u(t)\|_{1,p} \leq C$ and since by Sobolev's theorem $W^{1,p}(\mathbb{R}^2) \subset L^\infty(\mathbb{R}^2)$ for $p > 2$, it follows that $\|u(t)\|_{0,\infty} \leq C$ for $0 \leq t < T$.

Finally, since $S(t)$ is an isometry on $L^2(\mathbb{R}^2)$ it follows from (1.15) and (1.5) that

$$\|u(t)\|_{2,2} \leq \|S(t)u_0\|_{2,2} + \int_0^t \|S(t - s)F(u(s))\|_{2,2}\, ds$$

$$\leq \|u_0\|_{2,2} + C\int_0^t \|u(s)\|_{0,\infty}^2\|u(s)\|_{2,2}\, ds$$

which by Gronwall's inequality implies the boundedness of $\|u(t)\|_{2,2}$ on $[0, T[$ as desired. $\qquad \square$

Combining Lemma 1.3 with Lemma 1.4 yields our main result,

**Theorem 1.5.** *Let* $u_0 \in H^2(\mathbb{R}^2)$. *If* $k \geq 0$ *then the initial value problem*

$$\begin{cases} \dfrac{1}{i}\dfrac{\partial u}{\partial t} - \Delta u + k|u|^2 u = 0 \\ u(0, x) = u_0(x) \end{cases} \qquad (1.17)$$

*has a unique global solution* $u \in C([0, \infty[\,:\, H^2(\mathbb{R}^2)) \cap C^1([0, \infty[\,:\, L^2(\mathbb{R}^2))$.

In conclusion we make a few comments. First we note that the local solution of (1.17) exists, by Lemma 1.3, also without the restriction $k \geq 0$. We can actually obtain global existence also for $k < 0$ provided that $|k|\,\|u_0\|_{0,2}^2 < 2$ since this condition together with (1.12) and (1.9) imply that $\|u(t)\|_{1,2}$ is bounded on $[0, T[$ and as in the proof of Lemma 1.4 this implies the boundedness of $\|u(t)\|_{2,2}$.

Also, it is not difficult to show that the initial value problem (1.2) has local solutions for the more general $F(u)$ defined by $F(u) = k|u|^{p-1}u$ with $p \geq 1$. Moreover, it can be shown that for $k > 0$ the solutions of (1.2) with $F(u) = k|u|^{p-1}u$ are actually global solutions for every $p \geq 1$.

# 8.2. A Nonlinear Heat Equation in $\mathbb{R}^1$

Consider the following initial value problem

$$\begin{cases} \dfrac{\partial u}{\partial t} = \dfrac{\partial^2 u}{\partial x^2} + f(u) & 0 < x < 1, t > 0 \\[2mm] u(t,0) = u(t,1), u'_x(t,0) = u'_x(t,1) & t \geq 0 \\[2mm] u(0,x) = u_0(x). \end{cases} \tag{2.1}$$

This problem represents the heat flow in a ring of length one with a temperature dependent "source". In this section we will prove, under suitable conditions, the existence of local and global classical solutions of this initial value problem and study the asymptotic behavior of the global solutions as $t \to \infty$.

We start by introducing a convenient abstract frame. Let $X = C_p([0,1])$ be the space of all continuous real valued periodic functions having period 1 with the supremum norm $\|u\| = \max_{0 \leq x \leq 1}|u(x)|$. $X$ consists therefore of continuous functions on $[0,1]$ satisfying $u(0) = u(1)$. Let $A$ be the linear operator in $X$ defined by $D(A) = \{u : u, u', u'' \in X\}$ where $u'$ and $u''$ are the first and second derivatives of $u$ respectively and for $u \in D(A)$, $Au = u''$.

**Lemma 2.1.** *The operator $A$ defined above is the infinitesimal generator of a compact analytic semigroup $T(t)$, $t \geq 0$ on $X$.*

PROOF. Since the domain of $A$ contains all the trigonometrical polynomials (with period 1) it is dense in $X$ by the Weierstrass approximation theorem. Let $g \in X$ and $\lambda = \rho e^{i\vartheta}$ with $\rho > 0$ and $-\pi/2 < \vartheta < \pi/2$. Consider the boundary value problem

$$\begin{cases} \lambda^2 u - u'' = g \\ u(0) = u(1), \quad u'(0) = u'(1). \end{cases} \tag{2.2}$$

A direct computation shows that this problem has a solution $u$ which is given by,

$$u(x) = \frac{1}{2\lambda \sinh \dfrac{\lambda}{2}}$$

$$\cdot \left[ \int_0^x \cosh \lambda \left( x - y - \frac{1}{2} \right) g(y)\, dy + \int_x^1 \cosh \lambda \left( x - y + \frac{1}{2} \right) g(y)\, dy \right]$$

$$\tag{2.3}$$

and that this solution is unique. Denoting $\operatorname{Re} \lambda = \mu = \rho \cos \vartheta > 0$ and using the elementary inequalities

$$\left| \sinh \frac{\lambda}{2} \right| \geq \sinh \frac{\mu}{2}, \qquad \left| \cosh \lambda \left( x - y \pm \frac{1}{2} \right) \right| \leq \cosh \mu \left( x - y + \frac{1}{2} \right)$$

we find

$$|u(x)| \leq \frac{\|g\|}{2|\lambda| \sinh \frac{\mu}{2}}$$

$$\cdot \left[ \int_0^x \cosh \mu \left( x - y - \frac{1}{2} \right) dy + \int_x^1 \cosh \mu \left( x - y + \frac{1}{2} \right) dy \right]$$

$$= \frac{\|g\|}{\cos \vartheta |\lambda|^2}. \tag{2.4}$$

Fixing any $\pi/4 < \vartheta_0 < \pi/2$ we find that

$$\rho(A) \supset \Sigma(\vartheta_0) = \{ \lambda : |\arg \lambda| < 2\vartheta_0 \}$$

and

$$\| R(\lambda : A) \| \leq (\cos \vartheta_0 |\lambda|)^{-1} \qquad \text{for} \quad \lambda \in \Sigma(\vartheta_0).$$

From Theorem 2.5.2 it follows now that $A$ is the infinitesimal generator of an analytic semigroup $T(t)$, $t \geq 0$.

Since $T(t)$ is analytic it is continuous in the uniform operator topology for $t > 0$. This is an immediate consequence of the inequality

$$\| T(t + h) - T(t) \| \leq h \| AT(t) \| \leq \frac{C}{t} h \tag{2.5}$$

which holds for analytic semigroups for every $t > 0$ and $h \geq 0$. Furthermore, since for $\lambda \in \Sigma(\vartheta_0)$, $R(\lambda : A)$ maps $X$ into $D(A)$ such that bounded sets in $X$ are mapped into bounded sets in $D(A)$ which have also a uniform bound on their first derivative, it follows from the Arzela-Ascoli theorem that $R(\lambda : A)$ is a compact operator. From Theorem 2.3.2 it follows now that $T(t)$, $t \geq 0$ is a compact semigroup and the proof is complete.   $\square$

From Lemma 2.1 in conjunction with Theorems 6.2.1 and 6.2.2 we now have:

**Theorem 2.2.** *For every continuous real valued function $f$ and every $u_0 \in X$ there exists a $t_0 > 0$ such that the initial value problem (2.1) has a mild solution $u(t, x)$ on $[0, t_0[$ and either $t_0 = \infty$ or if $t_0 < \infty$ then $\limsup_{t \to t_0} \| u(t, x) \| = \infty$.*

If we assume further that $f$ is Hölder continuous then the mild solution given by Theorem 2.2 is a classical solution. In this case we have:

**Theorem 2.3.** *If $f$ is a Hölder continuous real valued function then for every $u_0(x) \in X$ there is a $t_0 > 0$ such that the initial value problem (2.1) has a* *classical solution* $u(t, x)$ *on* $[0, t_0[$ *and either* $t_0 = \infty$ *or if* $t_0 < \infty$ *then* $\limsup_{t \to t_0} \|u(t, x)\| = \infty$.

PROOF. From Theorem 2.2 it follows that the initial value problem (2.1) has a mild solution $u$ which by definition is continuous on $[0, t_0[ \times [0, 1]$. Therefore $t \to f(u(t; x))$ is continuous in $X$ and by Theorem 4.3.1 $u(t, x)$ is Hölder continuous. Since by our assumption, $f$ is Hölder continuous it follows that $t \to f(u(t, x))$ is Hölder continuous on $[0, t_0[$. But then Corollary 4.3.3 implies that $u$ is a classical solution of the initial value problem and the proof is complete.                                                              □

We turn now to the study of global solutions of the initial value problem (2.1) and start by noting that the conditions of Theorem 2.3 do not imply the existence of a global solution of (2.1). Indeed, choosing for example $f(s) = s^2$ and $u_0(x) \equiv 1$ it is easy to see that the unique solution of (2.1) in this case is $u(t, x) = (1 - t)^{-1}$ which blows up as $t \to 1$.

**Lemma 2.4.** *Let $f$ be continuous and let $u$ be a bounded mild solution of* (2.1) *on* $[0, \infty[$ *then the set* $\{u(t, x): t \geq 0\}$ *is precompact in $X$.*

PROOF. Let $\|u(t)\| \leq K$ for $t \geq 0$. The continuity of $f$ implies that $\|f(u(t))\| \leq N$ for some constant $N$. Let $T(t)$, $t \geq 0$ be the semigroup generated by $A$ and recall that, by Lemma 2.1, $T(t)$ is compact for $t > 0$. Let $0 < \varepsilon < 1$, $t \geq 1$ and set

$$u(t) = T(\varepsilon)u(t - \varepsilon) + [u(t) - T(\varepsilon)u(t - \varepsilon)] = u_\varepsilon(t) + v_\varepsilon(t).$$

The set $\{u_\varepsilon(t): t \geq 1\}$ is precompact in $X$ since $\{u(t - \varepsilon): t \geq 1\}$ is bounded and $T(\varepsilon)$ is compact. Also,

$$\|v_\varepsilon(t)\| = \left\| \int_{t-\varepsilon}^t T(t - s)f(u(s))\, ds \right\|$$

$$\leq \int_{t-\varepsilon}^t \|T(t - s)\|\, \|f(u(s))\|\, ds \leq \varepsilon MN$$

where $M = \sup\{\|T(t)\| : 0 \leq t \leq 1\}$. Therefore $\{u(t): t \geq 1\}$ is totally bounded i.e. precompact. Since $\{u(t): 0 \leq t \leq 1\}$ is compact as the continuous image of the interval $[0, 1]$ the result follows.                                        □

**Lemma 2.5.** *Let $f$ be Hölder continuous. If for some $u_0 \in X$ the initial value problem* (2.1) *has a bounded global solution $u(t, x)$ then there is a sequence $t_k \to \infty$ such that*

$$\lim_{t_k \to \infty} u(t_k, x) = \varphi(x) \tag{2.6}$$

*where $\varphi(x)$ is a solution of the boundary value problem*

$$\begin{cases} \varphi'' + f(\varphi) = 0 \\ \varphi(0) = \varphi(1), \qquad \varphi'(0) = \varphi'(1). \end{cases} \tag{2.7}$$

PROOF. Multiplying the equation

$$\frac{\partial u}{\partial t} = \frac{\partial^2 u}{\partial x^2} + f(u) \tag{2.8}$$

by $\partial u/\partial t$ and integrating over $x$ and $t$ yields

$$\int_0^T \int_0^1 \left| \frac{\partial u}{\partial t} \right|^2 dx\, dt + \frac{1}{2} \int_0^1 \left| \frac{\partial u}{\partial x}(T, x) \right|^2 dx - \int_0^1 F(u(T, x))\, dx$$

$$\leq \frac{1}{2} \int_0^1 \left| \frac{\partial u}{\partial x}(0, x) \right|^2 dx - \int_0^1 F(u(0, x))\, dx \tag{2.9}$$

where $F(s) = \int_0^s f(r)\, dr$. Since $|u(t, x)| \leq K$ for some constant $K$, we deduce from (2.9) that

$$\int_0^\infty \int_0^1 \left| \frac{\partial u}{\partial t} \right|^2 dx\, dt < \infty.$$

Therefore, there exists a sequence $t_l \to \infty$ for which $\lim_{t_l \to \infty}(\partial u(t_l, x)/\partial t) = 0$ a.e. on $[0, 1]$, or $(\partial u(t_l, x)/\partial t) \to 0$ in $L^2(0, 1)$. From Lemma 2.4 it follows that for a subsequence of $t_l$ which we denote by $t_k$, we have $\lim_{t_k \to \infty} u(t_k, x) = \varphi(x)$ uniformly for $0 \leq x \leq 1$. Therefore, $\lim_{t_k \to \infty} f(u(t_k, x)) = f(\varphi(x))$ uniformly in $x$ for $x \in [0, 1]$. Passing to the limit as $t \to \infty$ through the sequence $t_k$, in equation (2.8) in the sense of $L^2(0, 1)$ and using the closedness of the operator $Au = u''$ as an operator in $L^2(0, 1)$ we find $\varphi''(x) + f(\varphi(x)) = 0$ in $L^2(0, 1)$. Since $f(\varphi(x))$ is continuous, this equation holds in a classical sense. Furthermore, the periodicity conditions are satisfied by $\varphi(x)$ since they are satisfied by $u(t, x)$. $\square$

**Corollary 2.6.** If $f$ is Hölder continuous and $f(s) \neq 0$ for all $s \in \mathbb{R}$, then the initial value problem (2.1) has no bounded global solutions.

PROOF. If $f(s) \neq 0$ the boundary value problem (2.7) has no solution. Indeed, integrating the equation $\varphi'' + f(\varphi) = 0$ over $[0, 1]$ yields

$$\varphi'(1) - \varphi'(0) = \int_0^1 f(\varphi(s))\, ds \neq 0$$

and therefore the boundary conditions cannot be fulfilled. Thus by Lemma 2.5, no bounded solution of (2.1) can exist. $\square$

We conclude our discussion with the following result:

**Theorem 2.7.** If $f$ is Hölder continuous and $sf(s) < 0$ for all $s \neq 0$ then all solutions of the initial value problem (2.1) are bounded and moreover, all solutions of (2.1) tend to zero as $t \to \infty$.

PROOF. The boundedness of the solution and, even more, the estimate:

$$\max_{0 \leq x \leq 1} |u(t, x)| \leq \max_{0 \leq x \leq 1} |u(s, x)| \qquad \text{for} \quad t \geq s \tag{2.10}$$

are immediate consequences of the maximum principle. Therefore all solutions of the initial value problem (2.1) are bounded. Moreover from Lemma 2.5 we know that for some sequence $t_k \to \infty$, $u(t_k, x) \to \varphi(x)$ where $\varphi(x)$ is a solution of the boundary value problem (2.7). But the only solution of this boundary value problem is $\varphi \equiv 0$. This can be seen by multiplying $\varphi'' + f(\varphi) = 0$ by $\varphi$, integrating over $[0, 1]$ and obtaining

$$\int_0^1 |\varphi'|^2 \, dx \le 0$$

which implies $\varphi' \equiv 0$ and $\varphi = \text{const}$. However the only solution of $f(s) = 0$ is $s = 0$ and therefore $\varphi \equiv 0$. Thus we have

$$\lim_{t_k \to \infty} u(t_k, x) = 0. \tag{2.11}$$

Combining (2.10) and (2.11) yields $u(t, x) \to 0$ as $t \to \infty$.                         □

## 8.3. A Semilinear Evolution Equation in $\mathbb{R}^3$

Let $\Omega$ be a bounded domain with smooth boundary $\partial\Omega$ in $\mathbb{R}^3$ and consider the following nonlinear initial value problem

$$\begin{cases} \dfrac{\partial u}{\partial t} = \Delta u + \displaystyle\sum_{i=1}^3 u \dfrac{\partial u}{\partial x_i} & \text{in } ]0, T] \times \Omega \\[2mm] u(t, x) = 0 & \text{on } [0, T] \times \partial\Omega \\[2mm] u(0, x) = u_0(x) & \text{in } \Omega \,. \end{cases} \tag{3.1}$$

We will use the results of Section 6.3 to obtain a strong solution of the initial value problem (3.1) in $L^2(\Omega)$.

In this section we will denote by $(\, ,)$ and $\|\cdot\|$ the scalar product and norm in $L^2(\Omega)$. As in Section 7.2 we define an operator $A$ by

$$D(A) = H^2(\Omega) \cap H_0^1(\Omega), \qquad Au = -\Delta u \qquad \text{for } u \in D(A). \tag{3.2}$$

The operator $A$ is clearly symmetric and since $-A$ is an infinitesimal generator of a $C_0$ semigroup on $L^2(\Omega)$ (e.g. by Theorem 7.2.5) it follows that $A$ is self adjoint. Moreover, from Theorem 7.2.7 it follows that $-A$ is the infinitesimal generator of an analytic semigroup on $L^2(\Omega)$. Therefore we can use the results of Section 2.6 to define fractional powers of $A$. In particular we have for some $\delta > 0$,

$$(Au, u) = (A^{1/2}u, A^{1/2}u) = \|A^{1/2}u\|^2 = \|\nabla u\|^2 \ge \delta \|u\|^2 \tag{3.3}$$

where $\nabla u$ is the gradient of $u$ and the inequality is a consequence of Poincaré's inequality. The domain of $A$ consists of Hölder continuous functions. This follows from a version of Sobolev's imbedding theorem or can be shown directly as follows:

**Lemma 3.1.** $D(A)$ *consists of Hölder continuous functions with exponents* $\frac{1}{2}$ *and there is a constant $C$ such that*

$$|u(x_1) - u(x_2)| \le C\|Au\| \, |x_1 - x_2|^{1/2} \quad \text{for} \quad u \in D(A) \quad (3.4)$$

*where $x_i \in \Omega$, $i = 1, 2$ and $|x_1 - x_2|$ denotes the Euclidean distance between $x_1$ and $x_2$.*

PROOF. For $\varphi \in C_0^\infty(\Omega)$ we have the classical identity

$$\varphi(x) = C\int_\Omega \frac{\Delta\varphi(y)}{|x - y|} \, dy. \quad (3.5)$$

From (3.5) and the Cauchy-Schwartz inequality we deduce

$$|\varphi(x_1) - \varphi(x_2)|^2 \le C^2\left(\int_\Omega \Delta\varphi(y)\left(\frac{1}{|x_1 - y|} - \frac{1}{|x_2 - y|}\right) dy\right)^2$$

$$\le C^2\int_\Omega |\Delta\varphi|^2 \, dy \cdot \int_\Omega \left(\frac{1}{|x_1 - y|} - \frac{1}{|x_2 - y|}\right)^2 dy.$$

But,

$$\int_\Omega \left(\frac{1}{|x_1 - y|} - \frac{1}{|x_2 - y|}\right)^2 dy \le C|x_1 - x_2|$$

where $C$ is a constant depending only on $\Omega$. Therefore

$$|\varphi(x_1) - \varphi(x_2)| \le C\|A\varphi\| \, |x_1 - x_2|^{1/2}. \quad (3.6)$$

Approximating $u \in D(A)$ in $H^2(\Omega) \cap H_0^1(\Omega)$ by a sequence $\varphi_n \in C_0^\infty(\Omega)$ and passing to the limit yields (3.4) since $H^2(\Omega) \subset C(\overline{\Omega})$ by Theorem 7.1.2. $\qquad\square$

For functions $u$ in the domain of $A$ we will need the following estimate.

**Lemma 3.2.** *There is a constant $C$ such that*

$$\|u\|_{0,\infty}^4 \le C\|Au\|^3\|u\| \quad \text{for} \quad u \in D(A). \quad (3.7)$$

PROOF. First we note that by Theorem 7.1.2 $u \in D(A)$ is in $C(\overline{\Omega})$ and since $\partial\Omega$ is assumed to be smooth it also follows that $u$ vanishes on $\partial\Omega$. For $u \equiv 0$ (3.7) is trivial. Let $\|u\|_{0,\infty} = L > 0$. From Lemma 3.1 we have

$$|u(x_1) - u(x_2)| \le K|x_1 - x_2|^{1/2}$$

where $K = C\|Au\|$. Without loss of generality we assume that $|u(0)| = L$ and let $B_R$ be an open ball of radius $R = (L/K)^2$ around 0. In this ball we have

$$|u(x)| > |u(0)| - |u(x) - u(0)| \ge L - K|x|^{1/2} > L - K\frac{L}{K} = 0.$$

$$(3.8)$$

Since $u$ vanishes on $\partial\Omega$ we deduce from (3.8) that $B_R \subset \Omega$ and for $x \in B_R$

$$|u(x)| \geq L - K|x|^{1/2}. \tag{3.9}$$

Now,

$$\|u\|^2 \geq \int_{B_R} |u(x)|^2 \, dx \geq \int_{B_R} \left(L - K|x|^{1/2}\right)^2 dx$$

$$= 4\pi L^2 R^3 \int_0^1 \left(1 - \eta^{1/2}\right)^2 \eta^2 \, d\eta$$

$$= CL^2 R^3 = CL^8 K^{-6}$$

and (3.7) follows readily.                                                                 $\square$

**Lemma 3.3.** *For $\gamma > 3/4$ there is a constant $C$ depending only on $\gamma$ and $\Omega$ such that*

$$\|u\|_{0,\infty} \leq C\|A^\gamma u\| \qquad \text{for} \quad u \in D(A). \tag{3.10}$$

PROOF. Let $3/4 < \gamma < 1$. If $w = A^\gamma u$ then (3.10) is equivalent to $\|A^{-\gamma}w\|_{0,\infty} \leq C\|w\|$. In order to estimate $\|A^{-\gamma}w\|_{0,\infty}$ we use the definition of $A^{-\gamma}$ given by formula (6.4) of Section 2.6. So,

$$A^{-\gamma}w = \frac{\sin \pi\gamma}{\pi} \int_0^\infty t^{-\gamma}(tI + A)^{-1}w \, dt. \tag{3.11}$$

From (3.3) it follows that $\|A^{-1}\| \leq \delta^{-1}$ and that for every $t \geq 0$

$$\|(tI + A)^{-1}w\| \leq (t + \delta)^{-1}\|w\|. \tag{3.12}$$

Also since $-A$ is dissipative in $L^2(\Omega)$ we have

$$\|A(tI + A)^{-1}w\| \leq \|w\| \tag{3.13}$$

and since $(tI + A)^{-1}w \in D(A)$, Lemma 3.2 yields

$$\|(tI + A)^{-1}w\|_{0,\infty}^4 \leq C\|A(tI + A)^{-1}w\|^3\|(tI + A)^{-1}w\|. \tag{3.14}$$

Combining (3.11), (3.12), (3.13) and (3.14) yields

$$\|A^{-\gamma}w\|_{0,\infty} \leq C_1 \int_0^\infty t^{-\gamma}(\delta + t)^{1/4}\|w\| \, dt. \tag{3.15}$$

For $3/4 < \gamma < 1$, the integral in (3.15) converges and we have $\|A^{-\gamma}w\|_{0,\infty} \leq C\|w\|$. For $\gamma \geq 1$ the result follows from the result for $3/4 < \gamma < 1$ via the estimate $\|A^{-1}\| \leq \delta^{-1}$.                                                                 $\square$

We turn now to the nonlinear term of (3.1) and start with the following lemma.

**Lemma 3.4.** *Let*

$$f(u) = \sum_{i=1}^3 u \frac{\partial u}{\partial x_i}. \tag{3.16}$$

*If* $\gamma > 3/4$ *and* $u \in D(A)$ *then* $f(u)$ *is well defined and*

$$\|f(u)\| \leq C\|A^\gamma u\| \, \|A^{1/2} u\|. \tag{3.17}$$

*If* $u, v \in D(A)$ *then*

$$\|f(u) - f(v)\| \leq C(\|A^\gamma u\| \, \|A^{1/2} u - A^{1/2} v\| + \|A^{1/2} v\| \, \|A^\gamma u - A^\gamma v\|). \tag{3.18}$$

PROOF. Since $D(A) \subset H^2(\Omega)$ it follows from Sobolev's theorem (Theorem 7.1.2) that $u \in L^\infty(\Omega)$ and therefore $f(u) \in L^2(\Omega)$ and is thus well-defined. Moreover, from Lemma 3.3 we have

$$\|f(u)\| \leq \|u\|_{0,\infty} \|\nabla u\| \leq C\|A^\gamma u\| \, \|\nabla u\| = C\|A^\gamma u\| \, \|A^{1/2} u\|.$$

Also,

$$\|f(u) - f(v)\| \leq \|u\|_{0,\infty} \|\nabla(u - v)\| + \|u - v\|_{0,\infty} \|\nabla u\|$$

$$\leq C(\|A^\gamma u\| \, \|A^{1/2} u - A^{1/2} v\| + \|A^{1/2} v\| \, \|A^\gamma u - A^\gamma v\|).$$

$\square$

From (3.17) it follows that the mapping $f$ can be extended by continuity to $D(A^\gamma)$ and that (3.17) and (3.18) hold for every $u, v \in D(A^\gamma)$. Therefore the conditions of Theorem 6.3.1 are satisfied and we have:

**Theorem 3.5.** *The initial value problem* (3.1) *has a unique local strong solution for every* $u_0 \in D(A^\gamma)$ *with* $\gamma > 3/4$.

We note that from the results of Section 6.3 it follows in the same way as above that if

$$\|f(t_1, x) - f(t_2, x)\| \leq C|t_1 - t_2|^\beta \qquad 0 \leq \beta < 1$$

then the initial value problem

$$\begin{cases} \dfrac{\partial u}{\partial t} = \Delta u + \displaystyle\sum_{i=1}^{3} u \dfrac{\partial u}{\partial x_i} + f(t, x) & \text{in } ]0, T] \times \Omega \\[2mm] u(t, x) = 0 & \text{in } [0, T] \times \partial\Omega \\[2mm] u(0, x) = u_0(x) & \text{in } \Omega \end{cases} \tag{3.19}$$

has a unique local strong solution for every initial value $u_0(x) \in D(A^\gamma)$ with $\gamma > 3/4$.

# 8.4. A General Class of Semilinear Initial Value Problems

The present section is devoted to a general class of semilinear initial value problems which extends considerably the examples given in the previous two sections. The main tool that will be used is Theorem 6.3.1. In order to apply it we will have to use fractional powers of unbounded linear operators. We therefore start with some results concerning such fractional powers.

Recall that if $-A$ is the infinitesimal generator of an analytic semigroup in a Banach space $X$ and $0 \in \rho(A)$ we can define fractional powers of $A$ as we have done in Section 2.6. For $0 < \alpha \leq 1$, $A^\alpha$ is a closed linear operator whose domain $D(A^\alpha) \supset D(A)$ is dense in $X$. We denote by $X_\alpha$ the Banach space obtained by endowing $D(A^\alpha)$ with the graph norm of $A^\alpha$. Since $0 \in \rho(A)$, $A^\alpha$ is invertible and the norm $\| \ \|_\alpha$ of $X_\alpha$ is equivalent to $\|A^\alpha u\|$ for $u \in D(A^\alpha)$. Also, for $0 < \alpha < \beta \leq 1$, $X_\alpha \supset X_\beta$ and the imbedding is continuous.

Let $\Omega \subset \mathbb{R}^n$ be a bounded domain with smooth boundary $\partial \Omega$ and let

$$A(x, D) = \sum_{|\alpha| \leq 2m} a_\alpha(x) D^\alpha \tag{4.1}$$

be a strongly elliptic differential operator in $\Omega$. For the notations and pertinent definitions see Sections 7.1 and 7.2. For $1 < p < \infty$ we associate with $A(x, D)$ and operator $A_p$ in $L^p(\Omega)$ by

$$D(A_p) = W^{2m, p}(\Omega) \cap W_0^{m, p}(\Omega) \tag{4.2}$$

and

$$A_p u = A(x, D) u \qquad \text{for} \quad u \in D(A_p). \tag{4.3}$$

We have seen in Section 7.3 (Theorem 7.3.5) that $-A_p$ is the infinitesimal generator of an analytic semigroup on $L^p(\Omega)$. By adding to $A(x, D)$, and hence to $A_p$, a positive multiple of the identity we obtain an infinitesimal generator $-(A_p + kI)$ of an analytic semigroup, which is invertible. In the sequel we will tacitly assume that this has been done and thus assume directly that $A_p$ itself is invertible. From Theorem 7.3.1 we know the following a-priori estimate

$$\|u\|_{2m, p} \leq C(\|A_p u\|_{0, p} + \|u\|_{0, p}) \qquad \text{for} \quad u \in D(A_p).$$

Since we assume now that $A_p$ is invertible in $L^p(\Omega)$ it follows readily that $C\|u\|_{0, p} \leq \|A_p u\|_{0, p}$ for some constant $C > 0$ and therefore we have

$$\|u\|_{2m, p} \leq C\|A_p u\|_{0, p} \qquad \text{for} \quad u \in D(A_p). \tag{4.4}$$

Before we start describing some properties of the fractional powers of the operator $A_p$ we recall the well known Gagliardo-Nirenberg inequality.

**Lemma 4.1.** *Let $\Omega$ be a bounded domain in $\mathbb{R}^n$ with boundary $\partial \Omega$ of class $C^m$ and let $u \in W^{m, r}(\Omega) \cap L^q(\Omega)$ where $1 \leq r, q \leq \infty$. For any integer $j$, $0 \leq j < m$ and any $j/m \leq \vartheta \leq 1$ we have*

$$\|D^j u\|_{0, p} \leq C\|u\|_{m, r}^\vartheta \|u\|_{0, q}^{1 - \vartheta} \tag{4.5}$$

*provided that*

$$\frac{1}{p} = \frac{j}{n} + \vartheta \left( \frac{1}{r} - \frac{m}{n} \right) + (1 - \vartheta) \frac{1}{q} \tag{4.6}$$

*and $m - j - n/r$ is not a nonnegative integer. If $m - j - n/r$ is a nonnega-
tive integer (4.5) holds with $\vartheta = j/m$.*

The next lemma is our main working tool.

**Lemma 4.2.** *Let $1 < p < \infty$ and let $A_p$ be the operator defined above. For any
multi-index $\beta$, $|\beta| = j < 2m$ and any $j/2m < \alpha \leq 1$ we have*

$$\|D^\beta A_p^{-\alpha} u\|_{0,p} \leq C\|u\|_{0,p} \qquad for \quad u \in D(A_p). \tag{4.7}$$

PROOF. Set $B = D^\beta$. Since $|\beta| < 2m$ it is clear that $D(B) \supset D(A_p)$. From
the previous lemma we have

$$\|D^\beta u\|_{0,p} \leq C\|u\|_{2m,p}^{j/2m}\|u\|_{0,p}^{1-j/2m}. \tag{4.8}$$

Polarization of (4.8) together with the estimate (4.4) yield

$$\|D^\beta u\|_{0,p} \leq C\left(\rho^{-1+j/2m}\|A_p u\|_{0,p} + \rho^{j/2m}\|u\|_{0,p}\right) \tag{4.9}$$

for $\rho > 0$ and $u \in D(A_p)$. From Theorem 2.6.12 it follows now that
$D(B) \supset D(A_p^\alpha)$ for $j/2m < \alpha \leq 1$ i.e., $BA_p^{-\alpha}$ is bounded for these values of
$\alpha$ and the proof is complete.                                                  □

**Theorem 4.3.** *Let $\Omega \subset \mathbb{R}^n$ be a bounded domain with smooth boundary $\partial\Omega$
and let $A_p$ be as above. If $0 \leq \alpha \leq 1$ then*

$$X_\alpha \subset W^{k,q}(\Omega) \qquad for \quad k - \frac{n}{q} < 2m\alpha - \frac{n}{p}, \qquad q \geq p \tag{4.10}$$

$$X_\alpha \subset C^\nu(\overline{\Omega}) \qquad for \quad 0 \leq \nu < 2m\alpha - \frac{n}{p} \tag{4.11}$$

*and the imbeddings are continuous.*

PROOF. From Lemma 4.2 it follows readily that $X_\alpha \subset W^{j,p}(\Omega)$ provided
that $j < 2m\alpha$ and the imbedding is continuous. From Theorem 7.1.1 it
follows that $W^{j,p}(\Omega)$ is continuously imbedded in $W^{k,q}(\Omega)$ provided that
$k - n/q < j - n/p$ and (4.10) follows. From Sobolev's theorem (Theorem
7.1.2) it follows that $W^{j,p}(\Omega)$ is continuously imbedded in $C^\nu(\overline{\Omega})$ for
$0 \leq \nu < j - n/p$ and (4.11) follows.                                            □

We note in passing that Lemma 3.3 of the previous section is a special
case of Theorem 4.3 since it is a consequence of (4.10) taking $k = 0$, $q = \infty$,
$n = 3$, $p = 2$ and $m = 1$.

We turn now to the applications of Theorem 6.3.1. But rather than stating
and proving a very general result, we prefer to restrict ourselves to a simple
example in $\mathbb{R}^3$ with $p = 2$ and a second order operator, which contains
already most of the ingredients of the general case and then comment
(without proof) on more general results at the end of the section.

**Theorem 4.4.** *Let $\Omega$ be a bounded domain in $\mathbb{R}^3$ with smooth boundary $\partial\Omega$ and let $A(x, D)$ be a strongly elliptic operator given by*

$$A(x, D) = -\sum_{k,l=1}^{3} \frac{\partial}{\partial x_k} a_{k,l}(x) \frac{\partial}{\partial x_l}$$

*where $a_{k,l}(x) = a_{l,k}(x)$ are real valued and continuously differentiable in $\bar{\Omega}$. Let $f(t, x, u, p)$, $p \in \mathbb{R}^3$, be a locally Lipschitz continuous function of all its arguments and assume further that there is a continuous function $\rho(t, r) : \mathbb{R} \times \mathbb{R} \to \mathbb{R}^+$ and a real constant $1 \leq \gamma < 3$ such that*

$$|f(t, x, u, p)| \leq \rho(t, |u|)(1 + |p|^\gamma) \tag{4.12}$$

$$|f(t, x, u, p) - f(t, x, u, q)| \leq \rho(t, |u|)(1 + |p|^{\gamma-1} + |q|^{\gamma-1})|p - q| \tag{4.13}$$

$$|f(t, x, u, p) - f(t, x, v, p)| \leq \rho(t, |u| + |v|)(1 + |p|^\gamma)|u - v|. \tag{4.14}$$

*Then for every $u_0 \in H^2(\Omega) \cap H_0^1(\Omega)$ the initial value problem*

$$\begin{cases} \dfrac{\partial u}{\partial t} + A(x, D)u = f(t, x, u, \operatorname{grad} u) & \text{in } \Omega \\ u(t, x) = 0 & \text{on } \partial\Omega \\ u(0, x) = u_0(x) & \text{in } \Omega \end{cases} \tag{4.15}$$

*has a unique local strong solution in $L^2(\Omega)$.*

PROOF. We recall that with the strongly elliptic operator $A(x, D)$ we associate an operator $A$ in $L^2(\Omega)$ by $D(A) = H^2(\Omega) \cap H_0^1(\Omega)$ and $Au = A(x, D)u$ for $u \in D(A)$. From Theorem 7.3.6 it follows that $-A$ is the infinitesimal generator of an analytic semigroup on $L^2(\Omega)$ and from the strong ellipticity together with Poincaré's inequality it follows readily that $A$ is also invertible. From Theorem 4.3 it follows that if $\alpha > 3/4$ then $X_\alpha \subset L^\infty(\Omega)$ and if also $1/q > (5 - 4\alpha)/6$ then $X_\alpha \subset W^{1,q}(\Omega)$. Thus for $\max(3/4, (5\gamma - 3)/4\gamma) < \alpha < 1$ we have

$$X_\alpha \subset W^{1,2\gamma}(\Omega) \cap L^\infty(\Omega). \tag{4.16}$$

In order to apply Theorem 6.3.1 we have to show that the mapping

$$F(t, u)(x) = f(t, x, u(x), \nabla u(x)), \qquad x \in \Omega \tag{4.17}$$

is well defined on $\mathbb{R}^+ \times X_\alpha$ and satisfies a local Hölder condition there. From (4.12) and (4.16) we have for every $u \in X_\alpha$

$$\|F(t, u)\|_{0,2} \leq 2\rho(t, \|u\|_{0,\infty})\left(M^{1/2} + \|u\|_{1,2\gamma}^\gamma\right) \tag{4.18}$$

where $M$ is the measure of $\Omega$. Therefore $F$ is well defined on $\mathbb{R}^+ \times X_\alpha$. To show that $F$ satisfies a local Hölder condition we note that

$$\|F(t, u) - F(t, v)\|_{0,2}^2 \leq 2\int_0 |f(t, x, u, \nabla u) - f(t, x, u, \nabla v)|^2 \, dx$$

$$+ 2\int_\Omega |f(t, x, u, \nabla v) - f(t, x, v, \nabla v)|^2 \, dx$$

$$(4.19)$$

and estimate each of the two terms on the right of (4.19) separately. From (4.13) and (4.15) we have

$$\int_\Omega |f(t, x, u, \nabla u) - f(t, x, u, \nabla v)|^2 \, dx$$

$$\leq C \cdot \rho(t, \|u\|_{0,\infty})^2 \int_\Omega (1 + |\nabla u|^{2\gamma-2} + |\nabla v|^{2\gamma-2})|\nabla(u - v)|^2 \, dx$$

$$\leq C \cdot \rho(t, \|u\|_{0,\infty})^2 \left(M_1 + \|\nabla u\|_{0,2\gamma}^{2\gamma-2} + \|\nabla v\|_{0,2\gamma}^{2\gamma-2}\right) \|\nabla(u - v)\|_{0,2\gamma}^2$$

$$\leq L(\|u\|_\alpha, \|v\|_\alpha)\|u - v\|_{1,2\gamma}^2 \leq L(\|u\|_\alpha, \|v\|_\alpha)\|u - v\|_\alpha^2$$

where $\| \ \|_\alpha$ denotes the norm in $X_\alpha$ and $L$ is a constant depending on $\|u\|_\alpha$ and $\|v\|_\alpha$. To obtain the second inequality we used Hölder's inequality. The last inequality is a consequence of the continuous imbedding of $X_\alpha$ in $W^{1,2\gamma}(\Omega)$. Similarly for the second term we have by (4.14) and (4.16)

$$\int_\Omega |f(t, x, u, \nabla v) - f(t, x, v, \nabla v)|^2 \, dx$$

$$\leq C\rho(t, \|u\|_{0,\infty} + \|v\|_{0,\infty})^2 \int_\Omega (1 + |\nabla v|^{2\gamma})|u - v|^2 \, dx$$

$$\leq C\rho(t, \|u\|_{0,\infty} + \|v\|_{0,\infty})^2 \|u - v\|_{0,\infty}^2 \left(1 + \|v\|_{1,2\gamma}^{2\gamma}\right)$$

$$\leq L(\|u\|_\alpha, \|v\|_\alpha)\|u - v\|_\alpha^2$$

and therefore

$$\|F(t, u) - F(t, v)\|_{0,2} \leq L(\|u\|_\alpha, \|v\|_\alpha)\|u - v\|_\alpha \qquad (4.20)$$

and the existence of the strong local solution of (4.15) is a direct consequence of Theorem 6.3.1. $\qquad\qquad\qquad\qquad\qquad\qquad\qquad\qquad\qquad\qquad\quad\square$

Before continuing we note that Theorem 3.5 is a special case of Theorem 4.4 since $-\Delta$ is obviously strongly elliptic and $f(u, \nabla u) = u \cdot \nabla u$ certainly satisfies the conditions of Theorem 4.4. Furthermore, from Theorem 4.4 we also obtain an extension of the existence results of Section 8.2 for $\Omega \subset \mathbb{R}^3$ assuming however that $f$ is bounded and locally Lipschitz continuous in $\Omega$.

From Theorem 4.4 we obtain a local strong solution in the sense of $L^2(\Omega)$ of the initial value problem (4.15). This solution satisfies, by Theorem 4.4

$$u \in C([0, T_0[ : L^2(\Omega)) \cap C(]0, T_0[ : H^2(\Omega) \cap H_0^1(\Omega))$$

$$\cap C^1(]0, T_0[ : L^2(\Omega))$$

for some $T_0 > 0$. But in fact it is a classical solution of this initial value problem for $t > 0$. Indeed, since for $0 < t < T_0$, $u \in D(A) \subset C(\overline{\Omega})$ and, by Corollary 6.3.2, $t \to du/dt \in X_\alpha$ is locally Hölder continuous for $0 < t < T_0$ it follows that $(t, x) \to u(t, x)$ and $(t, x) \to (\partial/\partial t)u(t, x)$ are continuous on $0 < t < T_0$, $x \in \overline{\Omega}$. To show that $u$ is a classical solution of the equation it remains to show that $u(t, \cdot) \in C^2(\Omega)$. From the fact that for $0 < t < T_0$, $u(t, \cdot) \in D(A)$ we have $\nabla u \in W^{1, q_1}(\Omega) \subset L^{p_1}(\Omega)$ where $q_1 = 2$, $p_1 = 6/(3 - 2) = 6$. So, $Au = F(t, u) - du/dt \in L^{p_1/\gamma}(\Omega)$ by (4.12) whence by Theorem 7.3.1 $u \in W^{2, 6/\gamma}(\Omega)$ and therefore $\nabla u \in W^{1, q_2}(\Omega)$ with $q_2 = 6/\gamma > 2$. Repeating this process we find that $\nabla u \in W^{1/q_n}(\Omega)$ where $1/q_n = \gamma(1/q_{n-1} - 1/3)$. It is easy to check that after a finite number of steps (one step if $1 \leq \gamma < 2$) $q_n > 3$ and then $\nabla u(t, \cdot)$ is Hölder continuous in $\Omega$ and it follows that $F(t, u)$ is Hölder continuous in $\Omega$. Since $\alpha > 3/4$ and $(\partial/\partial t)u(t, \cdot) \in X_\alpha$ it follows that $(\partial/\partial t)u(t, \cdot)$ is Hölder continuous in $\Omega$. But then $Au = F(t, u) - du/dt$ is Hölder continuous in $\Omega$ and by a classical regularity theorem for elliptic equations it follows that $u(t, \cdot) \in C^{2+\delta}(\Omega)$ for some $\delta > 0$ that is, $u$ has second order Hölder continuous derivatives in $x$ and is thus the classical solution of (4.15).

We conclude this section with some comments on more general existence results. We assume that $A(x, D)$ is a strongly elliptic differential operator given by (4.1). We define an operator $A_p$ in $L^p(\Omega)$ by $D(A_p) = W^{2m, p}(\Omega) \cap W_0^{m, p}(\Omega)$ and $A_p u = A(x, D)u$ for $u \in D(A_p)$. By adding a positive multiple of the identity to $A_p$ we can assume as we will tacitly do that $A_p$ is invertible. From Theorem 7.3.5 it follows that $-A_p$ is the infinitesimal generator of an analytic semigroup on $L^p(\Omega)$. Let

$$F(t, u)(x) = f(t, x, u, Du, D^2 u, \ldots, D^{2m-1} u) \qquad (4.21)$$

where $D^j$ stands for any $j$-th order derivative. Assume that $f$ is a continuously differentiable function of all its variables and consider the initial value problem

$$\begin{cases} \dfrac{du}{dt} + A_p u = F(t, u) \\ u(0) = u_0 \end{cases} \qquad (4.22)$$

in $L^p(\Omega)$. From Theorem 4.3 it follows that if $1 - 1/2m < \alpha < 1$ and $p$ is sufficiently large, then $X_\alpha$ is continuously imbedded in $C^{2m-1}(\overline{\Omega})$. This implies that

$$\|F(t, A_p^{-\alpha} u) - F(s, A_p^{-\alpha} v)\|_{0, p} \leq C(|t - s| + \|u - v\|_{0, p}) \qquad (4.23)$$

where $C$ is a constant which depends on $\|D^j A^{-\alpha} u\|_{0, \infty}$, $\|D^j A^{-\alpha} v\|_{0, \infty}$ for $0 \leq j < 2m - 1$. Therefore if $p$ is large enough the conditions of Theorem 6.3.1 are satisfied and we have

**Theorem 4.5.** *Let $\Omega$ be a bounded domain in $\mathbb{R}^n$ with smooth boundary $\partial\Omega$ and let $A_p$ be the operator defined above. Let $F(t, u)$ be defined by (4.21) where $f$ is*

*a continuously differentiable function of all its variables with the possible exception of the x variables. If $p > n$ then for every $u_0 \in W^{2m,\,p}(\Omega) \cap W_0^{m,\,p}(\Omega)$ the initial value problem (4.22) has a unique local strong solution.*

If $p < n$ (as is the case in Theorem 4.4) the argument leading to Theorem 4.5 fails since for no $0 \le \alpha < 1$ $D^{2m-1}(A^{-\alpha}u) \in L^\infty(\Omega)$. In this case, in order to obtain an existence result one has to assume that the function $f$ satisfies some further conditions similar in nature to the estimates (4.12)–(4.14).

## 8.5. The Korteweg-de Vries Equation

In the present section we will use the results of Section 6.4 to obtain an existence theorem of a local solution of the Cauchy problem for the Korteweg-de Vries equation:

$$\begin{cases} u_t + u_{xxx} + uu_x = 0 & t \ge 0 \quad -\infty < x < \infty \\ u(0, x) = u_0(x). \end{cases} \tag{5.1}$$

Throughout this section we will assume that all functions are real valued, denote by $\int$ the integral over all of $\mathbb{R}$ and denote by $\hat{f}$ the Fourier transform of $f$.

For every real $s$ we introduce a Hilbert space $H^s(\mathbb{R})$ as follows; Let $u \in L^2(\mathbb{R})$ and set

$$\|u\|_s = \left( \int (1 + \xi^2)^s |\hat{u}(\xi)|^2 \, d\xi \right)^{1/2}. \tag{5.2}$$

The linear space of functions $u \in L^2(\mathbb{R})$ for which $\|u\|_s$ is finite is a pre-Hilbert space with the scalar product

$$(u, v)_s = \int (1 + \xi^2)^s \hat{u}(\xi) \bar{v}(\xi) \, d\xi. \tag{5.3}$$

The completion of this space with respect to the norm $\| \ \|_s$ is a Hilbert space which we denote by $H^s(\mathbb{R})$.

It is clear that $H^0(\mathbb{R}) = L^2(\mathbb{R})$. The scalar product and norm in $L^2(\mathbb{R})$ will be denoted by $(\,,)$ and $\| \ \|_0$. Furthermore, it is easy to check that the spaces $H^s(\mathbb{R})$ with $s = n$ coincide with the spaces $H^n(\mathbb{R})$, $n \ge 1$, as defined in Section 7.1 and the norms in the two different definitions are equivalent.

In the following lemma we collect some useful properties of the spaces $H^s(\mathbb{R})$.

**Lemma 5.1.** (i) *For $t \ge s$, $H^s(\mathbb{R}) \supset H^t(\mathbb{R})$ and $\|u\|_t \ge \|u\|_s$ for $u \in H^t(\mathbb{R})$.*
 (ii) *For $s > \frac{1}{2}$, $H^s(\mathbb{R}) \subset C(\mathbb{R})$ and for $u \in H^s(\mathbb{R})$*

$$\|u\|_\infty \le C\|u\|_s \tag{5.4}$$

*where $\|u\|_\infty = \sup\{|u(x)| : x \in \mathbb{R}\}$.*

PROOF. Part (i) is obvious from the definitions and the elementary inequality $(1 + \xi^2)^t \geq (1 + \xi^2)^s$ for $t \geq s$ and $\xi \in \mathbb{R}$.

From the Cauchy-Schwartz inequality we have,

$$|u(x)| = \left| \frac{1}{\sqrt{2\pi}} \int e^{ix\xi} \hat{u}(\xi) \, d\xi \right|$$

$$\leq \frac{1}{\sqrt{2\pi}} \left( \int \frac{d\xi}{(1 + \xi^2)^s} \right)^{1/2} \left( \int (1 + \xi^2)^s |\hat{u}(\xi)|^2 \, d\xi \right)^{1/2} = C\|u\|_s$$

so the integral defining $u$ in terms of $\hat{u}$ converges uniformly and $u$ is continuous. Moreover $\|u\|_\infty \leq C\|u\|_s$.                                                $\square$

Let $X = L^2(\mathbb{R}) = H^0(\mathbb{R})$ and $Y = H^s(\mathbb{R})$ with $s \geq 3$. We define an operator $A_0$ by $D(A_0) = H^3(\mathbb{R})$ and $A_0 u = D^3 u$ for $u \in D(A_0)$ where $D = d/dx$.

**Lemma 5.2.** $A_0$ is the infinitesimal generator of a $C_0$ group of isometries on $X$.

PROOF. $A_0$ is skew-adjoint i.e. $iA_0$ is self-adjoint or equivalently $(A_0 u, u) = 0$ for all $u \in D(A_0)$. This follows readily from

$$(A_0 u, u) = \int D^3 u \cdot u \, dx = - \int u \cdot D^3 u \, dx = - (A_0 u, u)$$

where the second equality is achieved by integration by parts three times. From Stone's theorem (Theorem 1.10.8) it follows that $A_0$ is the infinitesimal generator of a group of isometries on $X = L^2(\mathbb{R})$.                        $\square$

Next we define for every $v \in Y = H^s(\mathbb{R})$, $s \geq 3$, an operator $A_1(v)$ by: $D(A_1(v)) = H^1(\mathbb{R})$ and for $u \in D(A_1(v))$, $A_1(v)u = v\,Du$. We then have:

**Lemma 5.3.** For every $v \in Y$ the operator $-A(v) = -(A_0 + A_1(v))$ is the infinitesimal generator of a $C_0$ semigroup $T_v(t)$ on $X$ satisfying

$$\|T_v(t)\| \leq e^{\beta t} \tag{5.5}$$

for every $\beta \geq \beta_0(v) = c_0\|v\|_s$ where $c_0$ is a constant independent of $v \in Y$.

PROOF. We note first that since $v \in H^s(\mathbb{R})$, $Dv \in H^{s-1}(\mathbb{R})$ and since $s \geq 3$ it follows from Lemma 5.1 that $Dv \in L^\infty(\mathbb{R})$ and that $\|Dv\|_\infty \leq C\|Dv\|_{s-1} \leq C\|v\|_s$.

Now, for every $u \in H^1(\mathbb{R})$ we have

$$(A_1(v)u, u) = \int v\,Du \cdot u \, dx = \tfrac{1}{2} \int v\,Du^2 \, dx = -\tfrac{1}{2} \int Dv \, u^2 \, dx$$

$$\geq -\tfrac{1}{2}\|Dv\|_\infty \|u\|^2 \geq -c_0\|v\|_s\|u\|^2.$$

Therefore $-(A_1(v) + \beta I)$ is dissipative for all $\beta \geq \beta_0(v) = c_0\|v\|_s$. Since $A_0$ is skew-adjoint, $A_0 + A_1(v) + \beta I$ is also dissipative for $\beta \geq \beta_0(v)$. Moreover,

$$\|(A_1(v) + \beta I)u\| \leq \|v\,Du\| + \beta\|u\| \leq \|v\|_\infty\|Du\| + \beta\|u\|. \tag{5.6}$$

Using integration by parts it is not difficult to show that for every $u \in H^3(\mathbb{R})$ we have $\|Du\| \leq \|u\|^{2/3}\|D^3u\|^{1/3}$ and by polarization we obtain for every $\varepsilon > 0$,

$$\|Du\| \leq \varepsilon\|D^3u\| + C(\varepsilon)\|u\|. \tag{5.7}$$

Choosing $\varepsilon = 1/2\|v\|_\infty$ and substituting (5.7) into (5.6) yields

$$\|(A_1(v) + \beta I)u\| \leq \tfrac{1}{2}\|A_0 u\| + C\|u\| \qquad \text{for} \quad u \in D(A_0) \tag{5.8}$$

whence, by Corollary 3.3.3, $A_0 + A_1(v) + \beta I = A(v) + \beta I$ is the infinitesimal generator of a $C_0$ semigroup of contractions of $X$ for every $\beta \geq \beta_0(v)$. Therefore, $A(v)$ is the infinitesimal generator of a $C_0$ semigroup $T_v(t)$ satisfying (5.5). $\qquad \square$

We let now $B_r$ be the ball of radius $r > 0$ in $Y$ centered at the origin and consider the family of operators $A(v)$, $v \in B_r$. We want to show that this family satisfies the conditions of Theorem 6.4.6. Because of the special form of the family $A(v)$, $v \in B_r$, it follows that it suffices to prove the following three conditions:

($A_1$) The family $A(v)$, $v \in B_r$, is a stable family in $X$.
($A_2$) There is an isomorphism of $Y$ onto $X$ such that for every $v \in B_r$
$SA(v)S^{-1} - A(v)$ is a bounded operator in $X$ and

$$\|SA(v)S^{-1} - A(v)\| \leq C_1 \qquad \text{for all} \quad v \in B_r. \tag{5.9}$$

($A_3$) For each $v \in B_r$, $D(A(v)) \supset Y$, $A(v)$ is a bounded linear operator from $Y$ into $X$ and

$$\|A(v_1) - A(v_2)\|_{Y \to X} \leq C_2\|v_1 - v_2\|. \tag{5.10}$$

We note that ($A_1$) is the same as the condition ($\tilde{H}_1$) of Section 6.4 and ($A_2$) implies both ($\tilde{H}_2$) and ($\tilde{H}_5$) as can be easily seen from Lemma 5.4.4 and Theorem 5.4.6. The condition ($A_3$) implies ($\tilde{H}_3$) and ($\tilde{H}_4$) while ($\tilde{H}_6$) is satisfied since both $X$ and $Y$ are reflexive. Finally, if $\|u_0\|_s < r$ and $v \in B_r$ then

$$\|A(v)u_0\| \leq \|D^3u_0\| + \|v Du_0\|$$

$$\leq \|D^3u_0\| + \|v\|_\infty\|Du_0\|$$

$$\leq \|u_0\|_3(1 + r) \leq r(1 + r) = k \tag{5.11}$$

and condition (4.21) of Theorem 6.4.6 is also satisfied.

In order to prove that the family $A(v)$, $v \in B_r$, satisfies the conditions ($A_1$)–($A_3$) we need one more preliminary result. Let

$$\Lambda^s f = \frac{1}{\sqrt{2\pi}} \int e^{ix\xi}(1 + \xi^2)^{s/2}\hat{f}(\xi)\,d\xi. \tag{5.12}$$

It is not difficult to check that $\Lambda^s$ is an isomorphism of $Y = H^s(\mathbb{R})$ onto $X = L^2(\mathbb{R})$. For a given function $f \in L^2(\mathbb{R})$ let $M_f$ be the operator of multiplication by the function $f$, i.e. $M_f u = fu$. We then have:

**Lemma 5.4.** *Let $f \in H^s(\mathbb{R})$, $s > 3/2$ and let $T = (\Lambda^s M_f - M_f \Lambda^s)\Lambda^{1-s}$. Then $T$ is a bounded operator on $X = L^2(\mathbb{R})$ and*

$$\|T\| \leq C\|\operatorname{grad} f\|_{s-1}. \tag{5.13}$$

PROOF. The Fourier transform of $T$ is the integral operator with kernel $k(\xi, \eta)$ given by

$$\sqrt{2\pi}\, k(\xi, \eta) = \left((1 + \xi^2)^{s/2} - (1 + \eta^2)^{s/2}\right)\hat{f}(\xi - \eta)(1 + \eta^2)^{(s-1)/2}.$$

Since

$$\left|(1 + \xi^2)^{s/2} - (1 + \eta^2)^{s/2}\right| \leq s|\xi - \eta|\left((1 + \xi^2)^{(s-1)/2} + (1 + \eta^2)^{(s-1)/2}\right)$$

we have

$$\sqrt{2\pi}\,|k(\xi, \eta)| \leq s(1 + \xi^2)^{(s-1)/2}|\xi - \eta|\hat{f}(\xi - \eta)(1 + \eta^2)^{(1-s)/2}$$
$$+ s|\xi - \eta|\hat{f}(\xi - \eta) = k_1(\xi, \eta) + k_2(\xi, \eta).$$

To show that $T$ is bounded it suffices to show that the operators $T_1$ and $T_2$ with kernels $k_1(\xi, \eta)$ and $k_2(\xi, \eta)$ are bounded. Using the inverse Fourier transform we find that

$$T_1 = s\Lambda^{s-1}M_g\Lambda^{1-s}, \qquad T_2 = sM_g \tag{5.14}$$

where $M_g$ is the multiplication operator by the function $g$ for which $\hat{g}(\xi) = |\xi|\hat{f}(\xi)$. From Lemma 5.1 (ii) it follows that

$$\|g\|_\infty \leq C\|g\|_{s-1} \leq C\|\operatorname{grad} f\|_{s-1}. \tag{5.15}$$

Now,

$$\|T_1 u\| = s\|\Lambda^{s-1}M_g\Lambda^{1-s}u\| = s\|M_g\Lambda^{1-s}u\|_{s-1} \leq s\|g\|_{s-1}\|u\| \tag{5.16}$$

and

$$\|T_2 u\| = s\|gu\| \leq s\|g\|_\infty \|u\|. \tag{5.17}$$

Therefore both $T_1$ and $T_2$ are bounded operators in $X$. Combining (5.15) with (5.14) and (5.17) yields the desired estimate (5.13).    □

We now have:

**Lemma 5.5.** *For every $r > 0$, the family of operators $A(v)$, $v \in B_r$, satisfies the conditions $(A_1)$–$(A_3)$.*

PROOF. Let $r > 0$ be fixed. From Lemma 5.3 it follows that if $\beta \geq c_0 r$, $A(v)$ is the infinitesimal generator of a $C_0$ semigroup $T_v(t)$ satisfying $\|T_v(t)\| \leq e^{\beta t}$ and therefore $A(v)$, $v \in B_r$ is a stable family in $X$ (see Definition 6.4.1).

As we have mentioned above $S = \Lambda^s$ is an isomorphism of $Y = H^s(\mathbb{R})$ onto $X = L^2(\mathbb{R})$. A simple computation shows that for $u, v \in Y$ we have

$$\left(SA(v)S^{-1} - A(v)\right)u = \left(S(vD)S^{-1} - vD\right)u = \left(Sv - vS\right)S^{-1}Du$$

and therefore by Lemma 5.4

$$\|(SA(v)S^{-1} - A(v))u\| = \|(\Lambda^s M_v - M_v \Lambda^s)\Lambda^{1-s}\Lambda^{-1} Du\|$$
$$\leq \|(\Lambda^s M_v - M_v \Lambda^s)\Lambda^{1-s}\| \, \|\Lambda^{-1} Du\|$$
$$\leq C\|\mathrm{grad}\, v\|_{s-1}\|u\| \leq C\|v\|_Y\|u\|.$$

Since $Y$ is dense in $X$ it follows that $\|SA(v)S^{-1} - A(v)\| \leq C\|v\|_Y \leq Cr$ and $(A_2)$ is satisfied.

Finally, since $s \geq 3$ $D(A(v)) \supset Y$ for every $v \in Y$ and for $v \in B_r$

$$\|A(v)u\| \leq \|D^3 u\| + \|v\, Du\| \leq \|D^3 u\| + \|v\|_\infty \|Du\|$$
$$\leq (1 + C\|v\|_s)\|u\|_s \leq (1 + Cr)\|u\|_Y$$

and therefore $A(v)$ is a bounded operator from $Y$ into $X$. Moreover if $v_1, v_2 \in B_r$, $u \in Y$ then

$$\|(A(v_1) - A(v_2))u\| = \|(v_1 - v_2)\, Du\|$$
$$\leq \|v_1 - v_2\| \, \|Du\|_\infty \leq C\|v_1 - v_2\| \, \|u\|_Y$$

and the proof is complete. $\qquad\square$

From Lemma 5.5 it follows that the family $A(v)$, $v \in B_r$ satisfies the conditions $(A_1)$–$(A_3)$ stated above and therefore by the remarks following these conditions all the assumptions of Theorem 6.4.6 are satisfied, provided only that $r > \|u_0\|_s$. Consequently we have:

**Theorem 5.6.** *For every $u_0 \in H^s(\mathbb{R})$, $s \geq 3$ there is a $T > 0$ such that the initial value problem*

$$u_t + u_{xxx} + uu_x = 0 \qquad t \geq 0, \quad -\infty < x < \infty$$
$$u(0, x) = u_0(x)$$

*has a unique solution $u \in C([0, T] : H^s(\mathbb{R})) \cap C^1([0, T] : L^2(\mathbb{R}))$.*

# Bibliographical Notes and Remarks

The abstract theory of semigroups of linear operators is a part of functional analysis. As such it is covered to some extent by many texts of functional analysis. The most extensive treatise of the subject is the classical book of Hille and Phillips [1]. Other general references are the books of Butzer and Berens [1]. Davies [1], Dunford and Schwartz [1], Dynkin [1], Friedman [1], Ladas and Lakshmikantham [1], Kato [9], Krein [1], Martin [1], Reed and Simon [1], Riesz and Nagy [1], Rudin [1], Schechter [4], Tanabe [6], Walker [1], Yosida [7] and others.

A good introduction to the abstract theory as well as to some of its applications is provided by the lecture notes of Yosida [3], Phillips [7] and Goldstein [3].

The theory of semigroups of bounded linear operators developed quite rapidly since the discovery of the generation theorem by Hille and Yosida in 1948. By now, it is an extensive mathematical subject with substantial applications to many fields of analysis. Only a small part of this theory is covered by the present book which is mainly oriented towards the applications to partial differential equations. We mention here briefly some themes which are not touched at all in this book.

Most of the classical theory of semigroups of bounded linear operators on a Banach space has been extended to equi-continuous semigroups of class $C_0$ in locally convex linear topological spaces. The first work in this direction was done by L. Schwartz [1]. Most of the classical results of the theory were generalized to this case by K. Yosida [7]. Further results in a more general set up are given in Komatsu [2], Dembart [1], Babalola [1], Ouchi [1] and Komura [1].

The theory was also generalized to semigroups of distributions. The first results in this direction are due to J. L. Lions [1]; see also Chazarin [1], Da Prato and Mosco [1], Fujiwara [1] and Ushijima [1], [2].

In the present book we deal only with strongly continuous semigroups. Different classes of continuity at zero were introduced and studied in Hille-Phillips [1]. Some more recent results on semigroups which are not $C_0$ semigroups can be found in Oharu [1]. Oharu and Sunouchi [1], Miyadera, Oharu and Okazawa [1], Okazawa [2] and Miyadera [3].

The theory of semigroups of bounded linear operators is closely related to the solution of ordinary differential equations in Banach spaces. Usually, each "well-posed" linear autonomous initial value problem gives rise to a semigroup of bounded linear operators. The book of S. G. Krein [1] studies the theory of semigroups from this point of view. There are however interesting results on differential equations in Banach spaces which are not well posed. In this direction we mention the work of Agmon and Nirenberg [1]; see also Lions [2], Lax [1], Zaidman [1], Ogawa [1], Pazy [1], Maz'ja and Plamenevskii [1] and Plamenevskii [1].

As we have just hinted semigroups of operators are obtained as solutions of initial value problems for a first order differential equation in a Banach space. Most of the theory deals with a single first order equation. The reason for this is that higher order equations can be reduced to first order systems and then by changing the underlying Banach space one obtains a first order single equation. There are however results for higher order equations which cannot be obtained by such a reduction and there are other results in which it is just more convenient to treat the higher order equation directly. We refer the interested reader to S. G. Krein [1] Chapter 3 for a discussion of equations of order two. Further references are Fattorini [1], [2], Goldstein [2], [4], Sova [1], Kisynski [3], [4], Nagy [1], [2], [3], Travis and Webb [1], [2], Rankin [1] and others.

In recent years the theory of semigroups of bounded linear operators has been extended to a large and interesting theory of semigroups of nonlinear operators in Hilbert and Banach spaces. We mention here only a few general references to the subject; Benilan, Crandall and Pazy [1], Brezis [1], Barbu [2], Crandall [1], Yosida [7], Pazy [4], [8] and Pavel [3].

Before we turn to a somewhat more detailed bibliographical account on the material presented in this book we note that no attempt has been made to compile a complete bibliography even of those parts of the theory which are covered by the present book. Most references given are only to indicate sources of the material presented, or closely related topics, and sources for further reading. An extensive bibliography of the subject was compiled by J. A. Goldstein and will appear in a forthcoming book by him.

**Section 1.1.** The results on semigroups of bounded linear operators which are continuous in the uniform operator topology at $t = 0$, or equivalently, semigroups which are generated by bounded linear operators can be considered as results about the exponential function in a Banach algebra. This approach was taken by M. Nagumo [1] and K. Yosida [1], see also Hille-Phillips [1] Chapter V. The representation of uniformly continuous groups of operators as an exponential of a bounded operator was also obtained by D. S. Nathan [1].

**Section 1.2.** Most of the results of this section are standard and can be found in every text dealing with semigroups of linear operators e.g. all the texts mentioned at the beginning of these bibliographical notes.

The proof of Theorem 2.7 follows a construction of I. Gelfand [1]. Lemma 2.8 is an extension of a classical inequality (Example 2.9) of E. Landau. In the present form it is due to Kallman and Rota [1]. For the case of a Hilbert space, T. Kato [12] proved that if $T(t)$ is a semigroup of contractions then 2 is the best possible constant in (2.13). For general Banach spaces, the best possible constant seems to be unknown. More details on related inequalities are given in Certain and Kurtz [1], see also Holbrook [1].

**Section 1.3.** The main result of this section is Theorem 3.1 which gave the first complete characterization of the infinitesimal generator of a strongly continuous semigroup of contractions. This result was the starting point of the subsequent systematic development of the theory of semigroups of bounded linear operators. It was obtained independently by E. Hille [2] and K. Yosida [2]. Our proof of the sufficient part of the theorem follows the ideas of K. Yosida [2]. The bounded linear operator $A_\lambda$ appearing in this proof is called the *Yosida approximation* of $A$. Hille's proof is based on a direct proof of the convergence of the exponential formula

$$T(t)x = \lim_{n \to \infty} \left( I - \frac{t}{n} A \right)^{-n} x$$

for $x \in D(A^2)$, see e.g. Tanabe [6] Section 3.1.

**Section 1.4.** The results of this section for the special case where $X = H$ is a Hilbert space are due to R. S. Phillips [5]. The extension to the general case was carried out by Lumer and Phillips [1]. We note in passing that the characterization of the infinitesimal generator $A$ of a semigroup of contractions as an $m$-dissipative operator i.e. a dissipative operator for which the range of $\lambda I - A$, $\lambda > 0$ is all of $X$ plays an essential role in the theory of nonlinear semigroups.

**Section 1.5.** The main result of this section is Theorem 5.2 which gives a complete characterization of the infinitesimal generator of a $C_0$ semigroup of bounded linear operators and thus generalizes the Hille-Yosida theorem which was restricted to the characterization of the generator of a $C_0$ semigroup of contractions. Theorem 5.2 was obtained independently and almost simultaneously by W. Feller [1], I. Miyadera [1] and R. S. Phillips [2]. Our proof of the theorem is a simplification of Feller's proof.

Another way to prove the sufficient part of Theorem 5.2 is to prove Theorem 5.5 directly using a straightforward generalization of the proof of the sufficient part of the Hille-Yosida theorem, see e.g. Dunford-Schwartz [1] Chapter VIII.

**Section 1.6.** The study of semigroups of linear operators started actually with the study of groups of operators. The first results were those for groups generated by bounded linear operators (see Section 1.1). These works were followed by M. Stone [1] and J. von-Neumann [1]. Theorem 6.3 is due to E. Hille [1] and Theorem 6.6 is due to J. R. Cuthbert [1].

**Section 1.7.** The results about the inversion of the Laplace transform are standard. Better results for the inversion of the Laplace transform can be obtained by a somewhat more delicate analysis, see Hille-Phillips [1] Chapter II. The conditions of Theorem 7.7 imply actually that $A$ is the infinitesimal generator of an analytic semigroup (see Section 2.5). Usually one proves for such an $A$ directly, using the Dunford-Taylor operator calculus, that $U(t)$ defined by (7.26) is a semigroup of bounded linear operators and that $A$ is its infinitesimal generator, see e.g. Friedman [1] Part 2 Section 2. Instead, we prove that the condition (7.24) implies the conditions of Theorem 5.2 and $A$ is thus the infinitesimal generator of a $C_0$ semigroup.

**Section 1.8.** Theorem 8.1 is due to E. Hille [1]. In this context see also Dunford-Segal [1]. It is interesting to note that this paper of Dunford and Segal stimulated K. Yosida strongly and led him to the characterization of the infinitesimal generator of a $C_0$ semigroup of contractions, Yosida [2].

Theorem 8.3 is due to E. Hille (see Hille-Phillips [1]). The exponential formula given in Theorem 8.3 served as a base of Hille's proof of the characterization of the infinitesimal generator of a $C_0$ semigroup of contractions. This formula was also the starting point of the theory of semigroups of nonlinear contractions in general Banach spaces which started in 1971 by the fundamental result of Crandall and Liggett. The proof of Theorem 8.3 that we give here follows Hille-Phillips [1]. A different proof of a more general result is given in Section 3.5.

**Section 1.9.** The results of this section are based on Hille-Phillips [1] Chapter V and Kato [3]. See also Kato [9] Chapter 8.

**Section 1.10.** In the definition of the adjoint semigroup we follow Phillips [4], see also Hille-Phillips [1] Chapter XIV and K. Yosida [7] Chapter IX in which an extension, by H. Komatsu [2], of the results of Phillips to locally convex spaces is given.

A slightly different approach which leads however to the same strongly continuous semigroups is taken by Butzer and Berens [1] Chapter I. Theorem 10.8 is due to M. Stone [1] and was the first result concerning semigroups generated by an unbounded linear operator.

**Section 2.1.** The algebraic semigroup property, $T(t + s) = T(t)T(s)$, amplifies many topological properties of the semigroup $T(t)$. Theorem 1.1 which is due to K. Yosida [3] is one example of such an effect. Another example, Hille-Phillips [1] Chapter X, is;

**Theorem.** *If $T(t)$ is a semigroup of bounded linear operators which is strongly measurable on $]0, \infty[$ then it is strongly continuous on $]0, \infty[$. If moreover $T(t)$ is weakly continuous at $t = 0$ then $T(t)$ is a $C_0$ semigroup.*

**Section 2.2.** The results of this section cover most of the results of Chapter XVI of Hille-Phillips [1]. While the proofs of the results in Hille-Phillips [1] use the Gelfand representation theory, our proof of Theorem 2.3 is completely elementary and follows the approach taken in Hille [1]. Theorems 2.4, 2.5 and 2.6 also follow Hille [1].

A counter example to the converse of Theorem 2.6 is given in Hille-Phillips [1] (page 469), see also Greiner, Voigt and Wolff [1]. Further results on the spectral mapping theorem for $C_0$ semigroups of positive operators can be found in Greiner [1], Derdinger [1] and Derdinger and Nagel [1].

**Section 2.3.** Theorem 3.2 is due to P. D. Lax (see Hille-Phillips [1] Chapter X). Theorem 3.3 and Corollaries 3.4 and 3.5 are taken from Pazy [3]. Theorem 3.6 comes from Hille-Phillips [1] Chapter XVI but, while the proof there uses the Gelfand representation theory, our proof is elementary. Theorem 3.6 gives a necessary condition for an infinitesimal generator $A$ to generate a $C_0$ semigroup which is continuous in the uniform operator topology for $t > 0$. It seems that a full characterization of the infinitesimal generator of such semigroups in terms of properties of their resolvents is not known.

**Section 2.4.** Some early results on the differentiability of $C_0$ semigroups were obtained by E. Hille [3] and K. Yosida [4]. The full characterization of the infinitesimal generator of a differentiable semigroup, Theorem 4.7, is due to Pazy [3]. Theorem 4.7 was extended to semigroups of distributions by V. Barbu [1] and to semigroups of linear operators on locally convex spaces by M. Watanabe [1]. Corollary 4.10 is due to Yosida [4]. Theorem 4.11 and Corollaries 4.12 and 4.14 come from Pazy [5].

**Section 2.5.** Theorem 5.2 is due to E. Hille [1]. Our proof follows Yosida [4]. Theorem 5.3 is due to E. Hille [3]. Theorem 5.5 is taken from Crandall, Pazy and Tartar [1] while Theorem 5.6 is due to Kato [10]. Corollary 5.7 is due to J. Neuberger [1] and T. Kato [10]. Corollary 5.8 seems to be new. The uniform convexity of the underlying space or a similar condition is necessary since there are concrete examples of analytic semigroups of contractions for which $\lim_{t \to 0} \|I - T(t)\| = 2$, G. Pisier (private communication).

Related to the results of this section are also the deep results of A. Beurling [1] and M. Certain [1].

**Section 2.6.** Let $A$ be the infinitesimal generator of a $C_0$ semigroup. The fractional powers of $-A$ were first investigated by S. Bochner [1] and R. S. Phillips [1]. Later A. V. Balakrishnan [1], [2] gave a new definition of the fractional powers of $-A$ and extended the theory to a wider class of operators. About the same time several other authors contributed to this subject. Among them M. Z. Solomjak [1], K. Yosida [5], T. Kato [4], [5], [7], Krasnoselskii and Sobolevskii [1], J. Watanabe [1]. Subsequently, H. Komatsu gave a unified point of view in a series of papers Komatsu [3]–[7].

Our simplified treatment follows mainly Kato [4] and [5], see also Friedman [1] Part 2 Section 14 and Tanabe [6] Section 2.3.

**Section 3.1.** The results of this section are due to R. S. Phillips [2]. For related results see Hille-Phillips [1] Chapter XIII and Dunford-Schwartz [1] Chapter 8. Phillips [2] also started the study of properties of $C_0$ semigroups which are conserved under bounded perturbations (i.e. perturbations of the infinitesimal generator by a bounded operator). Among other results he showed that continuity in the uniform operator topology for $t > 0$ is conserved while the same property for $t > t_0 > 0$ is not conserved. The problem whether or not the differentiability for $t > 0$ of a semigroup $T(t)$ is conserved under bounded perturbations of its generator seems to be still open. For a result related to this problem see Pazy [3].

**Section 3.2.** Theorem 2.1 is due to E. Hille [1], see also T. Kato [9] Chapter 9 and Hille-Phillips [1] Chapter XIII. A related result is given in Da Prato [1].

**Section 3.3.** Corollary 3.3 was essentially proved by H. F. Trotter [2] for the case $\alpha < \frac{1}{2}$, see also Kato [9] Chapter 9. The general case of Corollary 3.3 with $\alpha < 1$ was proved by K. Gustafson [1]. Theorem 3.2 is a consequence of a more symmetric version of Corollary 3.3 proved in Pazy [9].

Theorem 3.4 was proved by P. Chernoff [2]. Corollary 3.5 is due to P. Chernoff [2] and N. Okazawa [1], it is a generalization of the result of R. Wüst [1] in Hilbert space.

**Section 3.4.** The main results of this section are due to H. F. Trotter [1]. J. Neveu [1] has proved the convergence theorem (Theorem 4.5) for the special case of semigroups of contractions independently. Convergence results of a similar nature are also given in T. Kato [9] and T. Kurtz [1], [2]. In Trotter [1] the proof that the limit of the resolvents $R(\lambda : A_n)$ of $A_n$ is itself a resolvent of some operator $A$ is not clear. This was pointed out and corrected by T. Kato [3]. In Theorem 4.5 the condition that $(\lambda_0 I - A)D$ is dense in $X$ assures that $\bar{A}$ (the closure of $A$) is an infinitesimal generator of a $C_0$ semigroup. A different necessary and sufficient condition for this is given in M. Hasegawa [1]. An interesting proof of Trotter's theorem was given by Kisynski [2].

Trotter [1], treats also the question of convergence of $C_0$ semigroups acting on different Banach spaces. Results of this nature are very useful in proving the convergence of solutions of certain difference equations to the solutions of a corresponding partial differential equation. An example of this type is given in Section 3.6 below.

Convergence in a Banach space, of semigroups which are not $C_0$ semigroups was studied by I. Miyadera [2] and Oharu-Sunouchi [1].

The convergence results were also extended to semigroups on locally convex spaces, see e.g. K. Yosida [7], T. Kurtz [2] and T. I. Seidman [1].

**Section 3.5.** Lemma 5.1 is a simple extension of Corollary 5.2 which is due to P. Chernoff [1]. Theorem 5.3 and Corollary 5.4 are also extensions of the results of Chernoff [1]. Corollary 5.5 is an extension of the Trotter product formula, Trotter [2]. With regard to the conditions of this formula see Kurtz and Pierre [1].

**Section 3.6.** The results of this section are relevant to the numerical solutions of partial differential equations. They are similar in nature to the results of Trotter [1] and Kato [9]. For results of similar nature see also Lax-Richtmyer [1] and Richtmyer-Morton [1].

**Section 4.1.** The initial value problem (1.1) in the Banach space $X$ is called an abstract Cauchy problem. The systematic study of such problems started with E. Hille [4]. The uniqueness theorem (Theorem 1.2) is due to Ljubic [1]. Theorem 1.3 is due to Hille [4], see also Phillips [3]. Sufficient conditions for the existence of a solution of (1.1) for a dense subset $D$ of $X$ (not necessarily equal to $D(A)$) of initial data are given in R. Beals [1].

A different way of defining a weak solution of (1.1) was given by J. Ball [1], see remarks to the next section.

**Section 4.2.** Definition 2.3 defines a mild solution of (2.1) if $A$ is the infinitesimal generator of a semigroup $T(t)$. J. M. Ball [1] defines a "weak solution" of the equation

$$\frac{du}{dt} = Au + f(t) \tag{E}$$

where $A$ is a closed linear operator on $X$ and $f \in L^1(0, T; X)$ as follows:

**Definition.** A function $u \in C([0, T]: X)$ is a weak solution of (E) on $[0, T]$ if for every $v^* \in D(A^*)$ the function $\langle u(t), v^* \rangle$ is absolutely continuous on $[0, T]$ and

$$\frac{d}{dt} \langle u(t), v^* \rangle = \langle u(t), A^*v^* \rangle + \langle f(t), v^* \rangle \qquad \text{a.e. on} \quad [0, T].$$

He then proves,

**Theorem** (Ball). *There exists for each $x \in X$ a unique weak solution $u$ of* (E) *on* $[0, T]$ *satisfying $u(0) = x$ if and only if $A$ is the infinitesimal generator of a $C_0$ semigroup $T(t)$ of bounded linear operators on $X$, and in this case $u$ is given by*

$$u(t) = T(t)x + \int_0^t T(t - s)f(s)\, ds, \qquad 0 \le t \le T.$$

Corollaries 2.5 and 2.6 are due to Phillips [2], see also T. Kato [9]. Theorem 2.9 is a straightforward generalization of Theorem 2.4.

**Section 4.3.** Theorem 3.1 is essentially due to A. Pazy [7]. There it was only proved that $u$ is Hölder continuous with exponent $\beta$ satisfying $\beta < 1 - 1/p$. The fact that the result is true for $\beta = 1 - 1/p$ is due to L. Veron. Theorem 3.2 comes from Crandall and Pazy [1]. Corollary 3.3, for the more general situation where $A$ depends on $t$ (see Chapter 5) was proved by H. Tanabe [2], P. E. Sobolevskii [4], E. T. Poulsen [1] and Kato [9]. Theorem 3.5 is due to Kato [9], see also Da Prato and Grisvard [1]. Optimal regularity conditions for this problem are given in E. Sinestrari [1].

**Section 4.4.** Theorem 4.1 was taken from Pazy [6]. It is a simple generalization of a previous result of R. Datko [1]. The idea of Example 4.2 is taken from Greiner, Voigt and Wolff [1]. Other examples of this sort are also given in Hille-Phillips [1] Chapter XXIII and Zabczyk [1].

A more general result than Theorem 4.3 was proved by M. Slemrod [1]. The same problem is also treated in Derdinger and Nagel [1] and Derdinger [1].

Theorem 4.5 was taken from S. G. Krein [1] Chapter 4.

**Section 4.5.** The results of this section are technical and they are brought here mainly as a preparation to the first sections of Chapter 5. In this section we follow closely the results of T. Kato [11], see also H. Tanabe [6], Chapter 4.

**Section 5.1.** The results of this section are completely elementary and their sole aim is to motivate the rest of the results of this chapter and to familiarize the reader with the notion and main properties of evolution-systems. The term "evolution-system" is not standard, some authors call it a propagator, others a fundamental solution and still others an evolution-operator.

**Section 5.2.** The results of this section follow those of Kato [11]. The notion of stability defined here is stronger than the usual one used in the theory of finite difference approximations. When $A(t)$ is independent of $t$ then the stability condition coincides with the condition of Theorem 1.5.3 and therefore we can renorm the space so that in the new norm $A$ generates a semigroup $T(t)$ satisfying $\|T(t)\| \le e^{\omega t}$. If $A(t)$ depends on $t$ but $D(A(t))$

is independent of $t$ and the operators $A(t)$ commute for $t \geq 0$ then it is not difficult to show that the stability of $A(t)$ implies that $X$ can be renormed so that in the new norm $\|S_t(s)\| \leq e^{\omega s}$ for every $t \in [0, T]$ where $S_t(s)$ is the semigroup generated by $A(t)$.

**Sections 5.3–5.5.** The first construction of an evolution system for the initial value problem (3.1) with unbounded operators $A(t)$ was achieved by T. Kato [1]. His main assumptions were that $D(A(t)) = D$ is independent of $t$ and that for each $t \geq 0$, $A(t)$ is the infinitesimal generator of a $C_0$ contraction semigroup on $X$ together with some continuity conditions on the family of bounded operators $A(t)A(s)^{-1}$. The main result of Kato [1] is essentially a special case of Theorem 4.8. In an attempt to extend the results of Kato [1] and especially to remove the assumption that $D(A(t))$ is independent of $t$, several authors constructed evolution systems under a variety of conditions, see e.g. Elliot [1], Goldstein [1], Heyn [1], Kato [2], Kisynski [1], Yosida [7], [6] and others.

Our presentation follows closely that of T. Kato [11], [13] with a simplification due to Dorroh [1], see also H. Tanabe [6] Chapter 4.

A different method of studying the evolution equations (3.1) directly in the space $L^p(0, T: X)$, using a sum of operators technique is developed in Da Prato and Iannelli [1], see also Da Prato and Grisvard [1] and Iannelli [1].

Finally we note that the special partitions needed for Remark 3.2 are constructed in the appendix of Kato [13] or else in Evans [1].

**Sections 5.6–5.7.** The first evolution systems in the parabolic case were constructed by H. Tanabe [1], [2], [3] and independently but by a similar method by P. E. Sobolevskii [4]. In these works it was assumed that $D(A(t))$ is independent of $t$. This assumption was somewhat relaxed by T. Kato [6] and P. E. Sobolevskii [1], [3] who assumed that $D(A(t)^\gamma)$ for some $0 < \gamma < 1$ is independent of $t$. Later, T. Kato and H. Tanabe [1], [2] succeeded in removing the assumption that $D(A(t))$ is independent of $t$. They replaced it by some regularity assumptions on the function $t \rightarrow R(\lambda: A(t))$. In this context higher differentiability of the solution is obtained if one assumes higher differentiability of $t \rightarrow R(\lambda: A(t))$ see Suryanarayana [1]. Assuming that the conditions hold in a complex neighborhood of $[0, T]$ one obtains solutions of (6.2) that can be extended to a complex neighborhood of $]0, T]$ see Komatsu [1], Kato Tanabe [2] and K. Masuda [1]. K. Masuda [1] showed further that in this particular situation the Kato-Tanabe conditions are also necessary for the existence of an evolution system.

In Section 5.6 and 5.7 we deal only with the case where $D(A(t))$ is independent of $t$. We follow Tanabe [2], Sobolevskii [4] and Poulsen [1], see also H. Tanabe [6] Chapter 5 where the case of variable $D(A(t))$ is also treated.

A different approach to the solution of the evolution equation (6.1) (with $D(A(t))$ independent of $t$) which is not based on a construction of an evolution system for (6.2) is given in Da Prato and Sinestrari [1].

**Section 5.8.** Theorem 8.2 is due to H. Tanabe [4] and Theorem 8.5 is due to Pazy [2].

A subject which is related to the asymptotic behavior of solutions of the evolution equation (8.1) and which has not been touched in this chapter is singular perturbations, see e.g. Tanabe [5] and Tanabe and Watanabe [1].

**Section 6.1.** Theorems 1.2, 1.4 and 1.5 are due to I. Segal [1], see also T. Kato [8]. An example in which $f$ is Lipschitz continuous but the mild solution of (1.1) is not a strong solution can be found in Webb [1]. Theorems 1.6 and 1.7 are simple but useful modifications of the previous results.

We note that the Lipschitz continuity of $f$ can be replaced by accretiveness and one still obtains, under suitable conditions, global solutions of the initial value problem (1.1) see e.g. Kato [8], Martin [1] Chapter 8 and the very general paper of N. Pavel [2].

**Section 6.2.** The results of this section are based on Pazy [7]. Examples in which (3.1) with $A = 0$ and $f$ continuous does not have solutions are given e.g. in Dieudonne [1] page 287 and J. Yorke [1]. It is known, in fact, that with $A = 0$ the initial value problem (3.1) has a local strong solution for every continuous $f$ if and only if $X$ is finite dimensional, Godunov [1].

The main existence result, Theorem 2.1, of this section was extended by N. Pavel [1] as follows:

**Theorem** (Pavel). *Let $D \subset X$ be a locally closed subset of $X$, $f:[t_0, t_1[ \to X$ continuous and let $S(t)$, $t \geq 0$ be a $C_0$ semigroup, with $S(t)$ compact for $t > 0$. A necessary and sufficient condition for the existence of a local solution $u:[t_0, T(t_0:x_0)] \to D$, where $t_0 < T(t_0:x_0) \leq t_1$, to (3.1) for every $x_0 \in D$ is*

$$\lim_{h \to 0} h^{-1} \operatorname{dist}\left(S(h)z + hf(t, z): D\right) = 0$$

*for all $t \in [t_0, t_1[$ and $z \in D$.*

**Section 6.3.** The main result of this section, Theorem 3.1, is motivated by the work of H. Fujita and T. Kato [1]. Similar and more general existence results of this type can be found in Sobolevskii [4], Friedman [1] Part 2 Section 16 and Kielhöfer [3].

The treatise of D. Henri [1] "Geometric theory of semilinear parabolic equations" contains along with an existence result similar to Theorem 3.1 an extensive study of the dependence of the solutions on the data, their asymptotic behavior and many interesting applications.

Results which are to some extent between those of this section and the previous one are given in Lightbourne and Martin [1] and in Martin [2]. In these results $f$ is assumed to be continuous (but not necessarily Lipschitz continuous) with respect to some fractional power of $A$ and $S(t)$, the semigroup generated by $-A$, is assumed to be compact for $t > 0$.

The existence results of the previous sections were stated for the autonomous case (i.e. $A$ independent of $t$) mainly for the sake of simplicity. They can be extended to the nonautonomous case as is actually done in Segal [1] and Prüss [1] for the results of Section 6.1, in Fitzgibbon [1] for those of Section 6.2 and in Sobolevskii [4], Friedman [1] and Kielhöfer [3] for those of Section 6.3.

Some asymptotic results for nonautonomous semilinear evolution equations are given in Nambu [1].

**Section 6.4.** The results of this section follow closely Kato [14], see also Kato [15]. A different method to treat similar equations was recently developed by Crandall and Souganidis [1].

**Section 7.1.** As we have already mentioned in the introduction, the present book's main aim is the applications of semigroup theory to partial differential equations. The purpose of this and the next chapter is to present some examples of such applications.

A detailed study of Sobolev spaces is given in Adams [1], other references are Necas [1], Friedman [1] and Lions-Magenes [1].

**Section 7.2–7.3.** In the applications presented in these sections we restrict ourselves, for the sake of simplicity, to the Dirichlet boundary conditions. All the results hold for more general boundary conditions see e.g. Agmon [1], Stewart [2], Tanabe [6] Section 3.8 Pazy [2] and others. The needed a-priori estimates for the elliptic operators with general boundary conditions are given in Agmon, Douglis and Nirenberg [1], Nirenberg [1], Schechter [1], [2], [3] and Stewart [2], see also Lions-Magenes [1].

Theorem 2.2 is due to L. Gårding [1], for a proof see e.g. Agmon [2], Friedman [1], Yosida [7]. The regularity of solutions of elliptic boundary value problems (Theorem 2.3) was proved for general boundary values and $1 < p < \infty$ by Agmon, Douglis and Nirenberg [1] and for the Dirichlet boundary values by Nirenberg [1], see also Agmon [2], Friedman [1] and Lions-Megenes [1].

Theorem 3.1 is due to Agmon, Douglis and Nirenberg [1], Theorems 3.2 and 3.5 are due to Agmon [1] and Theorem 3.7 to Stewart [1].

Another interesting example of an operator that generates an analytic semigroup is the classical Stokes operator. For details see Giga [1].

**Section 7.4.** In this section we follow the treatment of K. Yosida [3], [7] in which more general hyperbolic equations are also treated.

**Section 7.5.** A proof of the classical Hausdorff-Young theorem used in this section can be found e.g. in Stein and Weiss [1] Chapter V.

**Section 7.6.** Results similar to those presented in this section, with more general boundary conditions can be found in Tanabe [6] Chapter 5 and Friedman [1] Part 2 Sections 9, 10.

**Section 8.1.** The results of this section are due to Baillon, Cazenave and Figueira [1] and to Ginibre and Velo [1]. Our presentation follows that of Baillon et al. Related results can be found in Lin and Strauss [1], Pecher and von Wahl [1] and Haraux [1].

Theorem 1.5 is also true in a bounded domain $\Omega$ in $\mathbb{R}^2$. The local existence of the solution in this case is similar to the case on all of $\mathbb{R}^2$ while the global existence is more complicated since one cannot apply Sobolev's imbedding theorem in a straightforward way. To prove the global existence in this case a new interpolation-imbedding inequality is used, see Brezis and Gallouet [1].

**Section 8.2.** The results of this section follow closely Pazy [7].

**Sections 8.3–8.4.** In these two sections fractional powers of minus the infinitesimal generators of analytic semigroups are used to obtain, via the abstract results of Section 6.3, solutions of certain nonlinear initial value problems for partial differential equations.

The results of Section 8.3 follow rather closely the ideas of Fujita and Kato [1] in which the linear operator $A$ is more complicated than in our case. Lemma 3.3 is due to Fujita-Kato [1].

The results of Section 8.4 follow those of Sobolevskii [4] and Friedman [1]. Results of similar nature in Hölder spaces and for unbounded domains can be found in Kielhöfer [1], [2].

The Gagliardo-Nirenberg inequalities used in this section are proved e.g. in Friedman [1] Part 1 Sections 9, 10.

In certain cases global solutions can be obtained, usually using some further conditions, see e.g. Kielhöfer [3] and von Wahl [1], [2]. For the Navier-Stokes equations in $\mathbb{R}^2$ see Fujita-Kato [1] and Sobolevskii [2].

Finally we note that for the sake of simplicity we have chosen to take the linear operator $A$ to be independent of $t$. Similar results can be obtained when $A$ depends on $t$, see e.g. Friedman [1] and Kielhöfer [3].

**Section 8.5.** The results of this section follow one of many examples given in Kato [14]. For this particular example better results including a global existence theorem are given in Kato [16].

# Bibliography

R. A. Adams
  [1] *Sobolev spaces*, Academic Press, New York (1975).

S. Agmon
  [1] On the eigenfunctions and on the eigenvalues of general elliptic boundary value problems, *Comm. Pure Appl. Math.* **15** (1962) 119–147.
  [2] Elliptic boundary value problems, Van Nostrand (1965).

S. Agmon and L. Nirenberg
  [1] Properties of solutions of ordinary differential equations in Banach spaces, *Comm. Pure Appl. Math.* **16** (1963) 121–239.

S. Agmon, A. Douglis and L. Nirenberg
  [1] Estimates near the boundary for solutions of elliptic partial differential equations, *Comm. Pure Appl. Math* **12** (1959) 623–727.

V. A. Babalola
  [1] Semigroups of operators on locally convex spaces, *Trans. Amer. Math. Soc.* **199** (1974) 163–179.

J. B. Baillon, T. Cazenave, and M. Figueira
  [1] Equation de Schrödinger non lineaire, *C.R. Acad. Sc. Paris* **284 A** (1977) 869–872.

A. V. Balakrishnan
  [1] An operator calculus for infinitesimal generators of semi-groups, *Trans. Amer. Math. Soc.* **91** (1959) 330–353.
  [2] Fractional powers of closed operators and the semi-groups generated by them, *Pacific J. Math.* **10** (1960) 419–437.

J. M. Ball
  [1] Strongly continuous semigroups, weak solutions and the variation of constants formula, *Proc. Amer. Math. Soc.* **63** (1977) 370–373.

V. Barbu
  [1] Differentiable distribution semi-groups, *Anal. Scuola Norm. Sup. Pisa* **23** (1969) 413–429.
  [2] *Nonlinear semigroups and differential equations in Banach spaces*, Noordhoff Int. Publ. Leyden the Netherlands (1976).

R. Beals
  [1] On the abstract Cauchy problem, *J. Func. Anal.* **10** (1972) 281–299.

Ph. Benilan, M. Crandall and A. Pazy
  [1] *Nonlinear evolution governed by accretive operators* (a book to appear).

A. Beurling
  [1] On analytic extension of semi-groups of operators, *J. Func. Anal.* **6** (1970) 387–400.

S. Bochner
  [1] Diffusion equations and stochastic processes, *Proc. Nat. Acad. Sc. U.S.A.* **35** (1949) 368–370.

P. L. Butzer and H. Berens
  [1] *Semi-groups of operators and approximation*, Springer-Verlag New York (1967).

H. Brezis
  [1] *Operateurs maximaux monotone et semigroups de contractions dans les espaces de Hilbert*, Math. Studies **5**, North Holland (1973).

H. Brezis and T. Gallouet
  [1] Nonlinear Schrödinger evolution equations, *Nonlinear Anal. TMA* **4** (1980) 677–682.

M. Certain
  [1] One-parameter semigroups holomorphic away from zero, *Trans. Amer. Math. Soc.* **187** (1974) 377–389.

M. Certain and T. Kurtz
  [1] Landau-Kolmogorov inequalities for semigroups and groups, *Proc. Amer. Math. Soc.* **63** (1977) 226–230.

J. Chazarin
  [1] Problemes de Cauchy abstrait et applications a quelques problemes mixtes, *J. Func. Anal.* **7** (1971) 386–446.

P. Chernoff
  [1] Note on product formulas for operator semi-groups, *J. Func. Anal.* **2** (1968) 238–242.
  [2] Perturbations of dissipative operators with relative bound one, *Proc. Amer. Math. Soc.* **33** (1972) 72–74.

M. G. Crandall
  [1] An introduction to evolution governed by accretive operators, "Dynamical-Systems-An international Symposium."(L. Cesari, J. Hale, J. LaSalle, Eds.) Academic Press, New York (1976) 131–165.

M. G. Crandall and A. Pazy
  [1] On the differentiability of weak solutions of a differential equation in Banach space, *J. Math. and Mech.* **18** (1969) 1007–1016.

M. G. Crandall, A. Pazy, and L. Tartar
   [1] Remarks on generators of analytic semigroups, *Israel J. Math* **32** (1979) 363–374.

M. G. Crandall and P. E. Souganidis
   [1] Nonlinear evolution equations MRC technical Rep. 2352 (1982).

J. R. Cuthbert
   [1] On semi-groups such that $T_t - I$ is compact for some $t > 0$. *Z. Whar.* **18** (1971) 9–16.

G. Da Prato
   [1] Somma di generatori infinitesimali di semigruppi analitici, *Rend. Sem. Math. Univ Padova* **40** (1968) 151–161.

G. Da Prato and P. Grisvard
   [1] Sommes d'operateurs lineaires et equations differentielles operationnelles, *J. Math. Pure et Appl.* **54** (1975) 305–387.

G. Da Prato and M. Iannelli
   [1] On a method for studying abstract evolution equations in the Hyperbolic case, *Comm. in Partial Diff. Eqs.* **1** (1976) 585–608.

G. Da Prato and U. Mosco
   [1] Semigruppi distribuzioni analitici, *Ann. Scuola Norm. Sup. Pisa* **19** (1965) 367–396.

G. Da Prato and E. Sinestrari
   [1] Hölder regularity for non autonomous abstract parabolic equations, *Israel J. Math.* **42** (1982) 1–19.

R. Datko
   [1] Extending a theorem of A. M. Liapunov to Hilbert space, *J. Math. Anal. and Appl.* **32** (1970) 610–616.

E. B. Davies
   [1] *One-Parameter Semigroups*, Academic Press, London (1980).

B. Dembart
   [1] On the theory of semigroups of operators on locally convex spaces, *J. Func. Anal.* **16** (1974) 123–160.

R. Derdinger
   [1] Über das Spektrum positiver Generatoren, *Math. Z.* **172** (1980) 281–293.

R. Derdinger and R. Nagel
   [1] Der Generator stark stetiger Verbandhalbgruppen auf $C(X)$ und dessen Spektrum, *Math. Ann.* **245** (1979) 159–177.

J. Dieudonne
   [1] *Foundation of modern analysis*, Academic Press, New York (1960).

J. R. Dorroh
   [1] A simplified proof of a theorem of Kato on linear evolution equations, *J. Math. Soc. Japan* **27** (1975) 474–478.

N. Dunford and J. Schwartz
   [1] *Linear operators, Part I General theory*, Interscience, New York (1958).

N. Dunford and I. E. Segal
  [1] Semi-groups of operators and the Weierstrass theorem, *Bull. Amer. Math. Soc.* **52** (1946) 911–914.

E. B. Dynkin
  [1] *Markov Processes*, Vol. I. Springer-Verlag, Berlin (1965).

J. Elliot
  [1] The equations of evolution in a Banach space, *Trans. Amer. Math. Soc.* **103** (1962) 470–483.

C. L. Evans
  [1] Nonlinear evolution equations in an arbitrary Banach space, *Israel J. Math.* **26** (1977) 1–42.

H. Fattorini
  [1] Ordinary differential equations in linear topological spaces I. *J. Diff. Eqs.* **5** (1968) 72–105.
  [2] Ordinary differential equations in linear topological spaces II. *J. Diff. Eqs.* **6** (1969) 50–70.

W. Feller
  [1] On the generation of unbounded semi-groups of bounded linear operators, *Ann. of Math.* **58** (1953) 166–174.

W. E. Fitzgibbon
  [1] Semilinear functional differential equations in Banach space, *J. Diff. Eqs.* **29** (1978) 1–14.

A. Freidman
  [1] *Partial differential equations*, Holt, Reinhart, and Winston, New York (1969).

H. Fujita and T. Kato
  [1] On the Navier-Stokes initial value problem I. *Arch. Rat. Mech. and Anal.* **16** (1964) 269–315.

D. Fujiwara
  [1] A characterization of exponential distribution semi-groups, *J. Math. Soc. Japan* **18** (1966) 267–274.

L. Gårding
  [1] Dirichlet's problem for linear elliptic partial differential equations, *Math. Scan.* **1** (1953) 55–72.

I. Gelfand
  [1] On one-parameter groups of operators in normed spaces, *Dokl. Akad. Nauk SSSR.* **25** (1939) 713–718.

Y. Giga
  [1] Analyticity of the semigroup generated by the Stokes operator in $L_r$ spaces, *Math. Z.* **178** (1981) 297–329.

J. Ginibre and B. Velo
  [1] On a class of nonlinear Schrödinger equations, *J. Func. Anal.* **32** (1979) 1–71.

A. N. Godunov
  [1] On Peano's theorem in Banach spaces, *Func. Anal. and Appl.* **9** (1975) 53–55.

J. A. Goldstein
  [1] Abstract evolution equations, *Trans. Amer. Math. Soc.* 141 (1969) 159–185.

[2] Semi-groups and second order differential equations, *J. Func. Anal.* **4** (1969) 50–70.

[3] Semi-groups of operators and abstract Cauchy problems, Tulane Univ. Lecture notes (1970).

[4] The universal addability problem for generators of cosine functions and operators, *Houston J. Math.* **6** (1980) 365–373.

G. Greiner

[1] Zur Perron-Froebenius-Theorie stark stetiger Halbgruppen, *Math. Z.* **177** (1981) 401–423.

G. Greiner, J. Voigt and M. Wolff

[1] On the spectral bound of the generator of semigroups of positive operators, *J. Operator Theory* **5** (1981) 245–256.

K. Gustafson

[1] A perturbation lemma, *Bull. Amer. Math. Soc.* **72** (1966) 334–338.

A. Haraux

[1] Nonlinear evolution equations—Global behavior of solutions, Lecture notes in Math. 841 Springer-Verlag (1981).

M. Hasegawa

[1] On the convergence of resolvents of operators, *Pacific J. Math.* **21** (1967) 35–47.

D. Henry

[1] Geometric theory of semilinear parabolic equations, Lecture notes in Math. 840 Springer-Verlag (1981).

E. Heyn

[1] Die Differentialgleichung $dT/dt = P(t)T$ für Operatorfunktionen, *Math. Nach.* **24** (1962) 281–330.

E. Hille

[1] Representation of one-parameter semi-groups of linear transformations, *Proc. Nat. Acad. Sc. U.S.A.* **28** (1942) 175–178.

[2] *Functional analysis and semi-groups*, Amer. Math. Soc. Colloq. Publ. **Vol. 31,** New York (1948).

[3] On the differentiability of semi-groups of operators, *Acta Sc. Math.* (Szeged) **12** (1950) 19–24.

[4] Une generalization du probleme de Cauchy, *Ann. Inst. Fourier* **4** (1952) 31–48.

E. Hille and R. S. Phillips

[1] *Functional analysis and semi-groups*, Amer. Math. Soc. Colloq. **Publ. Vol. 31,** Providence R.I. (1957).

J. Holbrook

[1] A Kallman-Rota inequality for nearly Euclidean spaces, *Adv. in Math.* **14** (1974) 335–345.

M. Iannelli

[1] On the Green function for abstract evolution equations, *Bul. U.M.I.* **6** (1972) 154–174.

R. R. Kallman and G. C. Rota
[1] On the inequality $\|f'\|^2 \le 4\|f\| \cdot \|f''\|$, Inequalities, **Vol. II**. Academic Press New York (1970) 187–192.

T. Kato
[1] Integration of the equation of evolution in Banach space, *J. Math. Soc. Japan* **5** (1953) 208–234.
[2] On linear differential equations in Banach spaces, *Comm. Pure. Appl. Math.* **9** (1956) 479–486.
[3] Remarks on pseudo-resolvents and infinitesimal generators of semi-groups, *Proc. Japan Acad.* **35** (1959) 467–468.
[4] Note on fractional powers of linear operators, *Proc. Japan Acad.* **36** (1960) 94–96.
[5] Fractional powers of dissipative operators, *J. Math. Soc. Japan* **13** (1961) 246–274.
[6] Abstract evolution equations of parabolic type in Banach and Hilbert spaces, *Nagoya Math. J.* **19** (1961) 93–125.
[7] Fractional powers of dissipative operators II, *J. Math. Soc. Japan* **14** (1962) 242–248.
[8] Nonlinear evolution equations in Banach spaces, *Proc. Symp. Appl. Math.* **17** *Amer. Math. Soc.* (1965) 50–67.
[9] *Perturbation theory for linear operators*, Springer Verlag, New York (1966).
[10] A characterization of holomorphic semi-groups, *Proc. Amer. Math. Soc.* **25** (1970) 495–498.
[11] Linear evolution equations of "hyperbolic" type, *J. Fac. Sc. Univ. of Tokyo* **25** (1970) 241–258.
[12] On an inequality of Hardy, Littlewood and Polya, *Adv. in Math.* **7** (1971) 107–133.
[13] Linear evolution equations of "hyperbolic" type II. *J. Math. Soc. Japan* **25** (1973) 648–666.
[14] Quasi-linear equations of evolution with applications to partial differential equations, Lecture Notes in Math. 448, Springer Verlag (1975) 25–70.
[15] Linear and quasi-linear equations of evolution of hyperbolic type. C.I.M.E. II Ciclo (1976).
[16] On the Korteweg-de Vries equation, *Manuscripta Math.* **28** (1979) 89–99.

T. Kato and H. Tanabe
[1] On the abstract evolution equation, *Osaka Math. J.* **14** (1962) 107–133.
[2] On the analyticity of solutions of evolution equations, *Osaka J. Math* **4** (1969) 1–4.

H. Kielhöfer
[1] Halbgruppen und semilinear Anfangs-Randwertprobleme, *Manuscripta Math.* **12** (1974) 121–152.
[2] Existenz und Regularität von Lösungen semilinearer parabolicher Anfangs-Randwertprobleme, *Math. Z.* **142** (1975) 131–160.
[3] Global solutions of semilinear evolution equations satisfying an energy inequality, *J. Diff. Eqs.* **36** (1980) 188–222.

J. Kisynski
[1] Sur les operateures de Green des Problemes de Cauchy abstraits, *Studia Math.* **23** (1963/4) 285–328.

[2] A proof of the Trotter-Kato theorem on approximation of semi-groups, *Colloq. Math.* **18** (1967) 181–184.

[3] On second order Cauchy's problem in a Banach space, *Bull. Acad. Pol. Sc.* **18** (1970) 371–374.

[4] On cosine operator functions and one parameter groups of operators, *Studia Math.* **144** (1972) 93–105.

H. Komatsu

[1] Abstract analyticity in time and unique continuation property of solutions of a parabolic equation, *J. Fac. Sc. Univ. of Tokyo* **9** (1961) 1–11.

[2] Semi-groups of operators in locally convex spaces, *J. Math. Soc. Japan* **16** (1964) 230–262.

[3] Fractional powers of operators, *Pacific J. Math.* **19** (1966) 285–346.

[4] Fractional powers of operators II Interpolation spaces, *Pacific J. Math.* **21** (1967) 89–111.

[5] Fractional powers of operators III Negative powers, *J. Math. Soc. Japan* **21** (1969) 205–220.

[6] Fractional powers of operators IV Potential operators, *J. Math. Soc. Japan* **21** (1969) 221–228.

[7] Fractional powers of operators V Dual operators, *J. Fac. Sc. Univ of Tokyo* **17** (1970) 373–396.

T. Komura

[1] Semi-groups of operators in locally convex spaces, *J. Func. Anal.* **2** (1968) 258–296.

M. A. Krasnoselskii and P. E. Sobolevskii

[1] Fractional powers of operators defined on Banach spaces, *Dokl. Akad. Nauk SSSR* **129** (1959) 499–502.

S. G. Krein

[1] *Linear differential equations in Banach spaces*, Translations Amer. Math. Soc. **29**, Providence, RI (1971).

T. G. Kurtz

[1] Extensions of Trotter's operator semi-group approximation theorems, *J. Func. Anal.* **3** (1969) 111–132.

[2] A general theorem on the convergence of operator semigroups, *Trans. Amer. Math. Soc.* **148** (1970) 23–32.

T. G. Kurtz and M. Pierre

[1] A counterexample for the Trotter product formula, *MRC Tech. Rep.* **2091** (1980).

G. E. Ladas and V. Lakshmikantham

[1] *Differential equations in abstract spaces*, Academic Press, New York (1972).

P. D. Lax

[1] On the Cauchy's problem for hyperbolic equations and the differentiability of the solutions of elliptic equations, *Comm. Pure Appl. Math.* **8** (1955) 167–190.

P. D. Lax and R. D. Richtmyer

[1] Survey of the stability of linear finite difference equations, *Comm. Pure Appl. Math* **9** (1956) 267–293.

J. H. Lightbourne and R. H. Martin
  [1] Relatively continuous nonlinear perturbations of analytic semigroups, *Nonlinear Anal. TMA* **1** (1977) 277–292.

J. E. Lin and W. Strauss
  [1] Decay and Scattering of solutions of nonlinear Schrödinger equations, *J. Func. Anal.* **30** (1978) 245–263.

J. L. Lions
  [1] Les semi-groupes distributions, *Portugal. Math* **19** (1960) 141–164.
  [2] *Equations differentielles operationelles et problemes aux limites*, Springer Verlag, Berlin (1961).

J. L. Lions and E. Magenes
  [1] *Problemes aux limites non homogenes et applications*, Vol. I, II Dunod Paris (1968) Vol. III (1970).

Ju. I. Ljubic
  [1] Conditions for the uniqueness of the solution of Cauchy's abstract problem, *Dokl. Akad. Nauk SSSR* **130** (1960) 969–972.

G. Lumer and R. S. Phillips
  [1] Dissipative operators in a Banach space, *Pacific J. Math* **11** (1961) 679–698.

R. H. Martin
  [1] *Nonlinear operators and differential equations in Banach spaces*, John Wiley and Sons, New York (1976).
  [2] Invariant sets and a mathematical model involving semilinear differential equations, Nonlinear equations in abstract spaces, Proc. Inter. Symp. Univ. of Texas Arlington, Academic Press, New York (1978) 135–148.

K. Masuda
  [1] On the holomorphic evolution operators, *J. Math. Anal. Appl.* **39** (1972) 706–711.

V. G. Maz'ja and B. A. Plamenevskii
  [1] On the asymptotic behavior of solutions of differential equations in Hilbert space, *Math. USSR Izvestia* **6** (1972) 1067–1116.

I. Miyadera
  [1] Generation of a strongly continuous semi-group of operators, *Tohoku Math. J.* **4** (1952) 109–114.
  [2] Perturbation theory for semi-groups of operators (Japanese), *Sugaku* **20** (1968) 14–25.
  [3] On the generation of semigroups of linear operators, *Tohoku Math. J.* **24** (1972) 251–261.

I. Miyadera, S. Oharu and N. Okazawa
  [1] Generation theorems of semigroups of linear operators, *Publ. R.I.M.S. Kyoto Univ.* **8** (1972/3) 509–555.

M. Nagumo
  [1] Einige analytische Untersuchungen in linearen metrischen Rigen, *Japan J. Math.* **13** (1936) 61–80.

B. Nagy
  [1] On cosine operator functions in Banach spaces, *Acta Sc. Math.* **36** (1974) 281–289.

[2] Cosine operator functions and the abstract Cauchy problem, *Periodica Math. Hung.* **7** (1976) 15–18.

[3] Approximation theorems for cosine operator functions, *Acta Math. Acad. Sc. Hung.* **29** (1977) 69–76.

T. Nambu

[1] Asymptotic behavior of a class of nonlinear differential equations in Banach space, *SIAM J. Math. Anal.* **9** (1978) 687–718.

D. S. Nathan

[1] One parameter groups of transformations in abstract vector spaces, *Duke Math. J.* **1** (1935) 518–526.

J. Necas

[1] *Les methodes directes en theorie des equations elliptiques*, Masson and Cie Ed. Paris (1967).

J. W. Neuberger

[1] Analiticity and quasi-analiticity for one-parameter semi-groups, *Proc. Amer. Math. Soc.* **25** (1970) 488–494.

J. von-Neumann

[1] Über die analytische Eigenschaften von Gruppen linearer Transformationen und ihrer Darstellungen, *Math. Z.* **30** (1939) 3–42.

J. Neveu

[1] Theorie des semi-groupes de Markov, *Univ. of Calif. Publ. in Statistics* **2** (1958) 319–394.

L. Nirenberg

[1] On elliptic partial differential equations, *Ann. Scuola Norm. Sup. Pisa* **13** (1959) 1–48.

H. Ogawa

[1] Lower bounds for solutions of differential inequalities in Hilbert space, *Proc. Amer. Math. Soc.* **16** (1965) 1241–1243.

[2] On the convergence of semi-groups of operators, *Proc. Japan Acad.* **42** (1966) 880–884.

S. Oharu

[1] Semigroups of linear operators in Banach spaces, *Publ. R.I.M.S., Kyoto Univ.* **7** (1971) 205–260.

S. Oharu and H. Sunouchi

[1] On the convergence of semigroups of linear operators, *J. Func. Anal.* **6** (1970) 292–304.

N. Okazawa

[1] A perturbation theorem for linear contraction semi-groups on reflexive Banach spaces, *Proc. Japan Acad.* **47** (1971) 947–949.

[2] A generation theorem for semigroups of growth of order $\alpha$, *Tohoku Math. J.* **26** (1974) 39–51.

S. Ouchi

[1] Semi-groups of operators in locally convex spaces, *J. Math. Soc. Japan* **25** (1973) 265–276.

N. H. Pavel
[1] Invariant sets for a class of semi-linear equations of evolution, *Nonlinear Anal. TMA* **1** (1977) 187–196.
[2] Global existence for nonautonomous perturbed differential equations and flow invariance, *Preprint Series in Math* **22** (1981) INCREST, Bucharest.
[3] Analysis of some nonlinear problems in Banach spaces and applications, Univ. "Al. I. Cuza" Iasi Facultatea de Matematica (1982).

A. Pazy
[1] Asymptotic expansions of solutions of ordinary differential equations in Hilbert space, *Arch. Rat. Mech. and Anal.* **24** (1967) 193–218.
[2] Asymptotic behavior of the solution of an abstract evolution equation and some applications, *J. Diff. Eqs.* **4** (1968) 493–509.
[3] On the differentiability and compactness of semi-groups of linear operators, *J. Math. and Mech.* **17** (1968) 1131–1141.
[4] Semigroups of nonlinear contractions in Hilbert space, C.I.M.E. Varenna 1970 Ed. Cremonese (1971) 343–430.
[5] Approximation of the identity operator by semi-groups of linear operators, *Proc. Amer. Math. Soc.* **30** (1971) 147–150.
[6] On the applicability of Lyapunov's theorem in Hilbert space, *SIAM J. Math. Anal.* **3** (1972) 291–294.
[7] A class of semi-linear equations of evolution, *Israel J. Math* **20** (1975) 23–36.
[8] Semigroups of nonlinear contractions and their asymptotic behaviour, Non-linear analysis and mechanics, Heriot-Watt Symp. Vol. III, R. J. Knops Ed. Pitman Research notes in Math 30 (1979) 36–134.
[9] A perturbation theorem for linear *m*-dissipative operators, *Memorias de Matematica da U.F.R.J.* **131** (1981).

H. Pecher and W. von-Wahl
[1] Time dependent nonlinear Schrödinger equations, *Manuscripta Math.* **27** (1979) 125–157.

R. S. Phillips
[1] On the generation of semi-groups of linear operators, *Pacific J. Math.* **2** (1952) 343–369.
[2] Perturbation theory for semi-groups of linear operators, *Trans. Amer. Math. Soc.* **74** (1953) 199–221.
[3] A note on the abstract Cauchy problem, *Proc. Nat. Acad. Sci. U.S.A.* **40** (1954) 244–248.
[4] The adjoint semi-group, *Pacific J. Math.* **5** (1955) 269–283.
[5] Dissipative operators and hyperbolic systems of partial differential equations, *Trans. Amer. Math. Soc.* **90** (1959) 193–254.
[6] Dissipative operators and parabolic differential equations, *Comm. Pure Appl. Math.* **12** (1959) 249–276.
[7] Semi-groups of contraction operators, Equazioni Differenzio Astratte, C.I.M.E. (Rome) 1963.

B. A. Plamenevskii
[1] On the existence and asymptotic solutions of differential equations with unbounded operator coefficients in a Banach space, *Math. USSR Izvestia* **6** (1972) 1327–1379.

E. T. Poulsen
[1] Evolutionsgleichungen in Banach Räumen, *Math. Z.* **90** (1965) 286–309.

J. Prüss
  [1] On semilinear evolution equations in Banach spaces, *J. Reine u. Ang. Mat.* **303 / 4** (1978) 144–158.

S. M. Rankin
  [1] A remark on cosine families, *Proc. Amer. Math. Soc.* **79** (1980) 376–378.

M. Reed and B. Simon
  [1] *Methods of modern mathematical physics II: Fourier analysis, Self-adjointness,* Academic Press, New York (1975).

F. Sz. Riesz and B. Nagy
  [1] *Functional Analysis,* P. Ungar, New York (1955).

W. Rudin
  [1] *Functional analysis,* McGraw-Hill, New York (1973).

R. D. Richtmyer and K. W. Morton
  [1] *Difference methods for initial value problems,* Second ed. Interscience, New York (1967).

M. Schechter
  [1] Integral inequalities for partial differential operators and functions satisfying general boundary conditions, *Comm. Pure Appl. Math.* **12** (1959) 37–66.
  [2] General boundary value problems for elliptic partial differential equations, *Comm. Pure Appl. Math.* **12** (1959) 457–486.
  [3] Remarks on elliptic boundary value problems, *Comm. Pure Appl. Math.* **12** (1959) 561–578.
  [4] *Principles of functional analysis,* Academic Press, New York (1971).

L. Schwartz
  [1] Lectures on mixed problems in partial differential equations and the representation of semi-groups, *Tata Inst. Fund. Research* (1958).

I. Segal
  [1] Non-linear semi-groups, *Ann. of Math.* **78** (1963) 339–364.

T. I. Seidman
  [1] Approximation of operator semi-groups, *J. Func. Anal.* **5** (1970) 160–166.

E. Sinestrari
  [1] On the solutions of the inhomogeneous evolution equation in Banach space, *Rend. Acc. Naz. Lincei* **LXX** (1981).

M. Slemrod
  [1] Asymptotic behavior of $C_0$ semi-groups as determined by the spectrum of the generator, *Indiana Univ. Math. J.* **25** (1976) 783–792.

P. E. Sobolevskii
  [1] First-order differential equations in Hilbert space with variable positive definite self-adjoint operator, a fractional power of which has a constant domain of definition, *Dokl. Akad. Nauk SSSR* **123** (1958) 984–987.
  [2] On non-stationary equations of hydrodinamics for viscous fluid, *Dokl. Akad Nauk SSSR* **128** (1959) 45–48.
  [3] Parabolic equations in a Banach space with an unbounded variable operator, a fractional power of which has a constant domain of definition, *Dokl. Akad. Nauk SSSR* **138** (1961) 59–62.

[4] On the equations of parabolic type in Banach space, *Trudy Moscow Mat. Obsc* **10** (1961) 297–350, English trans. *Amer. Math. Soc. Trans.* **49** (1965) 1–62.

M. Z. Solomjak
[1] Applications of the semi-group theory to the study of differential equations in Banach spaces, *Dokl. Akad. Nauk SSSR* **12** (1958) 766–769.

M. Sova
[1] Probleme de Cauchy pour equations hyperboliques operationelles a coefficients constants non-bornes, *Ann. Scuola Norm. Sup. Pisa* **22** (1968) 67–100.

E. M. Stein and G. Weiss
[1] *Introduction to Fourier analysis on Euclidean spaces*, Princeton Univ. Press (1971).

B. Stewart
[1] Generation of analytic semigroups by strongly elliptic operators, *Trans. Amer. Math. Soc.* **199** (1974) 141–161.
[2] Generation of analytic semigroups by strongly elliptic operators under general boundary conditions. *Trans. Amer. Math. Soc.* **259** (1980) 299–310.

M. H. Stone
[1] On one-parameter unitary groups in Hilbert space, *Ann. Math.* **33** (1932) 643–648.

P. Suryanarayana
[1] The higher order differentiability of solutions of abstract evolution equations, *Pacific J. Math.* **22** (1967) 543–561.

H. Tanabe
[1] A class of the equations of evolution in a Banach space, *Osaka Math. J.* **11** (1959) 121–145.
[2] On the equations of evolution in Banach space, *Osaka Math. J.* **12** (1960) 363–376.
[3] Remarks on the equations of evolution in a Banach space *Osaka Math. J.* **12** (1960) 145–165.
[4] Convergence to a stationary state of the solution of some kind of differential equations in a Banach space, *Proc. Japan Acad.* **37** (1961) 127–130.
[5] Note on singular perturbation for abstract differential equations, *Osaka J. Math.* **1** (1964) 239–252.
[6] *Equations of evolution* (English translation) Pitman, London (1979).

H. Tanabe and M. Watanabe
[1] Note on perturbation and degeneration of abstract differential equations in Banach space, *Funkcialaj Ekvacioj* **9** (1966) 163–170.

C. Travis and G. Webb
[1] Cosine families and abstract nonlinear second order differential equations, *Acta Math. Acad. Sc. Hung.* **32** (1978) 75–96.
[2] Second order differential equations in Banach spaces, Nonlinear equations in abstract spaces, Ed. V. Lakshmikantham, Academic Press, New York (1978) 331–361.

H. F. Trotter
[1] Approximation of semi-groups of operators, *Pacific J. Math.* **8** (1958) 887–919.

[2] On the product of semi-groups of operators, *Proc. Amer. Math. Soc.* **10** (1959) 545–551.

T. Ushijima
  [1] Some properties of regular distribution semi-groups, *Proc. Japan Acad.* **45** (1969) 224–227.
  [2] On the generation and smoothness of semigroups of linear operators, *J. Fac. Sc. Univ. Tokyo* **19** (1972) 65–127.

W. von-Wahl
  [1] Instationary Navier-Stokes equations and parabolic systems, *Pacific J. Math.* **72** (1977) 557–569.
  [2] Semilinear elliptic and parabolic equations of arbitrary order, *Proc. Royal Soc. Edinburgh A* **78** (1978) 193–207.

J. A. Walker
  [1] *Dynamical systems and evolution equations*, Plenum Press, New York (1980).

J. Watanabe
  [1] On some properties of fractional powers of linear operators, *Proc. Japan Acad.* **37** (1961) 273–275.

M. Watanabe
  [1] On the differentiability of semi-groups of linear operators in locally convex spaces, *Sc. Rep. Niigata Univ.* **9** (1972) 23–34.

G. Webb
  [1] Continuous nonlinear perturbations of linear accretive operators in Banach spaces, *J. Func. Anal.* **10** (1972) 191–203.

R. Wüst
  [1] Generalization of Rellich's theorem of perturbations of (essentially) self adjoint operators, *Math. Z.* **119** (1971) 276–280.

J. Yorke
  [1] A continuous differential equations in Hilbert space without existence, *Funkcialaj Ekvacioj* **12** (1970) 19–21.

K. Yosida
  [1] On the group embedded in the metrical complete ring, *Japan J. Math.* **13** (1936) 61–80.
  [2] On the differentiability and representation of one parameter semi-groups of linear operators, *J. Math. Soc. Japan* **1** (1948) 15–21.
  [3] Lectures on semi-group theory and its applications to Cauchy's problem in partial differential equations, *Tata Inst. Fund. Research* (1957).
  [4] On the differentiability of semi-groups of linear operators, *Proc. Japan Acad.* **34** (1958) 337–340.
  [5] Fractional powers of infinitesimal generators and the analyticity of the semi-groups generated by them, *Proc. Japan Acad.* **36** (1960) 86–89.
  [6] Time dependent evolution equations in locally convex space, *Math. Ann.* **162** (1965) 83–86.
  [7] *Functional analysis* (6th edition) Springer Verlag (1980).

J. Zabczyk
  [1] A note on $C_0$ semigroups, *Bull. Acad. Polon. Sc.* (Math. Astr. Phys.) **23** (1975) 895–898.

S. D. Zaidman
  [1] *Abstract differential equations*, Pitman Research Notes in Math. **36** (1979).

# Index

# Applied Mathematical Sciences

cont. from page ii